Progress in
MACROCYCLIC CHEMISTRY

Progress in
MACROCYCLIC CHEMISTRY

VOLUME 1

Edited by

REED M. IZATT

Department of Chemistry

JAMES J. CHRISTENSEN

Department of Chemical Engineering

Brigham Young University
Provo, Utah

A WILEY-INTERSCIENCE PUBLICATION

JOHN WILEY & SONS, New York • Chichester • Brisbane • Toronto

Copyright © 1979 by John Wiley & Sons, Inc.

All rights reserved. Published simultaneously in Canada.

Reproduction or translation of any part of this work beyond that permitted by Sections 107 or 108 of the 1976 United States Copyright Act without the permission of the copyright owner is unlawful. Requests for permission or further information should be addressed to the Permissions Department, John Wiley & Sons, Inc.

Library of Congress Cataloging in Publication Data:

Main entry under title:
Progress in macrocyclic chemistry.

"A Wiley-Interscience publication."
Includes index.
1. Cyclic compounds. I. Izatt, Reed McNeil, 1926- II. Christensen, James J., 1931-
QD331.P75   547'.5   78-14354
ISBN 0-471-03477-0

Printed in the United States of America

10 9 8 7 6 5 4 3 2 1

CONTRIBUTORS

D. Ammann, Department of Organic Chemistry, Swiss Federal Institute of Technology, Zurich, Switzerland

R. Bissig, Department of Organic Chemistry, Swiss Federal Institute of Technology, Zurich, Switzerland

F. de Jong, Shell Research B. V., Amsterdam, The Netherlands

J. L. Dye, Department of Chemistry, Michigan State University, East Lansing, Michigan

S. Lindenbaum, Department of Pharmaceutical Chemistry, University of Kansas, Lawrence, Kansas

W. E. Morf, Department of Organic Chemistry, Swiss Federal Institute of Technology, Zurich, Switzerland

N. S. Poonia, Department of Chemistry, University of Indore, Indore, India

E. Pretsch, Department of Organic Chemistry, Swiss Federal Institute of Technology, Zurich, Switzerland

D. N. Reinhoudt, Department of Organic Chemistry, Twente University, Enschede, The Netherlands

J. H. Rytting, Department of Pharmaceutical Chemistry, University of Kansas, Lawrence, Kansas

W. Simon, Department of Organic Chemistry, Swiss Federal Institute of Technology, Zurich, Switzerland

L. A. Sternson, Department of Pharmaceutical Chemistry, University of Kansas, Lawrence, Kansas

PREFACE

Interest in synthetic multidentate macrocyclic compounds has continually increased since C. J. Pedersen reported the synthesis of the crown ethers in 1967. The interest in these compounds is broad, ranging from synthesis of new molecules to investigation of selective metal binding properties to their use in catalyzing synthetic organic reactions. This book is an attempt to bring together in-depth investigations in currently important areas of macrocyclic research. The chapters deal mainly with macrocyclic compounds (saturated polyethers and their derivatives) and macrobicyclic compounds (cryptates).

In editing this book we had in mind those readers interested in becoming acquainted with one or more currently important areas of macrocyclic chemistry. The book is primarily aimed at researchers and students in organic, physical, analytical, and inorganic chemistry, as well as those in chemical engineering. However, we hope that it will also be of interest to many in the areas of biology, biochemistry, and physiology.

Chapter 1, by Morf, Ammann, Bissig, Pretsch, and Simon, discusses the ion selectivity of neutral macrocyclic and nonmacrocyclic ligands in membranes, with emphasis on alkali, alkaline earth, and ammonium ions. The discussion stresses the capability of these ligands to behave as ion carriers to selectively extract ions from aqueous solutions and transport them across a membrane phase. The origin of ion selectivity is treated for ligands in water and waterlike solvents, and this selectivity is compared with the ion selectivity demonstrated by these same ligands when they are contained in membranes. The more important parameters are enumerated for the design and development of ion carriers that will have high selectivity for specific metal ions in a membrane system. Specific examples are given to illustrate how such ligands can be developed. Ion-selective electrodes based on using neutral macrocyclic carriers are described. Also included is extensive discussion of the origin of cation premselectivity in thick, neutral carrier liquid membranes and in biological bilayers.

The second chapter, by Dye, examines the metal complexation properties of crown and cryptand ligands in amine and ether solvents, and the new compounds that have been prepared in these solutions. It is pointed out that it is the ability of these ligands to isolate the cation from its counterion which provides a really new dimension to metal-solution chemistry. Dye elucidates how the crown ethers and cryptands offer three advantages in the study of metal solutions: first, the concentration of species can be increased and new solvents can be used; second, the stoichiometry of the solution can be controlled; and third, new solid compounds and types of solid salts can be obtained from solutions containing crown or cryptand ligands. Also described is the use of alkali metal NMR spectroscopy to study the thermodynamics and kinetics of exchange and to detect and explore the nature of alkali metal ions in solution. The chapter points out that this technique has been used successfully to confirm the existence of spherically symmetric alkali metal anions (M^-).

Chapter 3, by Poonia, elaborates on the various factors that determine the complexation of macrocyclic ligands with alkali and alkaline earth metal ions. A major thrust of the chapter is the introduction of some new general principles for describing a wide variety of macrocycle-metal interactions. Poonia explains the interaction of cyclic polyethers, cryptands, and cyclic antibiotics with metal ions, by considering the charge density of the cation and conformation energy of the ligand. He also discusses the role played by organic and inorganic anions in determining the stability of the complex and the possibility of solid compound formation and stoichiometry. Experimental evidence is presented to support the principles put forth in the chapter.

The fourth chapter, by Reinhoudt and de Jong, deals with the synthesis of crown ethers and related macrocycles having one or more bis(methylene)-aromatic or -heteroaromatic moieties. Compounds containing oxygen; nitrogen and oxygen; sulfur and oxygen; and nitrogen, sulfur, and oxygen functional groups are discussed. Included are tables of compounds synthesized which contain the formulas and ring structures of the macrocycles together with reaction yields, compound melting points, and appropriate literature references. This is followed by a discussion of the use of NMR spectroscopy and mass spectrometry to elucidate the structure of crown ethers. The chapter then discusses properties of the macrocycles and their capacity to form complexes with alkali and alkaline earth metals and ammonium and alkylammonium salts. Stability and association constants are given for complexes formed in polar and in apolar solvents. The kinetics of complexation with alkylammonium salts, as determined by NMR lineshape methods, is discussed. Free energies of activation and decomplexation of the complexes are also included.

Chapter 5, by Lindenbaum, Rytting, and Sternson, concludes the book by describing factors that contribute to the specific ion permeability of membranes mediated by ionophores. Included is a discussion of complexation and membrane-transport phenomena. Ionophores are defined, and various types are discussed with respect to their structural features and biological applications. Some physiological secretory responses associated with ionophores are reported. Stability constants are given for the interaction of alkali and alkaline earth cations with cryptates, amine cations with 18-crown-6, and biogenic amines with lasalocid (X-537A). Included is a list of cation selectivities of several ionophores. The theory of membrane-transport phenomena is discussed, and a theoretical model for carrier-mediated permeation presented. The role of ion-transport mediators in energy coupling is reviewed, including the various proposed mechanistic models.

We would like to acknowledge our indebtedness to all those who have aided in bringing this book to fruition.

R. M. IZATT
J. J. CHRISTENSEN

Provo, Utah
October 1978

CONTENTS

1. Cation Selectivity of Neutral Macrocyclic and Nonmacrocyclic Complexing Agents in Membranes ... 1

   W. E. MORF, D. AMMANN, R. BISSIG, E. PRETSCH, and W. SIMON

2. The Role of Crown and Cryptand Complexation of Cations in the Formation of Metal-Amine and Metal-Ether Solutions ... 63

   J. L. DYE

3. Multidentate Macromolecules: Principles of Complexation with Alkali and Alkaline Earth Cations ... 115

   N. S. POONIA

4. Crown Ethers and Related Macrocycles with Bis(methylene)aromatic or -Heteroaromatic Subunits: Their Synthesis and Complexation ... 157

   D. N. REINHOUDT and F. DE JONG

5. Ionophores – Biological Transport Mediators ... 219

   S. LINDENBAUM, J. H. RYTTING, and L. A. STERNSON

Index ... 255

CHAPTER ONE
CATION SELECTIVITY OF NEUTRAL MACROCYCLIC AND NONMACROCYCLIC COMPLEXING AGENTS IN MEMBRANES

W. E. MORF, D. AMMANN, R. BISSIG,
E. PRETSCH, and W. SIMON

Department of Organic Chemistry
Swiss Federal Institute of Technology
Zurich, Switzerland

1 Introduction	2
2 Origin of the Ion Selectivity of Electrically Neutral Complexing Agents in Homogeneous Systems	8
2.1 Simple Binding Sites for A Cations, 8	
2.2 Effect of Coordination Number, 11	
2.3 Polydentate Ligands, 12	
3 Ion Selectivity of Membranes Containing Lipophilic Neutral Carriers	15
4 Design Features of Membrane-Active Complexing Agents	22
5 Ion-Selective Membrane Electrodes Based on Neutral Carriers	31
6 Carrier-Mediated Ion Transport Through Membranes	35
6.1 Bulk Membranes, 39	
6.2 Bilayer Membranes; Biological Systems, 44	
7 Future Prospects	53
Acknowledgment	57
References	57

1 INTRODUCTION

Electrically neutral, lipophilic ion-complexing agents of rather small relative molar mass are known to behave as ionophores or ion carriers (1), having the capability to selectively extract ions from aqueous solutions into a hydrophobic membrane phase and to transport these ions across such barriers by carrier translocation.

Although Moore and Pressman's discovery of the effect of naturally occurring neutral carrier antibiotics in biological membrane systems dates to 1964 (2), their fundamental property, namely their role as highly selective complexing agents for alkali metal ions, was recognized only some two years later (3). Some of the molecules studied, such as valinomycin (**1**) and the macrotetrolides (**2-6**) (see Figure 1), show a striking differentiation between Na^+ and K^+. Similar ion-selective properties are mimicked by the synthetic macrocyclic polyethers (crown compounds **7**) first studied by Pedersen (4, 5) as well as by the macroheterobicyclic ligands first synthesized by Lehn (**8-13**) (6, 7). A wide range of remarkable ion selectivities for alkali and alkaline earth metal cations may be induced in membranes by nonmacrocyclic synthetic ion carriers (8, 9). Such molecules (**14-20**) were first introduced in 1972 (10) (see Ref. 11). Similar ligands (**21**) were prepared more recently by Vögtle (12) and others (13).

In the meantime a large number of macrocyclic and nonmacrocyclic neutral complexing agents for ions are accessible. In addition to carriers for group 1A and 2A cations, ionophores for transition metals (14) and for enantiomeric ammonium ions (15-17) (compounds **23-26**, discrimination by chiral recognition) have been described.

A ligand that behaves as an ionophore must meet the following requirements:

1. The carrier molecule should be composed of polar and nonpolar groups.
2. The carrier should be able to assume a stable conformation that provides a cavity, surrounded by the polar groups, suitable for the uptake of a cation, while the nonpolar groups form a lipophilic shell around the coordination sphere. These groups must ensure sufficiently large lipid solubility for ligand and complex. This is one reason that classical electrically charged complexing agents such as EDTA do not behave as carriers in membrane systems.
3. Among the polar groups of the ligand sphere, there should be preferably 5 to 8, but not more than 12, coordinating sites such as oxygen atoms.
4. High selectivities are achieved by locking the coordinating sites into a rigid arrangement around the cavity. Such rigidity can be enhanced by the presence of bridged structures, e.g. hydrogen bonds. Within one group of the periodic system, the cation that best fits into the offered cavity is preferred. Ideally, all cations should be forced into accepting the same given number of coordinating groups.
5. Notwithstanding requirement 4, the ligand should be flexible enough to allow a sufficiently fast ion exchange. This is possible only with a stepwise

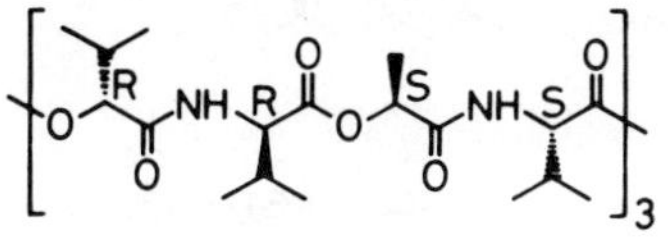

VALINOMYCIN 1

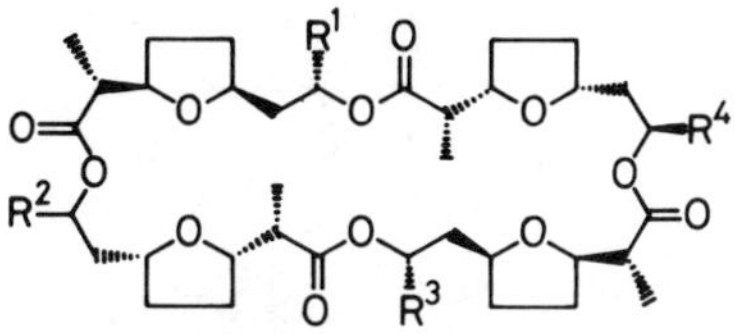

$R^1 = R^2 = R^3 = R^4 = CH_3$		NONACTIN	2
$R^1 = R^2 = R^3 = CH_3$	$R^4 = C_2H_5$	MONACTIN	3
$R^1 = R^3 = CH_3$	$R^2 = R^4 = C_2H_5$	DINACTIN	4
$R^1 = CH_3$	$R^2 = R^3 = R^4 = C_2H_5$	TRINACTIN	5
$R^1 = R^2 = R^3 = R^4 = C_2H_5$		TETRANACTIN	6

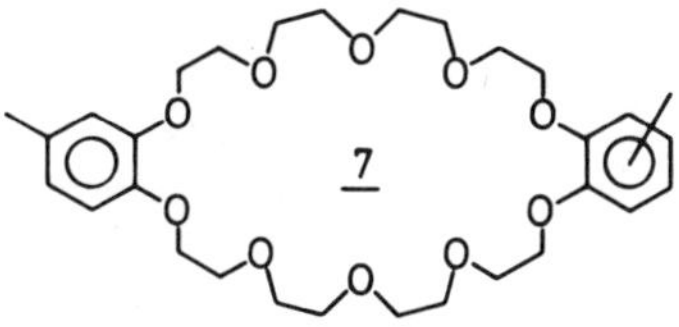

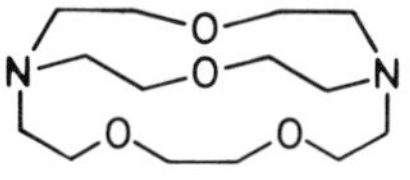

8

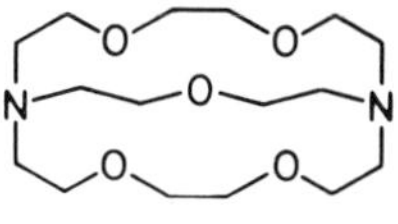

9

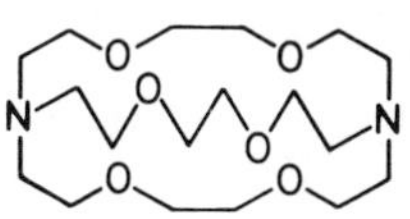

10

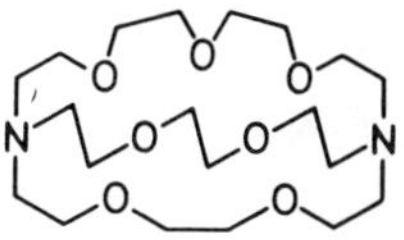

11

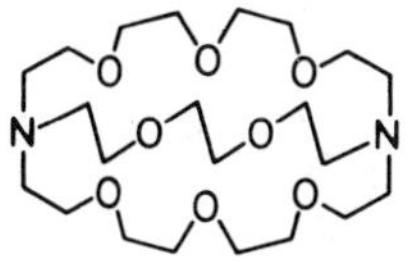

12

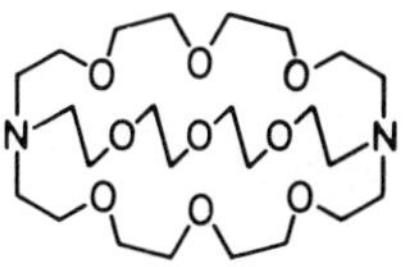

13

Figure 1 Ligands discussed in this work.

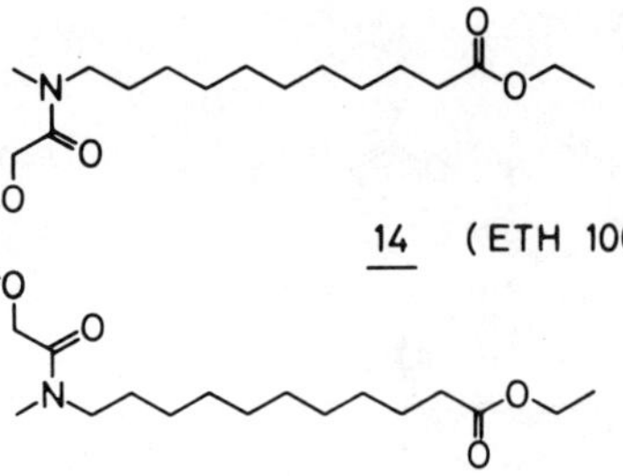

14 (ETH 1002)

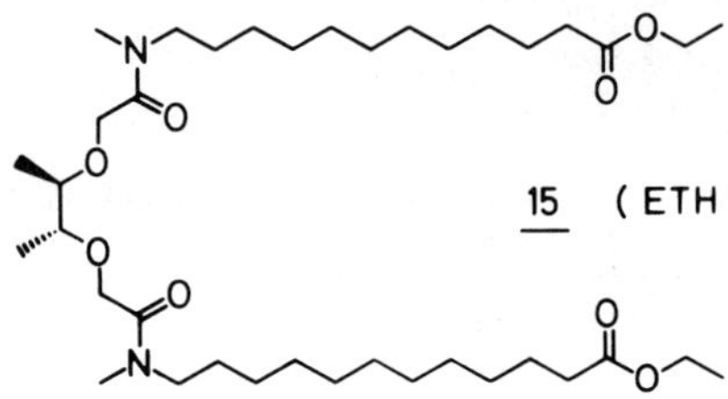

15 (ETH 1001)

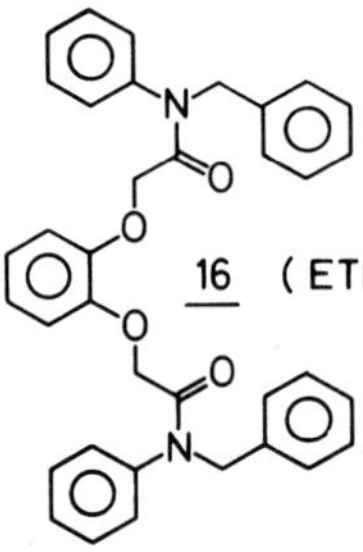

16 (ETH 157)

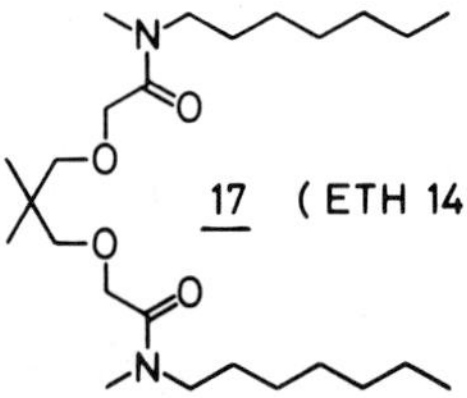

17 (ETH 149)

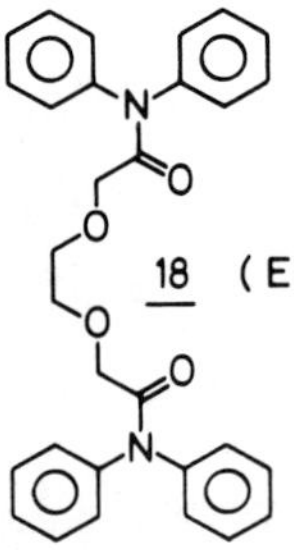

18 (ETH 67)

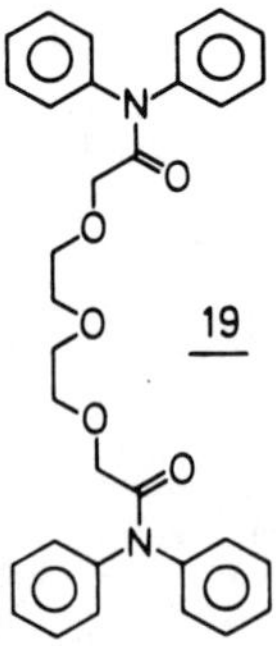

19 (ETH 231)

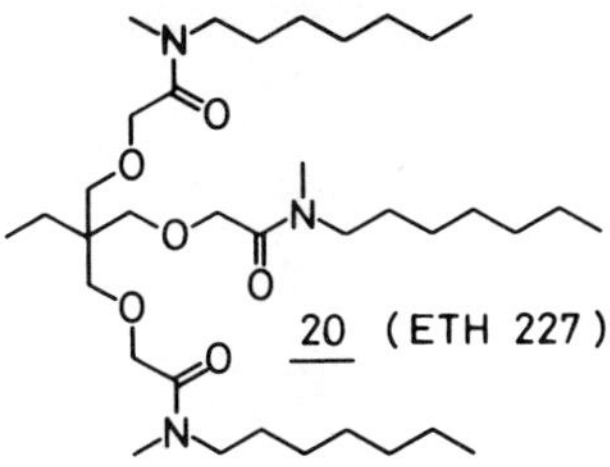

20 (ETH 227)

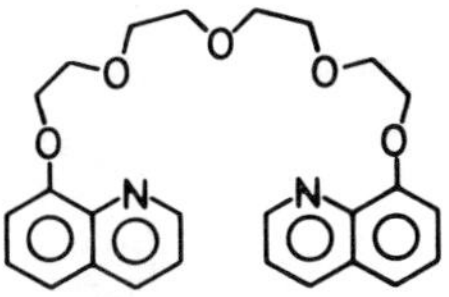

21

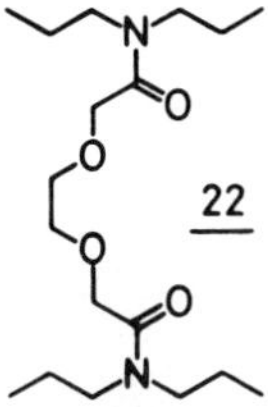

22 (ETH 65)

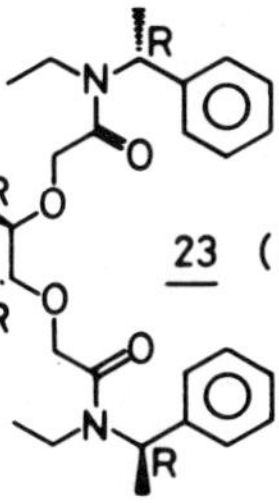

23 (ETH 1003)

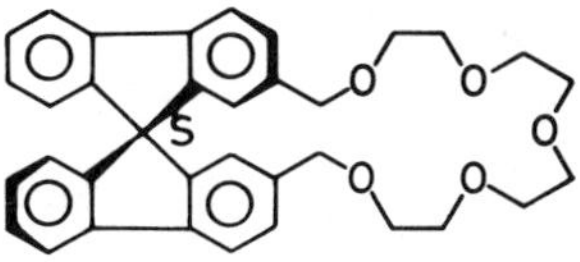

24

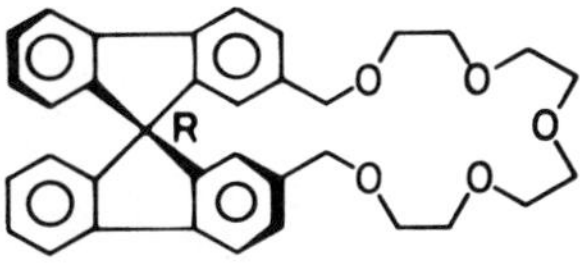

25

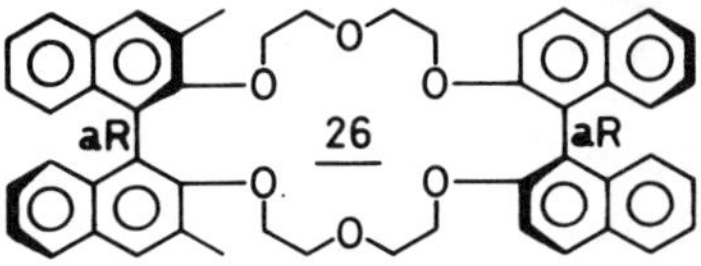

26

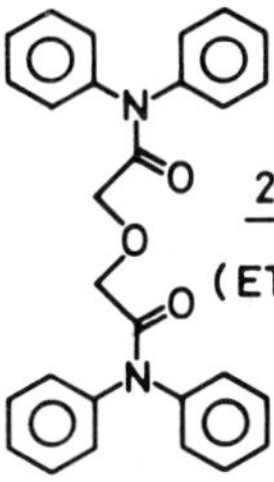

27 (ETH 1004)

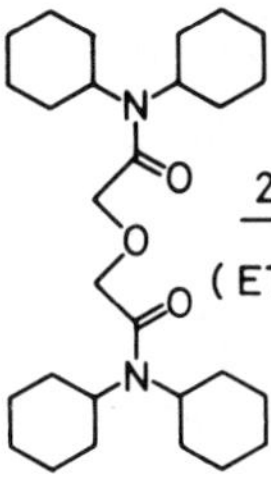

28 (ETH 129)

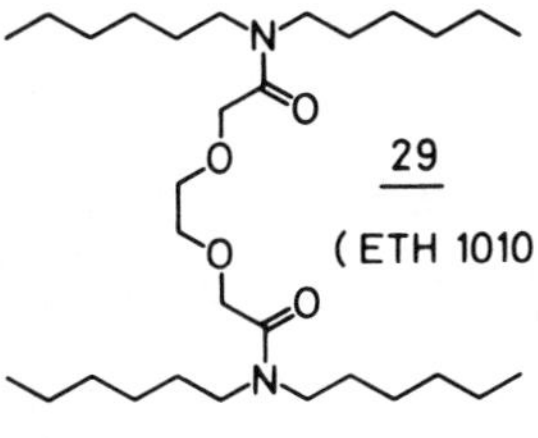

29 (ETH 1010)

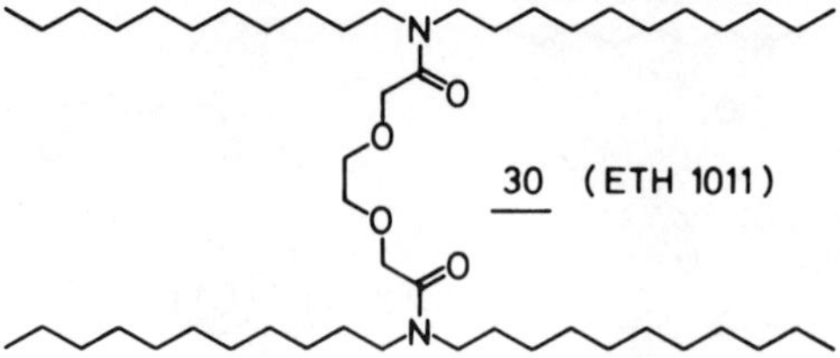

30 (ETH 1011)

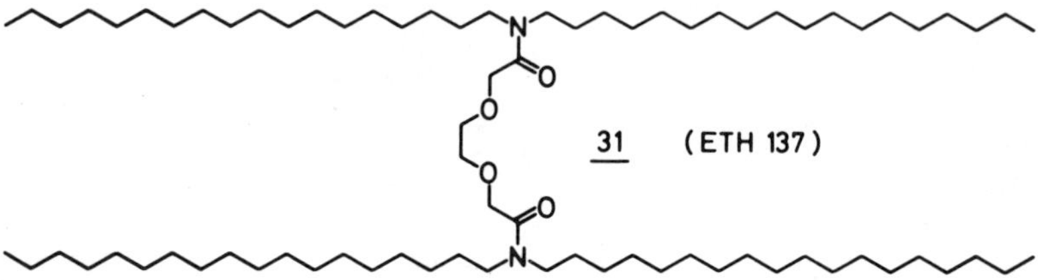

31 (ETH 137)

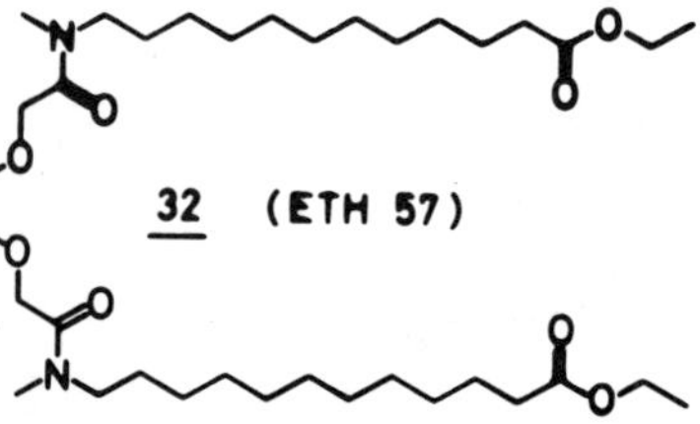

32 (ETH 57)

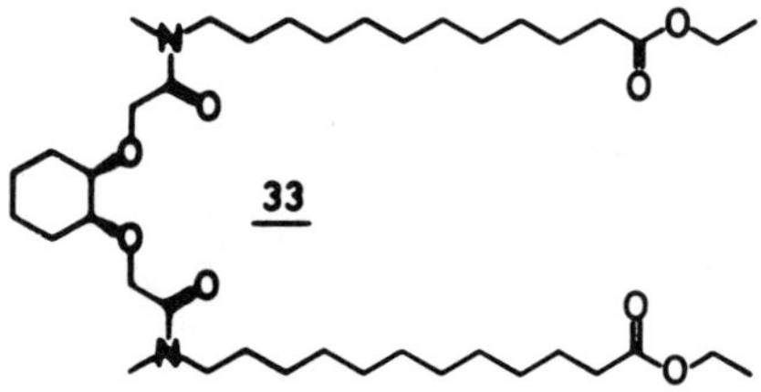

33

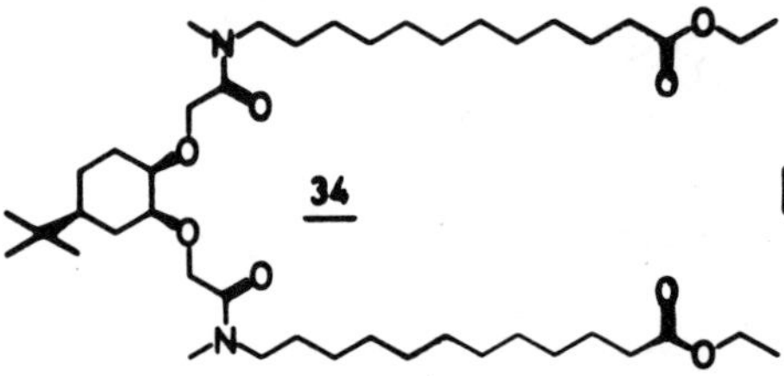

34

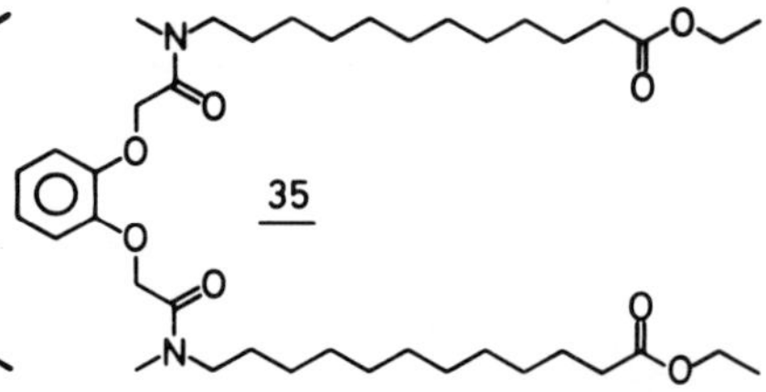

35

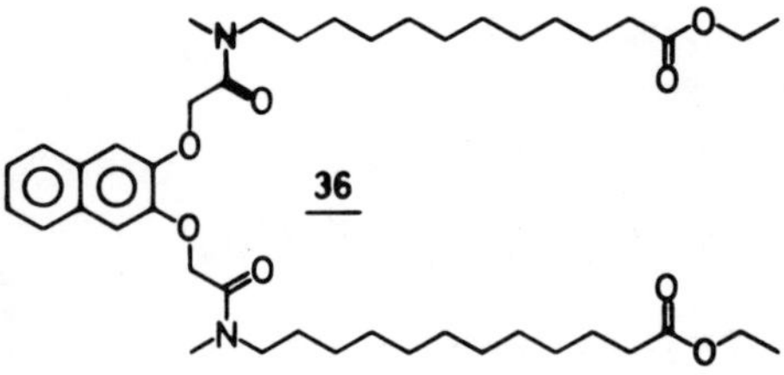

36

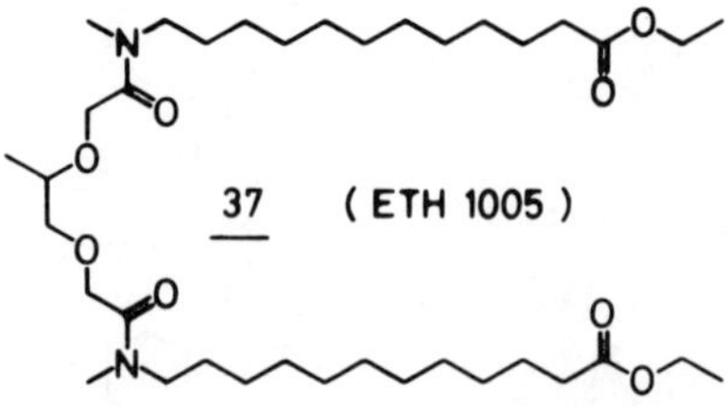

37 (ETH 1005)

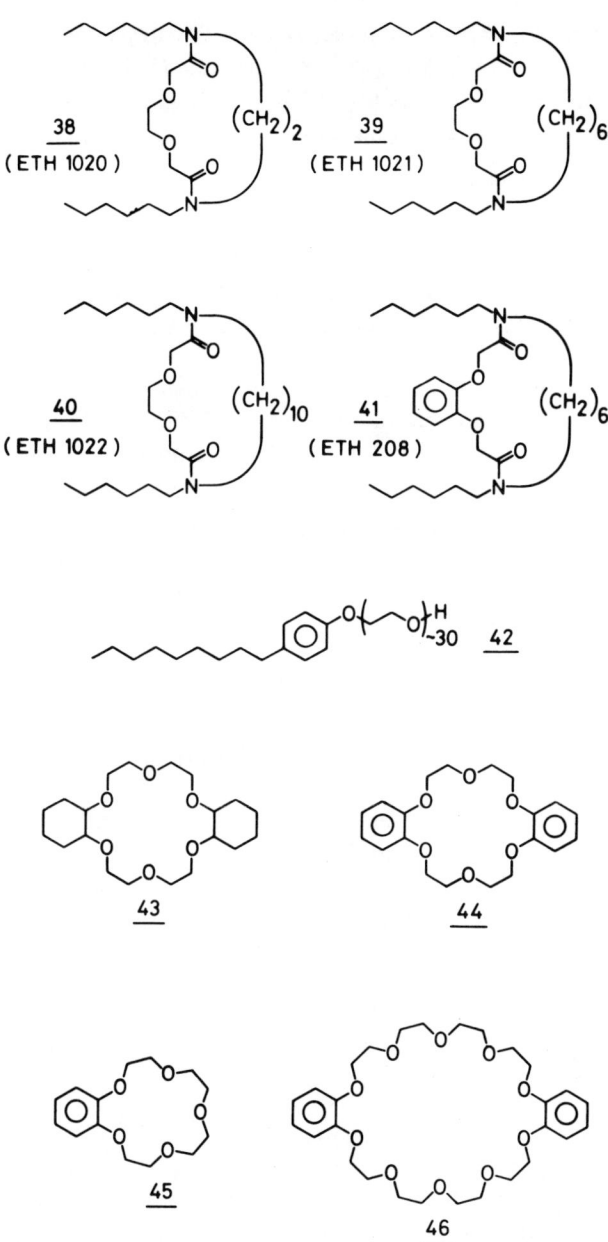

substitution of the solvent molecules by the ligand groups. Thus a compromise between stability (4) and exchange rate (5) has to be found.
6. To guarantee adequate mobility the overall dimensions of a carrier should be rather small but still compatible with high lipid solubility.

The underlying principles involved in points 1 to 6 are covered in somewhat greater detail in the following discussion.

2 ORIGIN OF THE ION SELECTIVITY OF ELECTRICALLY NEUTRAL COMPLEXING AGENTS IN HOMOGENEOUS SYSTEMS

In the present context lipophilic membranes in contact with aqueous solutions are of prime interest. To elucidate the origin of the ion selectivity of carriers in such systems, we focus first on the interaction of ions with ligands in water and waterlike media.

2.1 Simple Binding Sites for A Cations

Attractive binding sites are ligand atoms that are capable of competing with water molecules in the complexation of ions. An estimate of the effectiveness of different sites may be obtained by simple model calculations. For complexes of neutral ligands with group IA and IIA ions, the interaction between the ionic charge and the dipoles (permanent and induced) of the ligand groups, as well as contributions of the Lennard-Jones type resulting from repulsions between atoms, are of major importance.

The estimated energies of interaction of different group IA and IIA ions with one water molecule are given in Table 1, column 2. Calculated by considering the ion-dipole interactions, ion-induced dipole interactions, and contributions from repulsion (18, 19), these values are in excellent agreement with data obtained by much more sophisticated computational techniques (20) (column 3) as well as with experimental values (21) (column 4).

Through variation of molecular parameters such as dipole moment and polarizability of the binding sites, as well as the van der Waals radius of the ligand atoms, the interaction selectivity may be influenced considerably. This is demonstrated in Table 1 for 1:1 complexes of hypothetical molecules with cations. In columns 5 to 7 the increments of the interaction energies are given (relative to column 2) which result from changes in the dipole moment (column 5), the polarizability (column 6), and the radius (column 7) of the ligand site. The data indicate that an increase in the dipole moment or the polarizability, as well as a decrease in the radius of the ligand atom, increase the stability of the hypothetical complexes. The effect is especially large for small and multiply charged cations. Therefore small and polar binding sites generally prefer Ca^{2+} over Na^+ and Ba^{2+}, for example. An extreme situation is found for anionic sites (last column in Table 1), for which these effects are amplified.

Table 1 Interactions Between Binding Site Models and Cations

	Interaction Energy (kcal mole^{-1}) for the Complex Ion-H$_2$O			Change in Interaction Energy (kcal mole^{-1}) Obtained for			
Ion (radius, Å)	Calculated Values Using the Data of Table 2[a]	Values from Ref. 20	Values from Ref. 21	Increase in Dipole Moment by 0.5 D	Increase in Polarizability by 1 Å^3	Increase in Ligand Radius by 0.1 Å	Substitution of Dipole Moment by Charge $-e_0$
1	2	3	4	5	6	7	8
Li$^+$ (0.68)	−34.1	−34.1	−34.0	−6.8	−6.1	+3.8	−123
Na$^+$ (0.98)	−24.4	−23.2	−24.0	−5.2	−3.6	+2.3	−110
K$^+$ (1.33)	−17.5	−16.2	−17.9	−3.9	−2.0	+1.4	−98
Rb$^+$ (1.49)	−15.3		−15.9	−3.5	−1.6	+1.2	−93
Cs$^+$ (1.65)	−13.5		−13.7	−3.1	−1.3	+1.0	−89
Mg^{2+} (0.78)	−75.4			−12.4	−20.3	+8.9	−236
Ca^{2+} (1.06)	−54.1			−9.7	−12.5	+5.5	−214
Sr^{2+} (1.27)	−43.5			−8.2	−9.0	+4.0	−199
Ba^{2+} (1.43)	−37.4			−7.3	−7.1	+3.2	−190

[a] The basic formula (18) was $E = -\dfrac{x-2}{x}\dfrac{ze_0 p}{r^2} - \dfrac{x-4}{x}\dfrac{\alpha(ze_0)^2}{2r^4}$ with ionic charge ze_0, dipole moment p, polarizability α, ion-ligand distance r, and repulsion coefficient $x = 12$.

The data presented in Table 1 refer to interactions of one ion with one ligand molecule in the absence of other solvating species (gas phase) and are intended to give semiquantitative information. Some molecular parameters for a series of possible binding sites are compiled in Table 2. Unfortunately these single parameters cannot be varied independently in real molecules. In addition several binding sites are normally involved in the complexation of one cation. This means that the unambiguous isolation and experimental verification of the effect of any one molecular parameter on the observed behavior must remain an elusive goal.

Table 2 Molecular Parameters for Some Binding Sites (from Ref. 22)

Molecule	Permanent Dipole Moment (D)	Polarizability[a] ($Å^3$)	Van der Waals Radius of Ligand Atom (Å)
H_2O	1.85[b]	1.46[b]	1.38[b]
NH_3	1.47	2.26	1.50
H_2S	0.92	3.67	1.85
PH_3	0.55	4.28	1.90
CH_3-O-CH_3	1.30	0.65[c]	
$CH_3-NH-CH_3$	1.03	0.94[c]	
CH_3-S-CH_3	1.50	3.06[c]	
$CH_3-C(=O)-CH_3$	2.88	0.84[c]	
$CH_3-S(=O)_2-CH_3$	4.49		
$CH_3-C(=O)-O-CH_2CH_3$	1.78		
$CH_3-C(=O)-N(CH_3)-CH_3$	3.81		

[a] Calculated from molar refraction data.
[b] Values used in Table 1 (column 2) and in Ref. 18.
[c] Polarizability increment of ligand atom.

2.2 Effect of Coordination Number

The often observed variation in the coordination number of group IA and IIA cations is a consequence of the radius-ratio effect, which was introduced in discussions of ion packing in crystals (23, 24) and ionic hydration (18). If we consider, for example, coordination spheres of eight (cubic), six (octahedral), and four (tetrahedral) oxygen atoms, forming cavities for the uptake of a cation, we find minimal cavity radii of 1.0, 0.6, and 0.3 Å, respectively (see Table 3) as a result of the mutual repulsion between the oxygen atoms, which are assumed to behave as hard spheres of radius 1.40 Å. In agreement with Table 3, the reported coordination numbers are $\leqslant 6$ for Li^+ (18), $\leqslant 4$ for Be^{2+} (25, 26), $\leqslant 6$ for Mg^{2+} (24-26), and about 8 for Ca^{2+} (24) (for ionic radii see Table 1, column 1); they are also consistent with computed first-shell coordination numbers in water of 5.4 ± 0.7 for Li^+, 6.0 ± 1.1 for Na^+, and 7.2 ± 1.2 for K^+ (27). A similar situation is represented in Figure 2, where the influence of a fixed coordination number on the calculated free energy of hydration is illustrated. As a rule the interaction (e.g., $-\Delta G_H^0$) of a cation with a coordination sphere of n monodentate ligands increases with decreasing ionic radius until the cavity radius (given in Table 3) is reached. The calculated values of ΔG_H^0 in Figure 2 indicate that the most stable aquocomplexes of alkali metal ions and Mg^{2+} have a coordination number of 6, whereas the larger alkaline earth cations prefer $n = 8$.

Table 3 Radius of the Minimal Cavity Enclosed by n Oxygen Atoms[a] (from Ref. 23)

Coordination Number n (coordination geometry)	Radius r_m (Å)
2 (linear)	0.00
3 (triangular)	0.22
4 (tetrahedral)	0.31
4 (square)	0.58
5 (trigonal-bipyramidal)	0.58
5 (pyramidal)	0.64
6 (octahedral)	0.58
7 (symmetry C_{3v})	0.83
7 (pentagonal-bipyramidal)	0.98
8 (cubic)	1.02
9 (symmetry D_{3h})	1.02
12 (cubooctahedral)	1.40

[a] The radius of a coordinating oxygen atom is 1.40 Å.

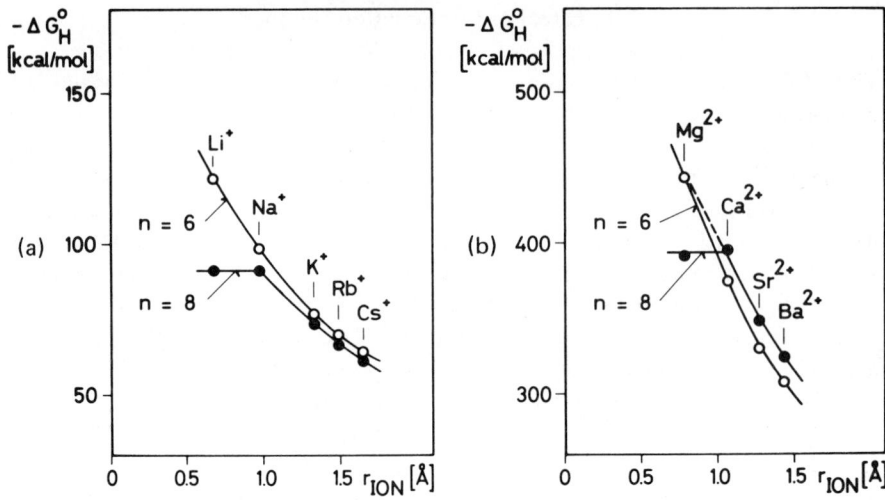

Figure 2 Calculated free energies of hydration ΔG_H^o for alkali ions (a) and alkaline earth ions (b). Values for cubic and octahedral coordination geometry are given as a function of the ionic radius (18).

If n could be fixed to 8 throughout, Mg^{2+} would be heavily destabilized relative to the situation with a free choice of n. For alkali ions a coordination with $n = 8$ would reject Li^+ and would also give some destabilization of the other ions (see Figure 2). These results clearly demonstrate that a coordination with oxygen and $n = 8$ is attractive for complexing Ca^{2+}.

Especially high selectivities are to be expected for polydentate ligands that offer a predetermined coordination sphere. In addition to the net effect of the coordination number and the binding properties of the ligand groups, steric interactions in the ligand skeleton may become important for the selection of an ion of a given size. Table 4 nicely demonstrates these trends. In accord with model calculations (see Figure 2 and Refs. 30 and 31), an increase of the coordination number results in a preference of alkaline earth over alkali metal cations of the same size. Simultaneously the radius of the selected cation increases. As a consequence of these trends it has not yet been possible to find neutral ion carriers with specificity for Mg^{2+} (30, 31).

2.3 Polydentate Ligands

It is well known that complexes of polydentate ligands possess enhanced stability over their unidentate counterparts (chelate effect) and that complexes of macrocyclic ligands are, as a rule, more stable than those of noncyclic polydentate ligands (macrocyclic effect) (32-38). According to Adamson (35), the

Table 4 Complex Formation Constants for Macroheterobicyclic Ligands in Aqueous Solution (from Ref. 28)

	Stability Constant ($\log K$) in H_2O							
Ligand	Li^+	Na^+	K^+	Rb^+	Cs^+	Mg^{2+}	Ca^{2+}	Sr^{2+}	Ba^{2+}
8	5.5	3.2	<2.0	<2.0	<2.0	2.5	2.5	<2.0	<2.0
9	2.5	5.4	3.95	2.55	<2.0	<2.0	6.95	7.35	6.3
10[a]	<2.0	3.9	5.4	4.35	<2.0	<2.0	4.4	8.0	9.5
11	<2.0	1.65	2.2	2.05	2.0	<2.0	~2.0	3.4	6.0
12	<2.0	<2.0	<2.0	<0.7	<2.0	<2.0	~2.0	~2.0	3.65
13	<2.0	<2.0	<2.0	<0.5	<2.0	<2.0	<2.0	<2.0	

[a]The crystal structures of the complexes with Na^+, K^+, and Cs^+ determined by Metz, Moras, and Weiss (29) show an eight coordination (two nitrogen and six oxygen atoms). In model calculations (30, 31) all eight coordinating atoms are assumed to be equal, at least in respect to the parameters used.

Table 5 Stability Constants [kg·mole^{-1}] of Ligands 15, 17, and 18 with Selected Alkali and Alkaline Earth Metal Cations as well as Ammonium Ion in Ethyl Alcohol (30°C) (from Refs. 39, 40)

Cation	Ligand 15		Ligand 17		Ligand 18	
	K_1	K_2	K_1	K_2	K_1	K_2
Li$^+$	6.0 (±1.0)·10	—	6.1(±0.9)·10	—	<1.5·10	—
Na$^+$	1.3(±0.2)·10^2	—	3.0(±0.5)·10	—	2.3(±0.20)·10^2	2.0(±0.2)·10
K$^+$	7.0(±1.0)·10	—	<10	—	1.6(±0.13)·10^2	4.0(±0.4)·10
Rb$^+$	8.0(±1.3)·10	—	<10	—	1.25(±0.10)·10^2	3.0(±0.3)·10
NH$_4^+$	<1.5·10	—	<10	—	<1.5·10	—
Mg^{2+}	9.9(±1.4)·10^2	—			2.40(±0.34)·10^2	—
Ca^{2+}	1.79(±0.44)·10^3	6.5(±2.1)·10	6.1(±1.3)·10^2	—	8.90(±3.10)·10^2	—
Sr^{2+}	4.15(±0.6)·10^3	2.60(±0.39)·10^3			$K_1 \cdot K_2 > 10^8$	
Ba^{2+}	2.7(±0.3)·10^3	6.90(±0.85)·10^2	1.4(±0.1)·10^3	2.3(±0.6)·10	1.10(±0.29)·10^4	2.3(±0.7)·10^3

chelate effect is largely a consequence of the asymmetry of the standard reference state. This has the following mathematical consequence:

$$\log K \ (n\text{-dentate}) = \log \beta_n \ (\text{unidentate}) + (n-1) \log 55.5 \qquad (1)$$

where K is the stability constant of the 1:1 cation-polydentate ligand complex in water, and β_n is the cumulative stability constant of the $1{:}n$ complex of the unidentate analogues (see also Refs. 35 and 38).

Stabilization effects beyond those discussed may be caused by a reduction in translational and/or rotational entropy of the free polydentate ligand.

In macrocyclic, and especially in macropolycyclic ligands, repulsions such as those between binding sites are already built in (32) and, in addition, optimal solvation of the free ligand sites may be prevented, further favoring formation of the complexes.

The existence of favored conformations of a multidentate ligand of a given constitution may allow the coordination shell (i.e., coordination number and cavity radius) to be predetermined, thus allowing selectivity even when non-macrocyclic molecules are used. Indeed, the noncyclic neutral carriers **14-20** and **22** (Figure 1) are capable of inducing extremely high selectivity for cations in certain membranes (see Section 4). Some stability constants determined in homogeneous systems for the nonmacrocyclic ligands **15, 17**, and **18** are given in Table 5.

3 ION SELECTIVITY OF MEMBRANES CONTAINING LIPOPHILIC NEUTRAL CARRIERS

The selectivity of complex formation of a given neutral carrier molecule in water or waterlike solvents is not necessarily identical to the ion selectivity of the corresponding carrier membrane, as in transport experiments. The problems involved have been discussed in detail elsewhere (31, 39, 41-44). For membranes incorporating electrically neutral carriers, the basic equilibriums at the membrane-solution interfaces are the following:

$$M^{zm} \ (\text{free, aqueous}) \xrightleftharpoons{K_m} M^{zm} \ (\text{complexed, membrane}) \qquad (2)$$

In addition to the pure complex formation reactions (4) discussed in Section 2, the distribution equilibriums (5) and (3) for complexes and free ligands become important:

$$S \ (\text{membrane}) \xrightleftharpoons{1/k_S} S \ (\text{aqueous}) \qquad (3)$$

$$M^{zm} \text{ (aqueous)} + nS \text{ (aqueous)} \underset{}{\overset{\beta^w_{ms,n}}{\rightleftarrows}} MS_n^{zm} \text{ (aqueous)} \tag{4}$$

$$MS_n^{zm} \text{ (aqueous)} \underset{}{\overset{k_{ms,n}}{\rightleftarrows}} MS_n^{zm} \text{ (membrane)} \tag{5}$$

Hence the overall partition coefficients K_m defined in Eq. 2 assume the form (41, 42 45-48)

$$K_m = \sum_n \beta^w_{ms,n} \, k_{ms,n} \left(\frac{c_s}{k_s}\right)^n \tag{6}$$

This relation follows from equilibrium (2) by the following steps:

1. Transfer of free carriers S (concentration c_s) from the membrane into the boundary layer of the outside solution (Eq. 3).
2. Formation of cationic complexes of the type MS_n^{zm} in the aqueous phase (Eq. 4).
3. Transfer of complexes into the membrane (Eq. 5).

The selectivity of a neutral carrier membrane toward cations of the same charge, I^{z+} and J^{z+}, is fundamentally determined by the ion-exchange reaction:

$$I^{z+} \text{ (complexed, membrane)} + J^{z+} \text{ (free, aqueous)} \underset{}{\overset{K_{ij}}{\rightleftarrows}}$$
$$J^{z+} \text{ (complexed, membrane)} + I^{z+} \text{ (free, aqueous)} \tag{7}$$

Indeed, a comparison of (7) with equilibriums of type (2) leads to

$$K_{ij} = \frac{K_j}{K_i} = \frac{a_i c_j}{a_j c_i} \tag{8}$$

where a_i and a_j denote the activities of free ions in the aqueous solution, and c_i and c_j are the total concentrations of ionic forms in the membrane phase. The ion-exchange equilibrium constant K_{ij} represents a measure for the preference of the membrane for the ion J^{z+} relative to the ion I^{z+} and is therefore called *selectivity factor* or *selectivity coefficient*. The selectivity coefficient K_{ij} introduced here may be obtained as follows (39, 43, 46-49):

$$K_{ij} = \frac{\sum_n \beta^w_{js,n} \, k_{js,n} \left(\dfrac{c_s}{k_s}\right)^n}{\sum_n \beta^w_{is,n} \, k_{is,n} \left(\dfrac{c_s}{k_s}\right)^n} \qquad (9)$$

It becomes evident that the cation selectivity of carrier-based membranes may depend on various factors: (a) the selectivity behavior of the carrier ligands used, which can be fully specified by the values $\beta^w_{ms,n}$, (b) the concentration of free ligands in the membrane (depending on the membrane composition), and (c) the extraction properties of the membrane solvent which are decisive for the magnitude of the ratios $k_{ms,n}/k_s^n$. The influence of the membrane solvent on K_{ij} may be estimated by using an electrostatic model (30, 31, 41, 47). According to such considerations the free energy of transfer of cationic complexes from water into the membrane (as compared to the uncomplexed ligands) is given by

$$-RT \ln \frac{k_{ms,n}}{k_s^n} = \Delta G_B \text{ (membrane)} - \Delta G_B \text{ (aqueous)} + \text{const} \qquad (10)$$

where $\Delta(\Delta G_B)$ represents the electrostatic contribution to the ionic partition coefficient $k_{ms,n}$ (Born term introduced in Refs. 18 and 30). This term is given by

$$\Delta G_B \text{ (membrane)} = N \frac{(z_m e)^2}{2 r_c} \left(\frac{1}{\epsilon} - 1\right) \qquad (11)$$

$$\Delta G_B \text{ (aqueous)} = N \frac{(z_m e)^2}{2 r_c} \left(\frac{1}{78.5} - 1\right) \qquad (12)$$

hence

$$\Delta(\Delta G_B) = N \frac{(z_m e)^2}{2 r_c} \left(\frac{1}{\epsilon} - \frac{1}{78.5}\right) \qquad (13)$$

where ϵ is the dielectric constant of the membrane solvent, $z_m e$ is the charge (in electrostatic units) and $2r_c$ is the overall diameter of the cationic forms, and N is Avogadro's number.

A very simple relation can be derived from Eq. 9 for neutral carriers that form predominantly 1:1 complexes with cations, as is the case for most of the natural ionophores known to date. Here the selectivity becomes independent of the ligand concentration. Since the dimensions and therefore the distribution coefficient of carrier complexes of given charge and stoichiometry are roughly independent of the nature of the central ion (see Eq. 13), the selectivity behavior can be approximately described by (31, 42-47, 50-53)

$$K_{ij} \approx \frac{\beta_{js}^w}{\beta_{is}^w} \qquad (14)$$

where the stability constants refer to the respective 1:1 complexes. Accordingly, the selectivity of corresponding neutral carrier membrane electrodes (K_{ij}^{Pot} values, see Section 5) among ions of the same charge is scarcely influenced by the ion-selective behavior of the membrane solvent used, but is mainly given by the complexation properties of the incorporated ligands. This is demonstrated in Figure 3 for a series of liquid-membrane electrodes based on the carrier antibiotic valinomycin. The correlations found attest a good agreement between the theoretical selectivity factors given in the simplified form (14) and the experimental values (see also Refs. 52-55). No such simple correlations are obtained, however, if cation-carrier complexes of different stoichiometries are involved (39, 43).

The selectivity of a neutral carrier membrane between divalent cations I^{2+} and monovalent cations J^+ is governed by the following equilibrium:

$$I^{2+} \text{(complexed, membrane)} + 2J^+ \text{(free, aqueous)} \rightleftharpoons$$
$$2J^+ \text{(complexed, membrane)} + I^{2+} \text{(free, aqueous)} \qquad (15)$$

By dissecting equilibrium (15) into fundamental reactions of type (2), we obtain the corresponding law of mass action:

$$\frac{K_j^2}{K_i} = \frac{a_i \, c_j^2}{a_j^2 \, c_i} \qquad (16)$$

where a_i and a_j are the activities of the free ions I^{2+} and J^+ in the aqueous solution, and c_i and c_j are the total concentrations of ionic forms of species I^{2+} and J^+ in the membrane phase.

In contrast to the case with I^{z+} and J^{z+}, the ion-exchange equilibrium (15) is heavily influenced by the concentration c of the anionic sites available in the membrane:

$$c = 2c_i + c_j \qquad (17)$$

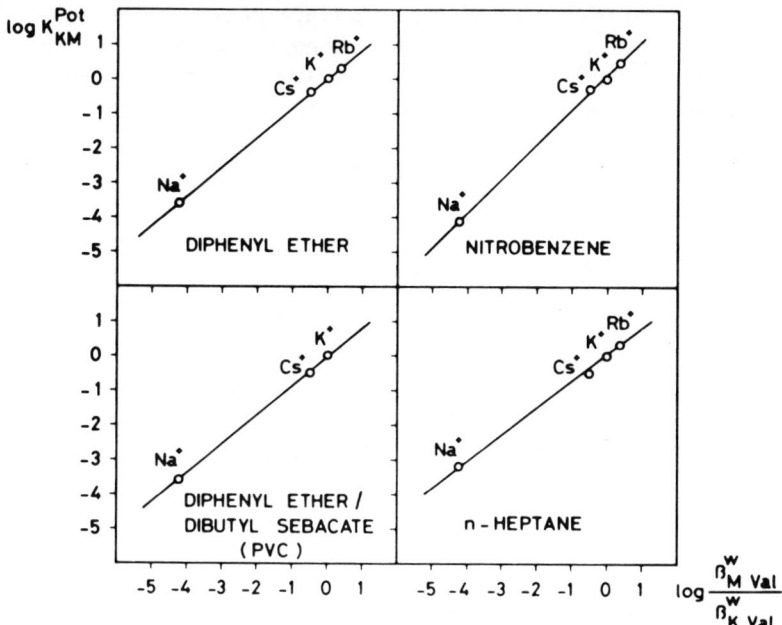

Figure 3 Correlation between theoretical and experimental selectivity factors for liquid-membrane electrodes based on the carrier valinomycin (structure **1** in Figure 1) in different membrane solvents (values taken from Refs. 31, and 56-60).

By defining the fractions of c that are occupied by each sort of cations as follows:

$$x_i = \frac{2c_i}{c} \tag{18a}$$

$$x_j = \frac{c_j}{c} \tag{18b}$$

a "standardized" equilibrium constant may be obtained for reaction (15) (41):

$$K_{ij} = \frac{K_j^2}{2c \cdot K_i} = \frac{a_i x_j^2}{a_j^2 x_i} \tag{19}$$

or

$$K_{ij} = \frac{\left(\sum_n \beta_{js,n}^w k_{js,n} (c_s/k_s)^n\right)^2}{2c \sum_n \beta_{is,n}^w k_{is,n} (c_s/k_s)^n} \tag{20}$$

As shown in Sections 5 and 6, K_{ij} represents the monovalent/divalent selectivity factor, which is accessible through potentiometric (K_{ij}^{Pot}) or transport (K_{ij}^{Tr}) studies. A discussion of this selectivity parameter is rather involved, since Eq. 20 obviously cannot be reduced to a simple form comparable to Eq 14. However, we may get the following general relationships when assuming one and the same stoichiometry, $1:n$, for all complexes in the membrane (see Eqs. 9, 10, 13, 14, and 20):

$$\frac{\partial \log K_{ij}}{\partial \log c_s} \approx (z_i - z_j)n \qquad (21)$$

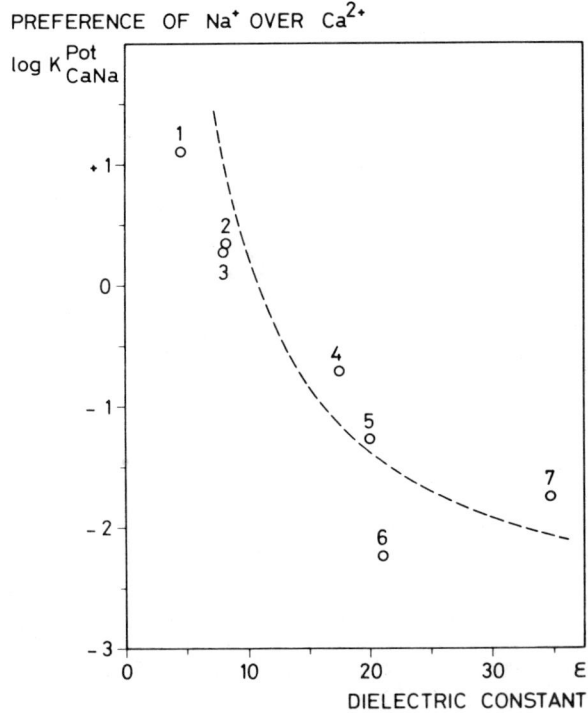

Figure 4 Dependence of the monovalent/divalent ion selectivity on the polarity (as described by the dielectric constant) of the membrane solvent used. The membrane solvents are: (1) dibutylsebacate; (2) tris(2-ethylhexyl)phosphate; (3) 1-decanol; (4) acetophenone; (5) 2-nitro-p-cymene; (6) p-nitroethylbenzene; (7) nitrobenzene. The selectivity factors were obtained from the EMF values measured in 0.1 M chloride solutions (31). The curve was calculated from Eq. 22 with $2r_c = 15$ Å.

3 Ion Selectivity of Membranes Containing Lipophilic Neutral Carriers

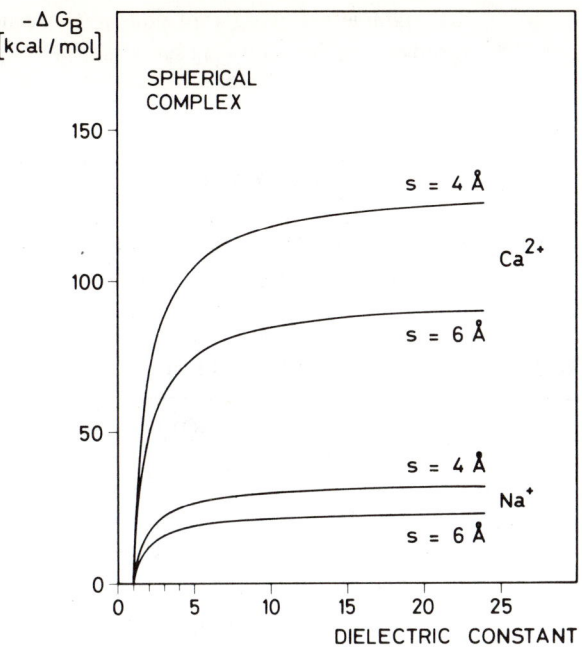

Figure 5 Free energy of the electrostatic interactions between cationic complex and membrane solvent. The values of ΔG_B were estimated for two metal ions of nearly the same size but of different charge, for two values of the ligand shell thickness s, and for a varying dielectric constant of the membrane medium (see Eq. 11, with $r_c = r_{ion} + s$).

$$\frac{\partial \log K_{ij}}{\partial (1/\epsilon)} \approx (z_i - z_j) z_i z_j \frac{243.3}{2 r_c (\text{Å})} \qquad (22)$$

The first expression predicts a certain variation of the monovalent/divalent cation selectivity with varying ligand concentration of the membrane. The second shows that the preference of a membrane for monovalent over divalent cations is efficiently improved when the polarity of the membrane solvent is reduced, and vice-versa. Such a trend was indeed observed for neutral carrier membranes in potentiometric measurements (Figure 4 and Refs. 41 and 61) and in electrodialysis experiments (41). The curve fitting the experimental points in Figure 4 was calculated according to Eq. 22. Equation 22 is a consequence of the Born term (11). This relation reveals another molecular parameter that is important in the discrimination between cations having the same radius but different charge. As Figure 5 indicates, small values of the average ligand-shell thickness s lead to a preferential uptake of, for example, Ca^{2+} relative to Na^+

into the membrane (other parameters being kept constant). Some implications of the rules derived are discussed in the following section.

4 DESIGN FEATURES OF MEMBRANE-ACTIVE COMPLEXING AGENTS

The effects of the more important design parameters may nicely be demonstrated by discussing the development of ionophore **15**; this carrier shows

Table 6 Lipophilicity Increments π_x Evolved from Hansch and Co-workers, (69) Research for Several Structural Fragments X

Structural Fragment X	π_x	Structural Fragment X	π_x
$-CH_3, -CH_2-$ (chain)	0.50	$-CH_2-SH$	0.27
$-CH_2-$ (cyclic)	0.41	$-CH_2-CS-NH_2$	-0.05
$-CH_2-$ (benzylic)	0.56	$-CH_2-SO-CH_2-$	-2.03
Chain branch	-0.20	$-CH_2-S-CH_2-$	0.95
$-CH_2-O-CH_2-$	0.03	$-CH_2-F$	0.33
$-CH_2-CO-O-CH_2-$	0.18	$-CH_2-Cl$	1.04
$-CH_2-CO-O-Ph$	1.49	$-CH_2-Br$	1.24
$-CH_2-CO-NH_2$	-1.46	$-CH_2-J$	1.69
$-CH_2-CO-NH-CH_2-$	-1.05	C_6H_6	2.13
$-CH_2-CO-N(CH_2-)(CH_2-)$	-0.77	$Ph-OCH_3$	2.11
		$Ph-CO-O-CH_2-$	2.12
		$Ph-CO-NH_2$	0.64
$-CH_2-CO-NH-Ph$	1.16	$Ph-CO-CH_2-$	1.58
$-CH_2-CO-N(Ph)(Ph)$	3.78	$Ph-COOH$	1.85
		$Ph-C\equiv N$	1.56
		$Ph-OH$	1.46
$-CH_2-CO-CH_2-$	-0.47	$Ph-NH_2$	0.90
$-CH_2-CO-Ph$	1.58	$Ph-N(CH_2-)(CH_2-)$	2.31
$-CH_2-COOH$	-0.17		
$-CH_2-C\equiv N$	-0.34		
$-CH_2-OH$	-0.66	$Ph-S-CH_2-$	2.74
$-CH_2-NH_2$	-0.57	$Ph-F$	2.27
$-CH_2-NH-CH_2-$	-0.15	$Ph-Cl$	2.84
$-CH_2-N(CH_2-)(CH_2-)$	0.27	$Ph-Br$	2.99
		$Ph-J$	3.25
		$Ph-NO_2$	1.85

4 Design Features of Membrane-Active Complexing Agents

extremely high selectivity for Ca^{2+} ions in membrane systems (11, 62, 63). Such 3,6-dioxaoctanedioic diamides form 1:2 Ca^{2+}-carrier complexes (39, 64, 65). As expected (see Section 2.2) calcium is coordinated by eight oxygen atoms (two ether groups and two amide groups per ligand) (64-66). The high polarity of the amide carbonyl coordinating sites ensures sufficient strength of interaction of the ligands with the cation. Simultaneously, highly polar binding sites lead to the preference of divalent over monovalent cations of the same size; small rather than large cations of the same charge are preferred by the ligand (Section 2.1). The arrangement of the four coordinating atoms in these ligands (14-16, 18, 19, 22, 23, 27-41) allows the formation of five-membered chelate rings in the complex (34). Although the ester carbonyl groups of the side chains (14, 15, 32-37) may act as coordinating sites under certain conditions (64, 66), a ^{13}C NMR spectroscopic investigation of the solvent polymeric membranes clearly showed that these groups do not participate in the coordination of the ions in the membrane phase (67, 68).

Table 7 Lipophilicities (log P) Calculated from the Increments Given in Table 6 for Various Ligands

Ligand (Figure 1)	Lipophilicity (log P)
1	−0.4
	2.0[a]
7	4.6
10	0.7
15	7.5
16	8.3
17	4.6
19	6.7
20	6.4
31	31.5
40	5.8
43	3.1

[a]Calculated from the increments given by Lehn (32). The difference in the log P values for ligand 1 demonstrates that especially for such rather complex ligand structures the absolute values are probably quite inaccurate. For ligands of similar structure, however, the Hansch parameters give a reasonable estimate of the relative ligand lipophilicities.

To ensure a long life of liquid membrane electrodes, the distribution of complex and ligand between aqueous solution and organic membrane phase should be in favor of the membrane. A rough estimate of the distribution of the ligand may be obtained by using lipophilicity increments π_x for various structural fragments X as described by Hansch and co-workers (69) (see Tables 6 and 7). The partition coefficient P of a compound in the system 1-octanol-water is given by

$$\log P = \Sigma \, \pi'_x \qquad (23)$$

It was shown that there is indeed a correlation between $\log P$ and the observed

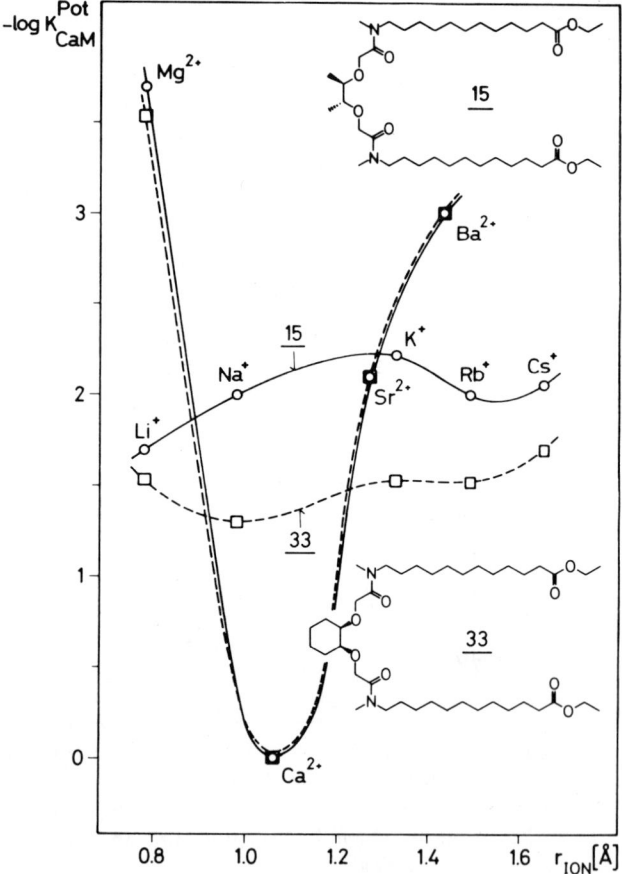

Figure 6 Influence of the average thickness of the ligand layer around the metal cation on the ion selectivity of liquid membrane electrodes.

4 Design Features of Membrane-Active Complexing Agents 25

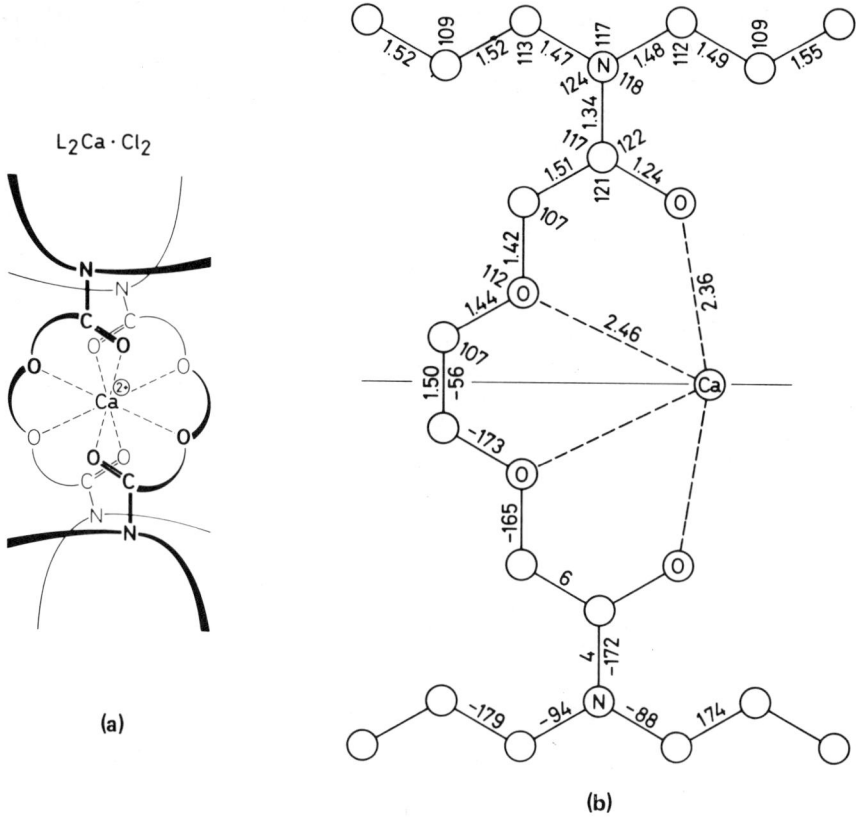

Figure 7 Molecular geometry for the Ca^{2+} complex of ligand **22** (crystallographic 222 symmetry). (*a*) Schematic representation of the complex; (*b*) bond lengths and angles (upper half), and torsion angles (lower half) in the complex (only one ligand is shown).

lifetime of a PVC liquid membrane electrode (70). For the investigated types of electrically neutral ligands, high lifetime of the membrane electrodes was achieved by using carrier molecules having log P values above about 4. Some compounds with extremely high lipophilicity, obtained by simply increasing the length of the hydrocarbon side chains in 3,6-dioxaoctanedioic diamide-type ligands [log $P > \sim 15$, cf. **30** (log P = 17.5), **31** (log P = 31.5)] did not induce cation selectivity in membrane electrodes (71). As these ligands and highly selective ligands (log P = 4 to 8) show similar complexing and ion-extracting properties, the lack of ion selectivity in membranes is probably caused by kinetic effects.

As the complexes of the ligands discussed are electrically charged, effects arising from the thickness s of the ligand shell around the cation (Eqs. 11-13 and Figure 5) become important. An increase of this parameter in the highly Ca^{2+}-selective ligand **15** causes a clear loss in the Ca^{2+} selectivity in respect to the monovalent ions (see Figure 6). As expected the selectivities among ions of the same charge are only slightly influenced (Figure 6). The nearly perfect fit of the nonsolvated Ca^{2+} ion in the cavity defined by two 3,6-dioxaoctanedioic diamides is confirmed by an X-ray analysis of a typical representative of this class of complexes (see Figure 7). The calcium ion-ligand atomic distances correspond nearly to the sum of the ionic radius of calcium (1.06 Å) and the van der Waals radius of the oxygen atoms (1.40 Å). The difference between the bulkiness of this calcium complex and that of the potassium complex of the carrier antibiotic valinomycin is clearly demonstrated by CPK models (see Figure 8). The X-ray analysis for the 1:3 (metal-ligand) complex of ligand **28** with $Ca(SCN)_2$ (72) has only recently been accomplished (73). All nine oxygen atoms of the three ligands take part in the coordination of the metal ion. The distances of the calcium ion from the ligand atoms are in the range of 2.40 to 2.55 Å. The six amide carbonyl groups lead to especially strong interactions with divalent and small ions (see Section 2.1). This may well be one of the reasons that, despite the bulkiness of this complex (see Figure 8), the calcium selectivity of this ligand is comparable to that of calcium carrier **15**.

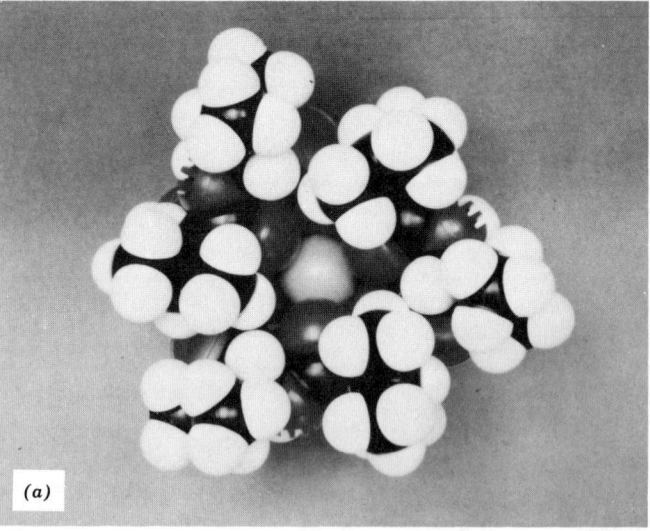

Figure 8 CPK models for (*a*) K^+ complex of valinomycin (**1**); (*b*) Ca^{2+} complex of ligand **22**; (*c*) Ca^{2+} complex of ligand **28**.

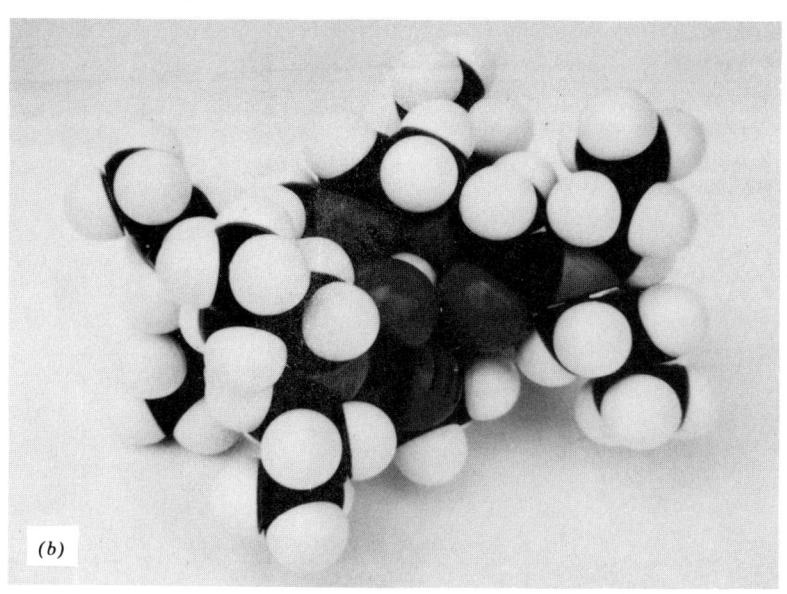

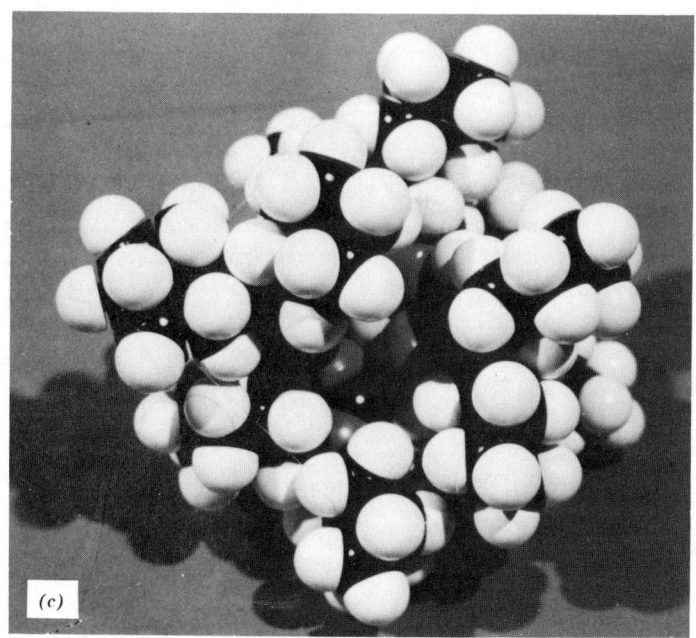

Figure 8 Continued

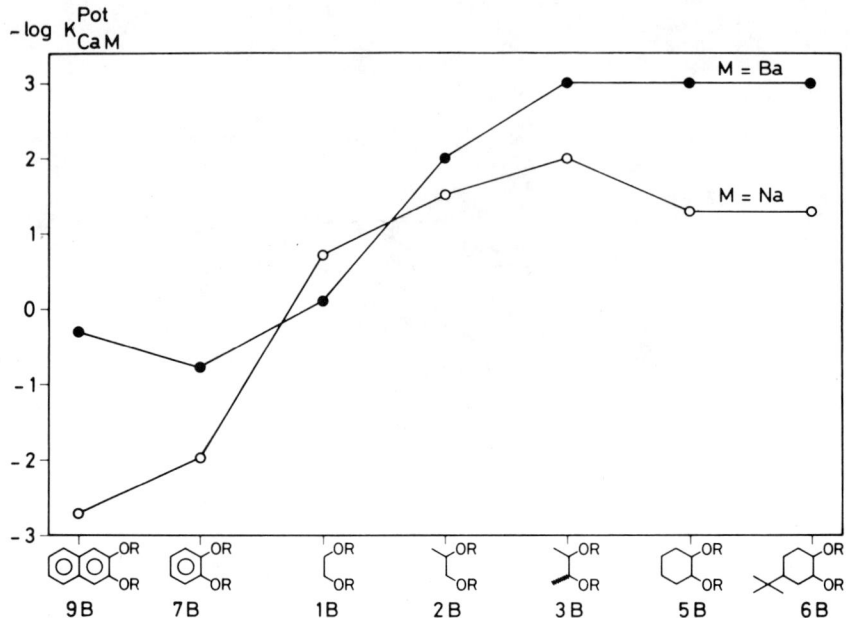

Figure 9 Influence of the constitution of ligands on the selectivity of the corresponding liquid membrane electrodes.

The design of a series of further ligands selective for Li^+, Na^+, and Ba^{2+} was achieved by a stepwise optimization of the ligand structure. Correlations of the type given in Figure 9 can be applied to this end. Unfortunately, every change of constitution of the ligand affects several molecular parameters simultaneously, such as dipole moment and polarizability of the ligand groups, conformation of the molecule, and ligand thickness. In Figure 9 selectivity factors for a number of compounds are plotted in order of decreasing polarity (decreasing with the expected polar substituent constant) (74, 75) of the substituents attached to the carbon atoms carrying the two ether oxygen atoms. For the molecules studied this order involves a change from a syn-periplanar to a synclinal arrangement of the ether oxygen atoms. According to model calculations (18, 31, 47) an increase of the dipole moment of the ligand groups* is expected to result in a selectivity enhancement for divalent relative to monovalent cations when the two are of the same radius (e.g., Ca^{2+}, Na^+) and for small relative to large cations of the same charge (e.g., Ca^{2+}, Ba^{2+}); these trends are obvious from Figure 9.

*This holds only for a given complex stoichiometry. It might be expected that the basicities of the ether oxygen atoms of the compounds shown in Figure 9 would increase from left to right (77).

4 Design Features of Membrane-Active Complexing Agents

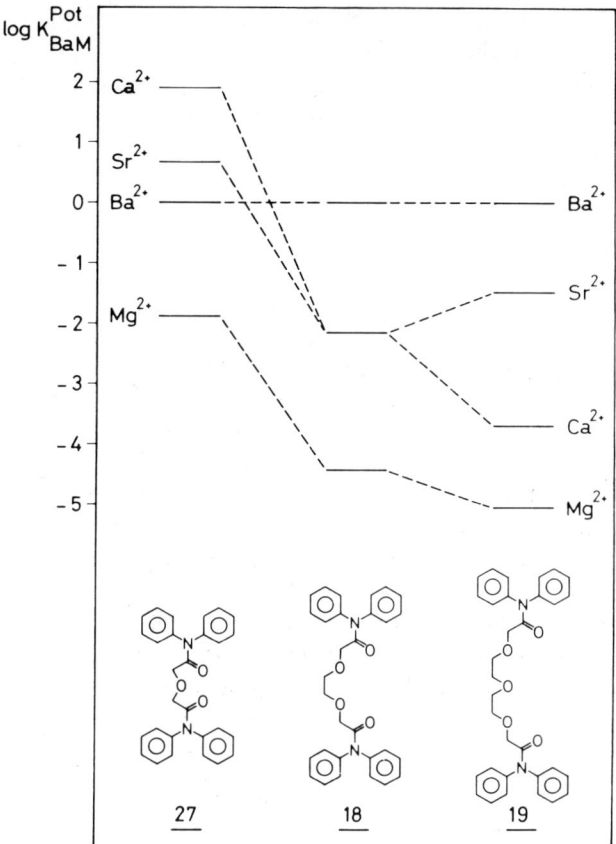

Figure 10 Influence of the number of ehtyleneoxide units between the diphenylamide groups on the ion selectivity of the corresponding liquid membrane electrodes.

Taking the specific molecular structures into account, this conclusion can be amplified as follows: Ligands with a high preference for Ca^{2+} over Na^+ and Ba^{2+} should conform either to one or to both of two specifications, that is, small polar substituent constant or synclinal arrangement of the ether oxygen atoms (15). Conversely, a high polar substituent constant along with a syn-periplanar arrangement (35) results in preference for Na^+ and Ba^{2+} over Ca^{2+} (76).

The deviations from this trend of the experimental data near both ends of the scale (36, 35, 33, 34) may be caused by the bulkiness of those ligands, as measured by the average thickness of the ligand layer around the complexed metal cation.

Figure 9 demonstrates clearly that even slight changes in the constitution of the ligand can shift the ion selectivities by orders of magnitude. Carrier **35**, which has a bulky phenylene substituent within the coordinating sites, and at the same time low dipole moments centered on the ether oxygen atoms, shows good selectivity for sodium over calcium. On the other hand a systematic study of the influence of the amide substituents on the ion selectivities of 3,6-dioxaoctanedioic diamides showed that *N*-benzyl-*N*-phenyl amides are attractive when sodium selectivity is sought. The combination of these two results led to one of the most successful neutral sodium carriers described so far, **16** (8).

N,*N*,*N'*,*N'*-Tetraphenyl-3,6-dioxaoctanedioic diamide (**18**) induces barium selectivity in ion selective membrane electrodes (see Figure 10). An improvement (ligand **19**) of this barium selectivity was achieved by the investigation of the influence of the number of ethylene oxide units between the diphenylamide groups (see Figure 10) (78).

Lithium-selective ligands were obtained by increasing the chain length between the ether groups of **14** by one carbon atom. The most promising representative of these 3,7-dioxanonanedioic diamides is **17**. This ligand forms only a 1:1 complex with lithium with ethyl alcohol as solvent (79). The reason for the

Table 8 Influence of the Ring Closure of Polydentate Ligands to Macrocyclic and Macrobicyclic Systems on the Stability of the Corresponding Complexes

System	Open-chain ligand	Macrocyclic/Macrobicyclic ligand
Cu^{2+} (H_2O, 25°C) (35)	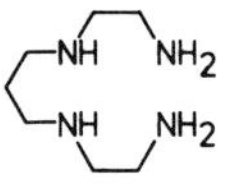 log K = 20.1	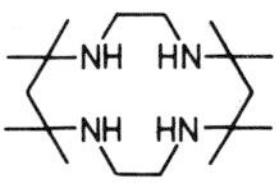 log K = 28.0, Δ = 7.9
K^+ (MeOH, 25°C) (85)	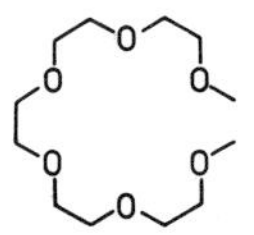 log K = 2.2	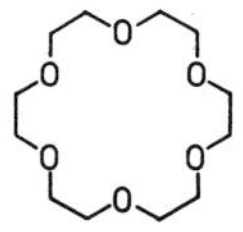 log K = 6.1, Δ = 3.9
Cu^{2+} [MeOH (80 wt. %), H_2O (20 wt. %), 25°C] (86)	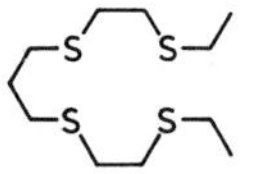 log K = 1.15	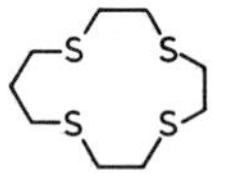 log K = 3.48, Δ = 2.33
Ba^{2+} (H_2O, 25°C) (6, 87)	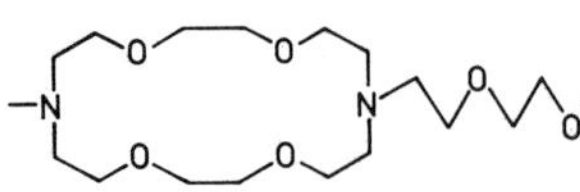 log K = 4.80	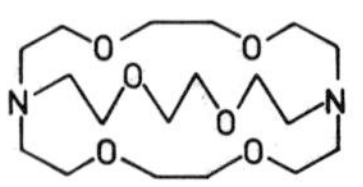 log K = 9.50, Δ = 4.95

somewhat surprising lithium ion selectivity may reside in the more favorable arrangement of the four coordinating oxygen atoms around the small lithium ion.

Carrier **20**, which is based on the skeleton of the lithium-selective ligands but contains two more potential coordinating sites, shows an especially strong discrimination of large alkali metal cations (K^+, Cs^+, Rb^+). Although this ligand is still slightly lithium selective, it is a strong candidate for clinical applications in sodium-selective membrane electrodes (e.g., blood serum measurements) (80).

It has been pointed out on several occasions that a ring closure of polydentate ligands (yielding macrocyclic ones) generally leads to a substantial increase of the stability of the complexes (see Table 8). A corresponding study with ligands of the types discussed here was carried out recently (81) (see Figure 11). The macrocyclic ligand **38** containing a 12-membered ring induces no selectivity behavior in membrane electrodes. When the ring size is increased a stepwise approach to the selectivity behavior of the corresponding noncyclic ligand is observed (see Figure 11). Ion-extraction studies confirm that the noncyclic ligand **29** appears to be a superior complexing agent for alkaline earth metal ions relative to the macrocyclic compounds **38-40** (Table 9). The arrangement of the two noncyclic ligand molecules around the Ca^{2+} ion, as determined by X-ray analysis (see Figure 7), is different from that of comparable small macrocyclic ligands (see **38**) for steric reasons. By increasing the ring size the same situation observed for noncyclic ligands is approached (see **40**).

5 ION-SELECTIVE MEMBRANE ELECTRODES BASED ON NEUTRAL CARRIERS

Most of the neutral carriers presented in the preceding sections may be used as ion-selective components in liquid membrane electrodes. A schematic diagram of a cell assembly that is suitable for the measurement of ion activities in aqueous sample solutions is shown in Figure 12. The potential difference (EMF) generated between the reference electrode and the ion-selective electrode largely results from the distribution of charged species between the sample solution and the membrane phase. Indeed the following equilibrium was found to be relevant [see also (2)]:

$$I^{z_i} \text{ (free, aqueous)} \xrightleftharpoons{K_i} I^{z_i} \text{ (complexed, membrane)} \qquad (24)$$

The partition coefficient introduced in Eq. 24 is approximately given by (41):

$$K_i = \frac{\tilde{c}_i}{\tilde{a}_i} \equiv \frac{c_i \exp(z_i F \psi_m / RT)}{a_i \exp(z_i F \psi_{aq} / RT)} \qquad (25)$$

Table 9 Extraction of Alkaline Earth Metal Picrates[a] in Methylene Chloride with the Ligands 15, 29, and 38-40

M^{2+}	Concentration of the Metal Picrate (mole l^{-1})	38 (5.7·10^{-3} mole l^{-1})	39 (1.94·10^{-3} mole l^{-1})	40 (2.05·10^{-3} mole l^{-1})	29 (2.05·10^{-3} mole l^{-1})	15 (1.5·10^{-3} mole l^{-1})
Mg^{2+}	6.3·10^{-4}	0.7%	0.4%	1.1%	2.5%	3%
Ca^{2+}	6.4·10^{-4}	0.7%	0.8%	9.6%	32.7%	38%[b]
Sr^{2+}	6.5·10^{-4}	0.8%	1.9%	7.3%	33.4%	15%
Ba^{2+}	6.6·10^{-4}	0.8%	3.2%	17.5%	48.8%	18%

[a]The percent extraction is given by 100%· (total concentration of picrate in CH$_2$Cl$_2$)/(total concentration of picrate in water).
[b]The concentration of Ca picrate used for this experiment was 6.5·10^{-4} mole l^{-1}.

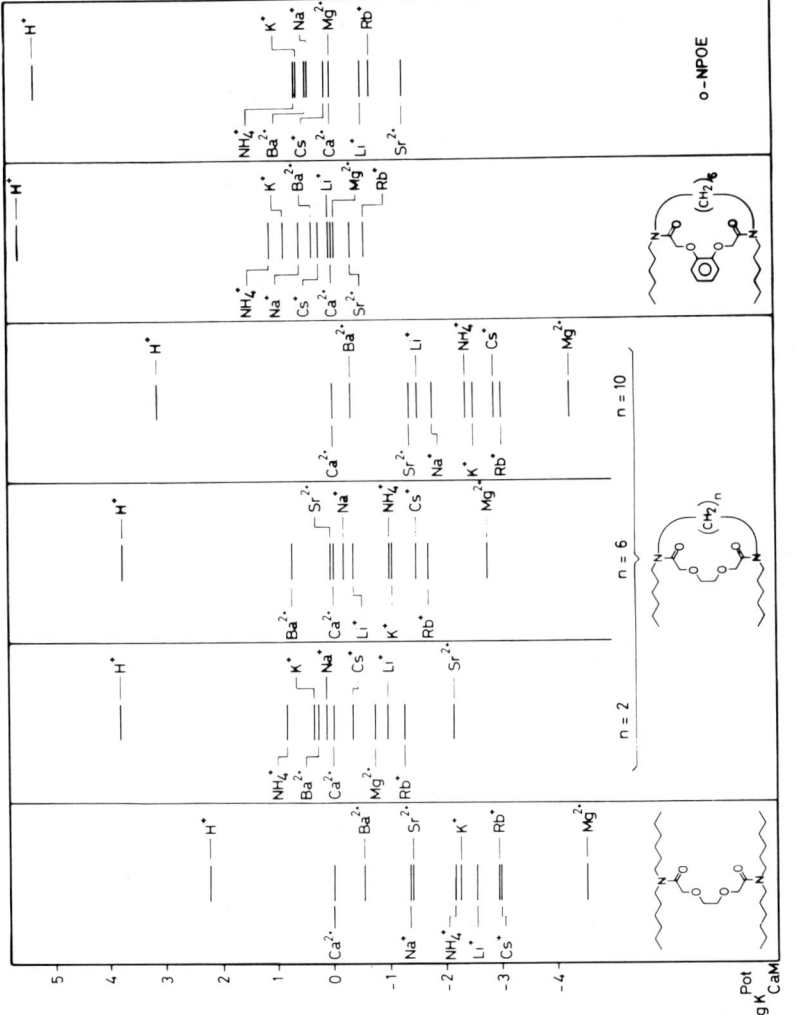

Figure 11 Ion selectivities observed for solvent polymeric membranes containing the ligands **29, 38, 39,** and **41**, and *o*-nitrophenyl octyl ether (*o*-NPOE) as plasticizer (0.1 M solutions of the chlorides, 25°C).

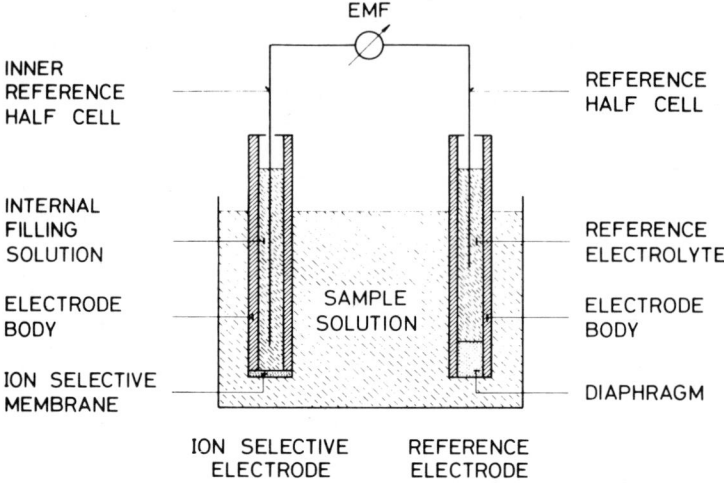

Figure 12 Schematic representation of a membrane electrode assembly.

where $\tilde{c}_i$ = electrochemical activity of complexes at the membrane surface
c_i = concentration of complexes at the membrane surface
ψ_m = electrical potential at the membrane surface
$\tilde{a}_i$ = electrochemical activity of free ions in the aqueous solution
a_i = activity of free ions in the aqueous solution
ψ_{aq} = electrical potential in the aqueous solution
R = gas constant
T = absolute temperature
F = Faraday constant

If one assumes a zero-current steady state and equal mobilities for all complexes in the membrane, the membrane potential (potential difference between the two solutions contacting the membrane) may be described directly in terms of the sample solution-membrane phase boundary:

$$E = \text{const} + (\psi_m - \psi_{aq}) \qquad (26)$$

With Eqs. 25 and 26 one can write

$$E = E_i^\circ + \frac{RT}{z_i F} \ln \frac{a_i}{x_i} \qquad (27)$$

$$x_i = \frac{|z_i| c_i}{\Sigma |z_i| c_i} \qquad (28)$$

where E_i° is equal to the membrane potential for a sample solution containing only one sort of cation (i.e., $x_i = 1$) having the activity $a_i = 1$. By using Eqs. 8, 19, 25, 27, and 28, the unknown membrane parameter x_i may be eliminated. For systems with two ions of the same charge z, one obtains the well-known Nicolsky equation (88, 89):

$$E = E_i^\circ + \frac{RT}{zF} \ln[a_i + K_{ij}^{\text{Pot}} a_j] \qquad (29)$$

The symbol K_{ij}^{Pot} is used instead of K_{ij} to indicate potentiometrically determined selectivity parameters.

A different relationship (53) is derived for cells with a divalent and a monovalent sample cation, I^{2+} and J^+, respectively (a similar equation is given elsewhere (90)):

$$E = E_i^\circ + \frac{RT}{F} \ln \left(\sqrt{a_i + \frac{1}{4} K_{ij} a_j^2} + \sqrt{\frac{1}{4} K_{ij} a_j^2} \right) \qquad (30)$$

This theoretical expression is slightly different from the more conventional formula, which is an empirical extension of the simple Nicolsky relation (29):

$$E = E_i^\circ + \frac{RT}{2F} \ln(a_i + K_{ij}^{\text{Pot}} a_j^2) \qquad (31)$$

According to Eqs. 29-31 a simple method for determining K_{ij}^{Pot} is the comparison of EMF values measured on cells containing one sample cation only. When this separate solution technique is used, the K_{ij}^{Pot} values determined from Eq. 31 become equivalent to K_{ij} defined by (30). EMF measurements therefore offer a simple method for assessing ion selectivities of neutral carriers in membranes. For a series of relevant membrane systems such selectivities are given in Table 10. The very high selectivities induced by some carriers in membranes suggest attractive applications of neutral carrier membrane electrodes in analytical and clinical chemistry (46, 47, 91-96). An extensive review on this subject is given elsewhere (97).

6 CARRIER-MEDIATED ION TRANSPORT THROUGH MEMBRANES

Most of the manifold remarkable effects of ionophores in biological as well as in artificial membrane systems are strictly a consequence of their ion-transport properties. This is nicely demonstrated in Figure 13, where the action of the carrier antibiotic monactin (3 in Figure 1) on the oxidative phosphorylation in

Table 10 Reported Selectivity Factors log K_{MN}^{Pot} for Liquid Membrane Electrodes Based on Neutral Complexing Agents

Ligand of the Membrane System N =	Li^+	Na^+	K^+	Rb^+	Cs^+	NH_4^+	H^+	Mg^{2+}	Ca^{2+}	Sr^{2+}	Ba^{2+}
$M = Li^+$											
Ligand **17** (79)	0	−1.3	−2.2	−2.5	−2.7	−1.4	−0.1	−3.8	−3.3	−3.4	−3.8
$M = Na^+$											
Ligand **16** (8)	−1.5	0	−0.3	−0.8	−1.1	−0.8	−0.3	−3.2	−2.9	−2.2	−2.4
Ligand **20** (80)	0.5	0	−2.4	−2.4	−2.5	−2.3	−0.7	−2.8	−0.5	−1.0	−1.4
$M = K^+$											
Ligand **1** (8)		−5.5	0	0.5	+0.4	−1.8	−5.0	−5.3	−4.6	−3.3	−3.4
Ligand **7** (82)	−2.3[a]	−2.4[b]	0	−0.1[a]	−0.6[a]	−1.2[a]	—	−4.0[a]	−3.5[a]	—	−4.0[a]
		−2.7[c]									
$M = NH_4^+$											
Ligand **2, 3** (8)	−2.7	−2.7	−0.8	−1.3	−2.4	0	−1.6	−5.0	−4.5	−4.0	−4.5
$M = Ca^{2+}$											

Ligand 15 (62)	−2.8	−5.0[c] −6.1[d]	−5.2[c]	−4.0	−4.0	−5.1	−0.1[c] −4.4	−5.1[c]	0	−2.1	−3.2
Ligand 28 (72)	−1.6[a]	−1.5[a]	−1.6[a]	−1.5[a]	−1.4[a]	−1.7[a]	−1.2[a]	−3.6[a]	0	−1.3[a]	−3.3[a]
M = Sr^{2+}											
Ligand 42[e] (83)	−2.7	−2.7	−2.1	—	2.3	−2.7	−3.3	−3.2	−2.7	0	2.5
M = Ba^{2+}											
Ligand 42[f] (84)	−3.7	−3.7	−2.1	—	—	−3.2	−3.7	−4.0	−4.0	−2.7	0
Ligand 19 (78)	−3.2	−2.4	−1.6	−2.0	−2.4	−2.3	−1.3	−5.1	−3.7	−1.5	0

Reported selectivities as obtained from the separate solution method using 10^{-1} M solutions, except for

[a] Separate solution method, 10^{-2} M solutions.
[b] Fixed interference method, no metal buffers. Background concentration of the interfering ions: 10^{-2} M.
[c] Fixed interference method, no metal buffers. Background concentration of the interfering ions: 10^{-1} M.
[d] Fixed interference method, metal-buffered solutions. Background concentration of the interfering ions: 10^{-1} M.
[e] Sr^{2+} complex of ligand 42 (83).
[f] Ba^{2+} complex of ligand 42 (84).

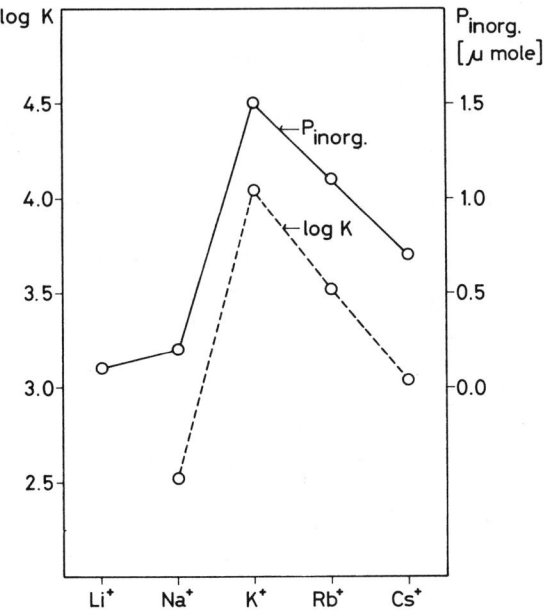

Figure 13 Comparison of the ATPase activity of monactin in presence of alkali metal cations (98) with the complex formation with different cations (31).

mitochondria is correlated with the selectivity of this molecule toward the alkali metal ions present, represented here by the stability constants of the complexes formed. In addition to the pronounced selectivity of carrier membranes versus different cations, the differentiation of such systems between cations and anions is of special significance. As a rule all the analytically relevant carrier membrane types discussed in the preceding section are capable of producing permselectivity for cations. This means that an exclusive transport of the cations of the sample solution is achieved by applying an electric field across the membrane, which fact is observable by a cation transference number of nearly 1.0 (see Table 11 and Refs. 42, 46, 48, 99, and 100). Analogously the potentiometric behavior of the same systems also shows no influence by sample anions such as chloride. Correspondingly the EMF relationships presented in Section 5 do not take into account any contribution by anions, but constitute a description of ideally cation-sensitive membranes.

Different mechanisms have been proposed to explain the origin of cation permselectivity of neutral-carrier-based membranes. The applicability and the consequences of these models depend heavily on the dimensions of the membrane. Therefore, we treat macroscopic and microscopic membranes separately.

6 Carrier-Mediated Ion Transport Through Membranes

6.1 Bulk Membranes

The origin of cation permselectivity in thick neutral-carrier liquid membranes has been unraveled only recently (99). By using radioactive-labeling techniques, it was shown that the carrier-induced extraction of cations into the membrane phase leads simultaneously to the generation of an equivalent number of anionic sites. These "immobilized" anions in the membrane are distinct from the dominating anions in the aqueous solutions (usually chloride). The latter species were found to be virtually excluded from the membrane phase, and therefore to induce practically no contribution to the electrical current across the membrane (see Table 11) or to the zero-current membrane potential (see Section 5 and Refs. 46-48, 99, and 100).

A theoretical approach to the ion-transport behavior of thick, electroneutral carrier membranes (41, 42) assumes constancy of the concentration c of anionic sites existing within the membrane phase. If the fraction of c occupied by the complexes of a cation I is denoted by x_i, the total flux of ionic forms of species I across the membrane may be approximated by

$$J_i = x_i \frac{Dc}{d} \phi \tag{32}$$

which is valid for an idealized cation-permselective membrane under symmetrical conditions (42). In Eq. 32 D is the mean diffusion coefficient of cations in the membrane, d is the membrane thickness, and ϕ is the dimensionless transmembrane potential, measured in units of $RT/F = 25.7$ mV (at $25°C$). This leads to the following expressions for the electrical current density j and for the cation transference numbers t_i, respectively:

$$j \equiv F \, \Sigma z_i J_i = \Sigma z_i x_i \cdot F \frac{Dc}{d} \phi \tag{33}$$

$$t_i \equiv \frac{F z_i J_i}{j} = \frac{z_i x_i}{\Sigma z_i x_i} \tag{34}$$

$$\Sigma t_i = 1 \tag{35}$$

For systems with only one sort of cation ($x_i = 1$) we immediately get $t_i = 1$, and Eq. 33 predicts ohmic behavior* of the membrane, the conductance being roughly independent of the nature of the cation (42, 48). This agrees with exper-

---

*Saturation of the current, which occurs only at voltages on the order of several volts, is caused by finite back diffusion of free carriers in the membrane (42).

Table 11 Transport Numbers of Cation Permselective Neutral Carrier Membranes (from Ref. 99)

	Membrane Composition			Electrolytes		
Cation Studied	Ligand (wt %)	Solvent[c] (wt %)	Matrix (wt %)	Anode Compartment	Cathode Compartment	Transference Number for Cations Studied[a]
Ca^{2+}	14; 3	o-NPOE; 65	PVC; 32	10^{-3} M $CaCl_2$	10^{-3} M KCl	0.99 ± 0.08
Ca^{2+}	14; 3	DBS; 65	PVC; 32	10^{-3} M $CaCl_2$	10^{-3} M KCl	1.00 ± 0.105
Ca^{2+}	14; 3	o-NPOE; 65	PVC; 32	5×10^{-4} M $CaCl_2$ 5×10^{-4} M $MgCl_2$	10^{-3} M KCl	0.99 ± 0.08
Ca^{2+}	14; 3	o-NPOE; 65	PVC; 32	5×10^{-4} M $CaCl_2$ 5×10^{-4} M NaCl	10^{-3} M KCl	0.99 ± 0.02
Ca^{2+}	14; 3	o-NPOE; 65	PVC; 32	10^{-4} M $CaCl_2$	10^{-4} M KSCN	0.995 ± 0.025
Li^+	17; 5.8	TEHP; 62.8	PVC; 31.4	10^{-2} M LiCl	10^{-2} M KCl	0.97 ± 0.11
Li^+	17; 5.8	TEHP; 62.8	PVC; 31.4	10^{-3} M LiCl	10^{-3} M KCl	1.02 ± 0.21
Li^+	17; 5.8	TEHP; 62.8	PVC; 31.4	10^{-4} M LiCl	10^{-4} M KCl	0.98 ± 0.10
K^+	1; 3	DPP; 67	PVC; 30	10^{-2} M KCl	10^{-2} M HCl	1.08 ± 0.07
K^+	1; 5	—	Silicone rubber; 95	10^{-2} M KCl	10^{-2} M HCl	1.1 ± 0.15
K^+	1; 5	—	Silicone rubber; 95	10^{-2} M KCl	10^{-2} M $HClO_4$	1.1 ± 0.15
K^+	1; 1	DOA; 66	PVC; 33	9×10^{-4} M KCl	9×10^{-4} M KCl	1.02 ± 0.04
$PEAH^{+b}$	24/25; 1	DOA; 65	PVC; 34	4×10^{-3} M PEAHCl	4×10^{-3} M PEAHCl	0.95 ± 0.04

[a] Measured on the specified membrane in the region of ohmic behavior of the current-voltage curve (at low voltage); 95% confidence limits.
[b] PEAH[+]: ^{14}C-α-phenylethylammonium cation.
[c] o-NPOE: o-nitrophenyloctyl ether; DBS: dibutyl sebacate; TEHP: tris(2-ethylhexyl)phosphate; DPP: dipentylphthalate; DOA: dioctyl adipate.

imental observations on 1.5 mm thick valinomycin-heptane liquid membranes for which a conductance ratio of $K^+/Na^+ \approx 2$ was reported (58). Such behavior is in striking contrast to the findings for lipid bilayer membranes, however, where the zero-current conductance is a direct measure of their ion selectivity (see Section 6.2).

A pronounced selectivity in the cation transport of thick carrier-based membranes can be observed by studying the transference number of different ions that are permeating simultaneously. For two cations of the same charge, I^{z+} and J^{z+}, the following ratio of transference numbers is obtained:

$$\frac{t_i}{t_j} = \frac{x_i}{x_j} \tag{36}$$

Recalling the equilibrium constant for the fundamental ion-exchange reaction (7) at the membrane-solution interfaces:

$$K_{ij} = \frac{K_j}{K_i} = \frac{a_i x_j}{a_j x_i} \tag{37}$$

one arrives at the following theoretical result (41-43, 46-48):

$$\frac{t_i}{t_j} = \frac{a_i}{K_{ij}^{Tr} a_j} \ ; \quad t_i + t_j = 1 \tag{38}$$

The symbol K_{ij}^{Tr} is introduced here to clearly indicate selectivity parameters determined by ion transport studies. In Figures 14 and 15, such K_{ij}^{Tr} values found in electrodialysis experiments on carrier membranes are correlated with the corresponding selectivity coefficients K_{ij}^{Pot} obtained potentiometrically. The same type of solvent-polymeric membrane was used for each experiment (42, 101, 102). Although widely different methods were applied to determine the ion selectivity, the agreement between the two sets of data is surprising, corroborating the membrane model outlined in this section. Accepting the fundamental commensurability of potentiometric and ion-transport selectivities, one can finally formulate the EMF response of carrier membrane electrodes to cations of the same charge z as follows (see Eqs. 29 and 38):

$$E = E_i^\circ + \frac{RT}{zF} \ln \frac{a_i}{t_i} \tag{39}$$

It becomes evident that specificity of membranes for one given ion is strictly exhibited only if the transference of any interfering ions becomes negligible. An

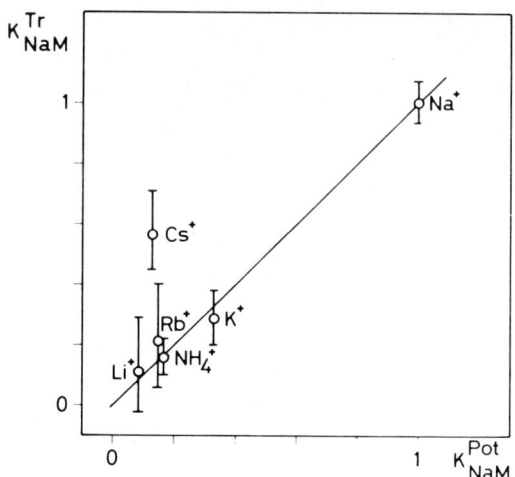

Figure 14 Transport selectivities K_{NaM}^{Tr} (transference number ratios for $a_{Na} = a_M$) and potentiometric selectivities K_{NaM}^{Pot} of PVC membranes based on a neutral Na^+ carrier (46-48). The experimental values for different ions M^+ are taken from Ref. 101.

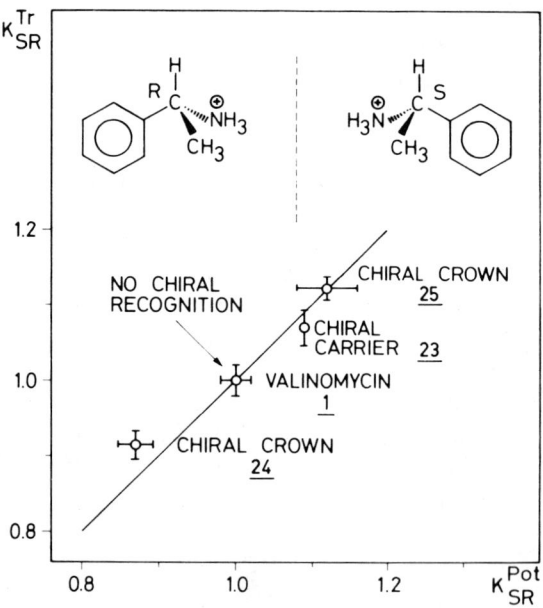

Figure 15 Transport selectivity and potentiometric selectivity between (R)- and (S)-phenylethylammonium ions for different enantiomer-selective carrier PVC membranes (48, 102).

6 Carrier-Mediated Ion Transport Through Membranes

example is given in Table 11, where the transference number for Ca^{2+} of membranes with ligand **14** remains at the value 0.99 even in the presence of Mg^{2+} or Na^+, which agrees with the potentiometric Ca^{2+} specificity of similar membranes presented in Section 5.

For cation-permselective membranes in contact with mixed solutions of two cations I^{2+} and J^+, the transference number ratio is given by

$$\frac{t_i}{t_j} = \frac{2x_i}{x_j} \tag{40}$$

$$x_i + x_j = 1 \tag{41}$$

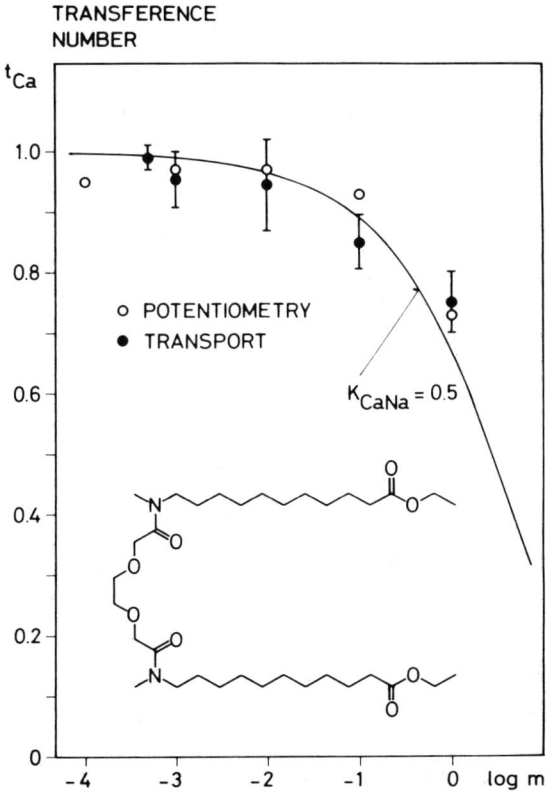

Figure 16 Transference number for Ca^{2+}, as obtained in the electrodialysis of Ca^{2+} and Na^+ (both at concentration m) through a Ca^{2+}-carrier membrane (42, 103). Full circles: experimental values; solid line; theoretical curve according to Eq. 43; open circles: values calculated from potentiometric selectivity data.

The monovalent/divalent ion selectivity was introduced in Section 3 in the following form:

$$K_{ij} = \frac{K_j^2}{2cK_i} = \frac{a_i x_j^2}{a_j^2 x_i} \tag{42}$$

Combination of Eqs. 40-42 then leads to the relation (41-43, 103)

$$\frac{t_i}{t_j} = \sqrt{\frac{a_i}{\frac{1}{4} K_{ij}^{Tr} a_j^2} + 1} - 1; \quad t_i + t_j = 1 \tag{43}$$

which can be inserted into Eq. 30 to give an expression analogous to (39):

$$E = E_i^\circ + \frac{RT}{2F} \ln \frac{a_i (2 - t_i)}{t_i} \tag{44}$$

Figure 16 shows results of the simultaneous transport of Ca^{2+} and Na^+ across PVC membranes based on the neutral ligand **14** (the first synthetic Ca^{2+} carrier, see Figure 1). The agreement between theory and experiment—electrodialysis as well as potentiometry—is excellent. It becomes evident that one given carrier membrane may exhibit specificity for divalent cations in relatively diluted aqueous solutions, and increasing selectivity for monovalent cations in highly concentrated solutions. The Ca^{2+} selectivity documented in Figure 16 turns out to be poor, however, when it is compared to the specifications of more recent carrier membrane systems discussed in the foregoing sections.

6.2 Bilayer Membranes; Biological Systems

Primitive biological membranes as well as their artificial counterparts are usually formed from two monomolecular layers of lipids, the resulting membrane thickness being on the order of only 5-10 nm (104). The concept of such bilayers differs significantly from bulk systems, since here electroneutrality need no longer be maintained in the membrane interior. Thus a preferential uptake of cations into the membrane may occur if these species are solubilized by lipophilic ligand shells, whereas hydrophilic anions are rejected. Finally, permselectivity in bilayer membranes can be explained simply in terms of the lipid solubilities of the ions involved (48, 91, 105-107).

The carrier-mediated ion-transport behavior of bilayer membranes was investigated extensively by the groups of Eisenman and Läuger (for a review, see Refs. 105-107). A quantitative description of the electrical properties of such

6 Carrier-Mediated Ion Transport Through Membranes

membrane systems has become available through the fundamental contributions by Hladky (108) and Ciani (109, 110). A generalized theory has been worked out recently (111) and is summarized below. For simplicity we introduce here the following assumptions:

1. The bilayer membrane separates two aqueous solutions of identical composition.
2. The carrier concentration is rather high, allowing a symmetrical distribution of free carriers with respect to the center of the membrane for the steady state.
3. The permeating ionic species are ion-carrier complexes with a fixed $1:n_i$ stoichiometry.

Then the general flux equation for an ion I of charge z reads (for a detailed explanation of the symbols used, see Ref. 111)

$$J_i = \frac{2k_i c_i \sinh(z\phi/2)}{F(\phi) + 2w_i \cosh(P_3 z\phi)} \tag{45}$$

$$c_i = \frac{\vec{k}_i}{\overleftarrow{k}_i} a_i c_s^{n_i} e^{-z\phi_0} \tag{46}$$

$$w_i = \frac{k_i}{\overleftarrow{k}_i} e^{z\phi_0/2} \tag{47}$$

where k_i = rate constant of ion transfer through the membrane interior (depending on the membrane thickness and on the shape of the free energy barrier for ion translocation)
$\vec{k}_i$ = rate constant of ion transfer from the aqueous solution into the membrane (rate of the complexation reaction)
$\overleftarrow{k}_i$ = rate constant of ion transfer out of the membrane (rate of the decomplexation reaction)
c_i = equilibrium concentration of ionic $1:n_i$ complexes within the membrane, at free energy minima near the interfaces
c_s = equilibrium concentration of free carrier ligands within the membrane
a_i = activity of free ions in the aqueous solutions
ϕ = reduced transmembrane potential (applied voltage, measured in units RT/F)
ϕ_0 = reduced potential difference between membrane and aqueous solutions for $\phi = 0$

$4P_3 - 1$ = fraction of the voltage drop ϕ occurring across the membrane interior ($P_3 \leq 0.5$)

The function $F(\phi)$ depends on the assumptions made in respect to the free energy profile within the membrane. For a trapezoidal shape of the free energy barrier, whose width at the top spans a fraction $2P_2$ of the membrane thickness, it follows that, according to the model by Ciani, Eisenman, and Krasne (109-111),

$$F(\phi) = \frac{\sinh(P_2 z \phi)}{P_2 z \phi} \qquad (48)$$

On the other hand, if the membrane interior is considered as a series of N sharp activation barriers, we obtain the solution (111)

$$F(\phi) = \frac{1}{N} \sum_{n=1}^{N} \exp\left[\frac{N+1-2n}{2N}(4P_3-1)z\phi\right] \qquad (49)$$

It can be shown that Eqs. 48 and 49 constitute equivalent descriptions. Thus for bulk membranes we usually find $P_2 = P_3 = 0.5$ and $N \to \infty$, respectively, hence

$$F(\phi) = \frac{\sinh(z\phi/2)}{z\phi/2} \qquad (50)$$

Since k_i assumes very low values in the case of thick membranes, it holds that $w_i \cong 0$ ["equilibrium domain" (106)], leading to an expression corresponding to Eq. 32:

$$J_i = k_i c_i z \phi \qquad (51)$$

In contrast, if the membrane interior can be treated as a single sharp free energy barrier (112), $P_2 \to 0$ and $N = 1$, respectively, and

$$F(\phi) = 1 \qquad (52)$$

The corresponding flux equation for $w_i = 0$ then reads

$$J_i = 2 k_i c_i \sinh \frac{z\phi}{2} \qquad (53)$$

Current-voltage behavior according to Eq. 53 was indeed observed for the transport of several lipophilic anions across dioleolyllecithin bilayers (see Figure 17).

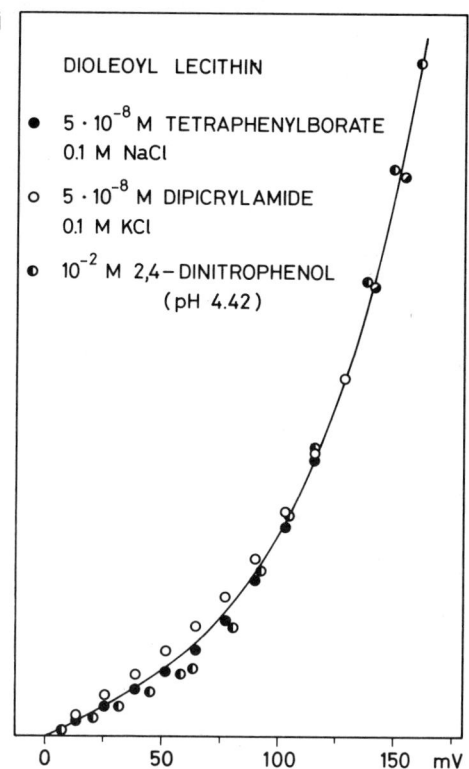

Figure 17 Current-voltage characteristics for dioleolyl lecithin bilayer membranes in the presence of different lipophilic anions (25°C). The experimental points are taken from conductance or current data reported by Läuger and co-workers (105, 113); the scales of current densities j were transformed to yield a common point at 128.5 mV ($\emptyset$ = 5). The solid line was calculated according to Eq. 53.

On the other hand Eq. 53 and its analogue for $w_i > 0$ (112) give only a poor correlation with experimental results when they are applied to carrier-mediated cation transport in bilayer membranes. This fact led Eisenman and co-workers (109, 110, 114) to the creation of a trapezoid-barrier concept for the cation translocation. An interpretation of the experimental data, as obtained for different carrier-bilayer systems, was achieved by inserting typical values of

$$P_2 \cong 0.35 \quad \text{and} \quad P_3 \cong 0.46 \tag{54}$$

into Eq. 45 and 48. A deeper understanding of these remarkable findings follows from the multibarrier model given here (see also Ref. 111). By treating the interior of the bilayer as a couple of barriers (here $N = 2$) of the Eyring type, we deduce from Eq. 49:

$$F(\phi) = \cosh\left[(4P_3 - 1)\frac{z\phi}{4}\right] \quad (55)$$

This result turns out to be practically identical to Eisenman's term (48):

$$F(\phi) = \frac{\sinh(P_2 z\phi)}{P_2 z\phi}$$

in the usual voltage range 0-150 mV (for $z = 1$) if it holds that

$$P_2 \cong 0.42\,(4P_3 - 1) \quad (56)$$

Obviously such a relation between P_2 and P_3 is in fact fulfilled for the carrier-mediated cation transport in lipid bilayers, as may be seen from the experimental values (54). Finally, the corresponding flux equation reads (111)

$$J_i = \frac{2k_i c_i \sinh(z\phi/2)}{\cosh(P_3 z\phi - z\phi/4) + 2w_i \cosh(P_3 z\phi)} \quad (57)$$

This expression may be replaced by the following approximation for $P_3 \approx 0.5$ and $w_i \approx 0$ ("equilibrium domain"):

$$J_i = 4k_i c_i \sinh\frac{z\phi}{4} \quad (58)$$

which is a result intermediate to the former cases (51) and (53).

Current-voltage behavior according to Eq. 57 is illustrated in Figure 18 for the transport of K^+ across phosphatidylserine membranes in the presence of the carrier antibiotics valinomycin **1** and monactin **3**. Figure 19 shows current-voltage behavior according to (58) for the transport of Ca^{2+} across lecithin bilayers modified by the synthetic ionophore **15**. Whereas the current-voltage curves for K^+ monactin and Ca^{2+} show the normal "hyperbolic" (110) shape predicted by Eq. 58, the characteristic for K^+ valinomycin is of the "saturation" type, which indicates that here the reaction kinetics at the membrane-solution interfaces becomes rate limiting ("kinetic domain" with $w_i > 0$). Further results are discussed in Ref. 110, where the term w_i/P_2 is used instead of the symbol w_i used here.

A readily accessible estimate of the cation transport selectivity of carrier-bilayer systems is obtained from the study of zero-current membrane con-

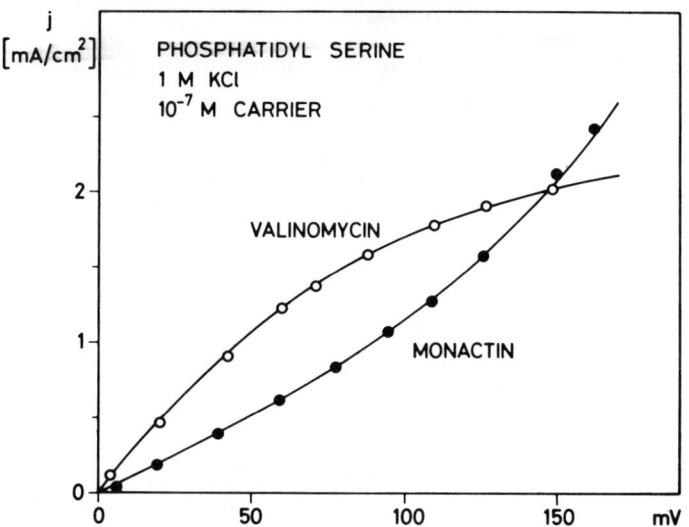

Figure 18 Current-voltage characteristics for phosphatidyl serine bilayer membranes in the presence of valinomycin (1) or monactin (3). The aqueous solution contained 1 M KCl (25°C). The experimental points are taken from Figure 10 in Ref. 115. The solid lines are theoretical curves according to Eq. 57 (valinomycin; $P_3 = 0.5$ and $w_i = 0.46$) and Eq. 58 (monactin; $P_3 = 0.5$ and $w_i = 0$).

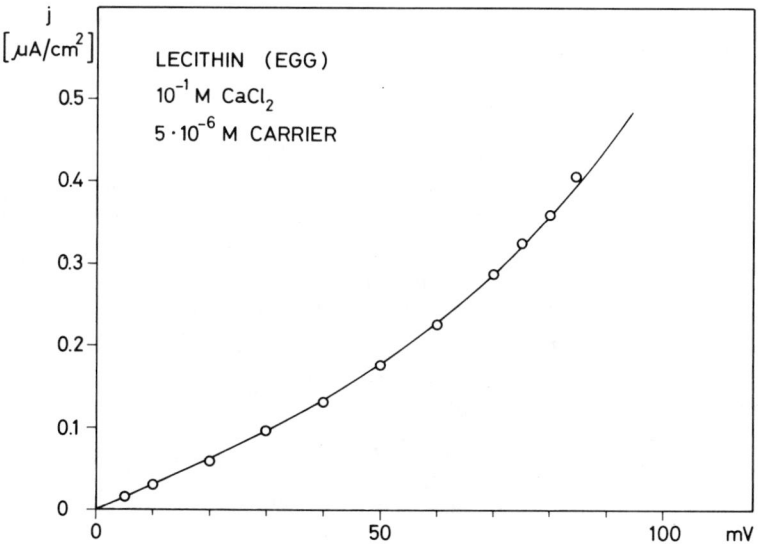

Figure 19 Current-voltage characteristics for lecithin bilayer membranes in the presence of the synthetic Ca^{2+} carrier **15**. The aqueous solution contained 0.1 M $CaCl_2$ and 5 μM ligand (25°C) (116). The theoretical curve was drawn according to Eq. 58.

ductances G^0 (106, 107). When only one kind of permeating ion is present, this parameter is obtained from Eq. 45 as follows:

$$G_i^0 = \lim_{\phi \to 0} \frac{zFJ_i}{(RT/F)\phi} = \frac{(zF)^2}{RT} \frac{k_i c_i}{1 + 2w_i} \tag{59}$$

or, after insertion of Eq. 46,

$$G_i^0 = \frac{(zF)^2}{RT} \frac{k_i(\vec{k}_i/\overleftarrow{k}_i)}{1 + 2w_i} a_i c_s^{n_i} e^{-z\phi_0} \tag{60}$$

This result is formally independent of the parameters P_3 and P_2 or N, although the latter are included implicitly in the terms k_i and w_i (111). In agreement with experimental evidence from lipid bilayers, Eq. 60 leads to the following conclusions:

The zero-current membrane conductance is

1. Proportional to the outside activity a_i of permeating ions (105-107, 112, 115).
2. Proportional to the n_ith power of the carrier concentration c_s (106, 107, 112, 115).
3. Heavily influenced by the surface potential ϕ_0 of the bilayer, which is mainly given by the nature of the lipid (107, 117).*

It should be noted that the second conclusion allows an easy determination of the stoichiometry of cation-carrier complexes in bilayer lipid membranes. Thus the formation of 1:1 complexes was confirmed for the natural ionophores **1-5** (106, 107, 112, 115). For certain crown compounds (**43** and an alkylated homolog) evidence has been given for 1:2 and even 1:3 complexes with alkali metal ions (107). The synthetic carrier **15** and a related one were found to yield predominantly 1:2 complexes with Ca^{2+} (116) and Na^+ (118), respectively, whereas **17** appears to form both 1:1 and 1:2 complexes with Li^+ (119).

Assuming identical conditions for the transport of two cations I^{z+} and J^{z+}, we can use Eq. 60 to obtain the following conductance ratio:

$$\frac{G_j^0}{G_i^0} = \frac{k_j}{k_i} \frac{\vec{k}_j/\overleftarrow{k}_j}{\vec{k}_i/\overleftarrow{k}_i} \frac{1 + 2w_i}{1 + 2w_j} \tag{61}$$

*In contrast, the boundary potentials at thick electroneutral membranes reflect to a large degree the activities and extractabilities of the ions present in the aqueous solutions, as shown in Section 5. Hence the G^0 values obtained on bulk membranes do not represent a measure of the ion selectivity of such systems (42, 48, 58).

Since the translocation rate constants k_i and k_j are nearly the same for "isosteric" complexes, the first factor in (61) approximates unity. The second term is found to be equivalent to K_j/K_i, that is, the equilibrium selectivity of the carrier membrane. For the equilibrium domain we finally get the simple relation (46-48, 106, 107, 110)

$$\frac{G_j^0}{G_i^0} = \frac{K_j}{K_i} = K_{ij}^{\text{Tr}} \qquad [w_i, w_j \ll 1] \qquad (62)$$

For rather hypothetical systems where the rate of cation translocation (k) is much higher than the effective rate of decomplexation $(\overleftarrow{k} \cdot e^{z\phi_0/2})$, the corresponding transport selectivity is (42, 106)

$$\frac{G_j^0}{G_i^0} = \frac{\vec{k}_j}{\vec{k}_i} \qquad [w_i, w_j \gg 1] \qquad (63)$$

Correlations between so-called permeability ratios, as obtained from con-

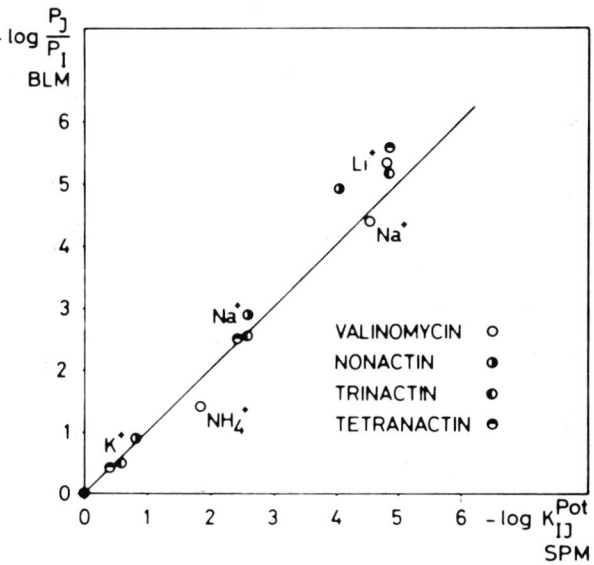

Figure 20 Comparison of the selectivities of neutral-carrier-modified solvent polymeric membranes (SPM) and bilayer membranes (BLM). The permeability ratios P_J/P_I (at "equilibrium" as far as available) fulfilled for the glyceryl dioleate BLMs are taken from Figure 10 and 11 in Ref. 110. Values for the SPMs were obtained using 0.1 M solutions of the aqueous chlorides and membranes of the composition 33.1 wt % polyvinyl chloride, 66.2 wt. % dioctyl adipate, and 0.7 wt. % carrier. For the macrotetrolides I^{z+} was NH_4^+; for valinomycin I^{z+} was K^+.

ductance data or potentiometric measurements on lipid bilayers, and theoretical selectivities calculated from extraction data were presented by Eisenman and co-workers (107, 110) (see also Figure 20). It was shown that Eq. 62 offers an adequate description for most carrier-bilayer systems. Distinct effects of interfacial kinetics were observed, however, for bilayers formed from glyceryl dioleate and related lipids. From the experimental data reported by Eisenman's group (110), we get a surprising correlation between the ratios of kinetic parameters w_j/w_i and the corresponding equilibrium selectivities P_j/P_i for alkali metal ions and different carriers (see Figure 21):

$$\frac{w_j}{w_i} \approx K_{ij}^{\mathrm{Tr}} \triangleq \frac{P_j}{P_i} \tag{64a}$$

or

$$\frac{k_j/\overleftarrow{k_j}}{k_i/\overleftarrow{k_i}} \approx \frac{k_j(\overrightarrow{k_j}/\overleftarrow{k_j})}{k_i(\overrightarrow{k_i}/\overleftarrow{k_i})} \tag{64b}$$

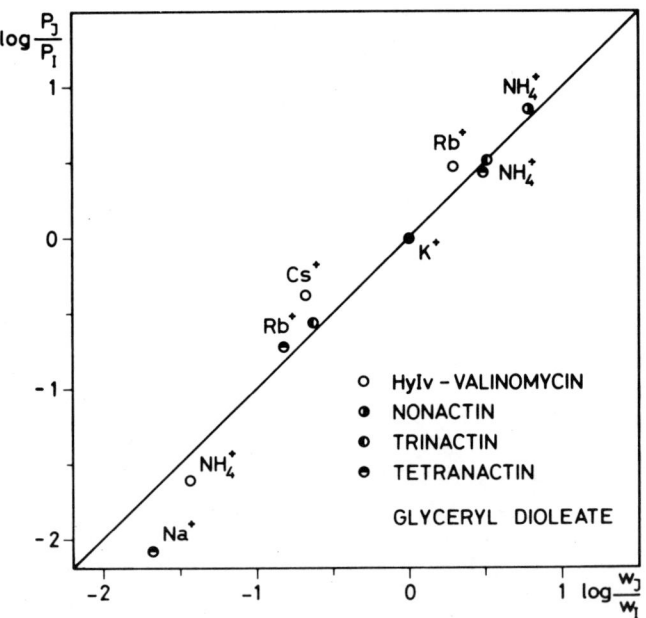

Figure 21 Correlation between the ratios of equilibrium permeabilities (P_J/P_I) and the ratios of kinetic parameters (w_J/w_I) for different ion-carrier combinations in glyceryl dioleate bilayers. The experimental values are taken from Ref. 110 ($I^{z+} = K^+$). The carrier, hydroxyisovalerate-valinomycin, is a homolog of **1** containing exclusively isopropyl substituents (see Figure 1).

This means nothing less than that the rate of the complexation reaction at the membrane-solution interface is nearly the same for all alkali ions (for a given carrier-bilayer system):

$$\vec{k}_i \approx \vec{k}_j \tag{65}$$

Thus, if substantial kinetic limitations at the interfaces come into play, a serious loss in the ion selectivity will occur, as is evident from Eqs. 63 and 65. Such a kinetic effect was predicted earlier (42), but is here corroborated by experimental evidence.

Finally, we may conclude that the complexation and decomplexation steps in the carrier-mediated ion transport through membranes should be relatively fast processes. This is strictly a prerequisite for both a high rate of and a high selectivity in ion permeation, which obtains in many biological membrane systems.

7 FUTURE PROSPECTS

In this chapter the ion selectivity of neutral macrocyclic and nonmacrocyclic complexing agents in membranes is discussed with emphasis on alkali, alkaline earth, and ammonium ions. Preliminary results indicate that carriers are feasible for transition metals, UO_2^{2+}, and even for certain anions. In addition to the analytical techniques discussed here, which are based on solvent polymeric membranes, separations by chromatographic methods using neutral complexing agents have been found to be attractive (120-123). Such separations have already been successful, especially for enantiomers (120). So far amazingly little effort has been invested in the separation of isotopes (124).

Although the present state of the art in ligand design is encouragingly successful, it is to a large extent a trial and error procedure. For the ideal tailoring of carrier molecules of a given selectivity an all-encompassing fundamental calculation of the free energy of interaction between host (ionophore) and guest (ion) molecules is necessary. Ab initio calculations make such estimations possible for modestly large molecules (20, 27, 125, 126). Models based on ab initio computations have become available only very recently (27, 127) for calculating interactions between large molecules. In a joint effort with E. Clementi and his group, interaction studies are in progress for ionophores and model compounds with different ions. An example of the interaction of ligand **28** with Na^+ is given in Figure 22. We hope that such computations will bring a more detailed understanding of the factors affecting ion selectivity and will lead to more direct, computer-aided ligand design.

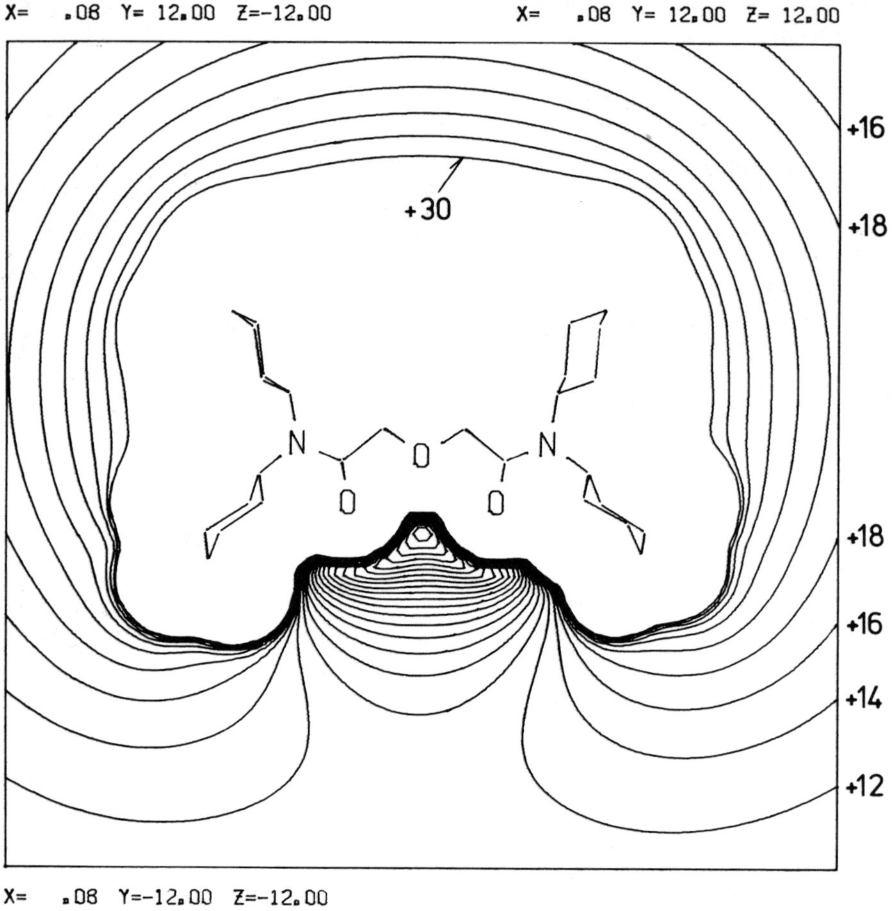

X= .08 Y= 12.00 Z=-12.00   X= .08 Y= 12.00 Z= 12.00

X= .08 Y=-12.00 Z=-12.00

Figure 22 Energy maps for the interaction of Na$^+$ with ligand 28. Distances (Na-O) of 224-230 pm were calculated for the equilibrium position of the ion. The corresponding interaction energies are close to −25.3 kcal mole^{-1} (−105.8 kJ mole^{-1}). The interval between contours is 2 kcal mole^{-1}.

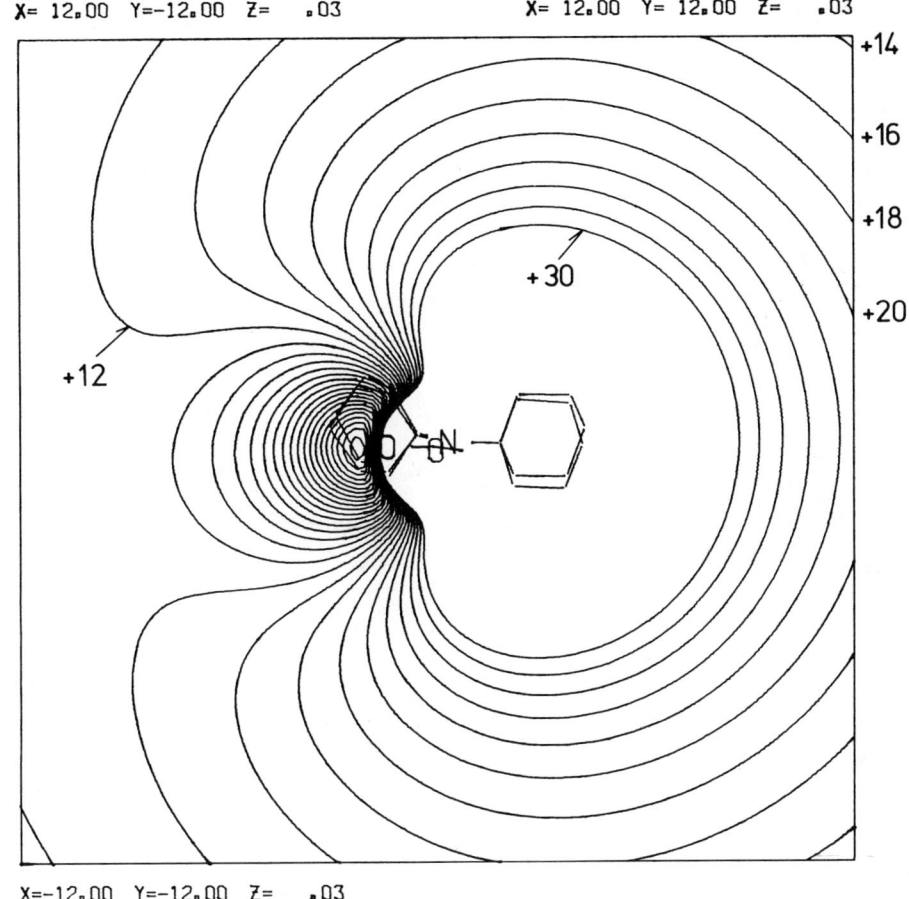

Figure 22 Continued

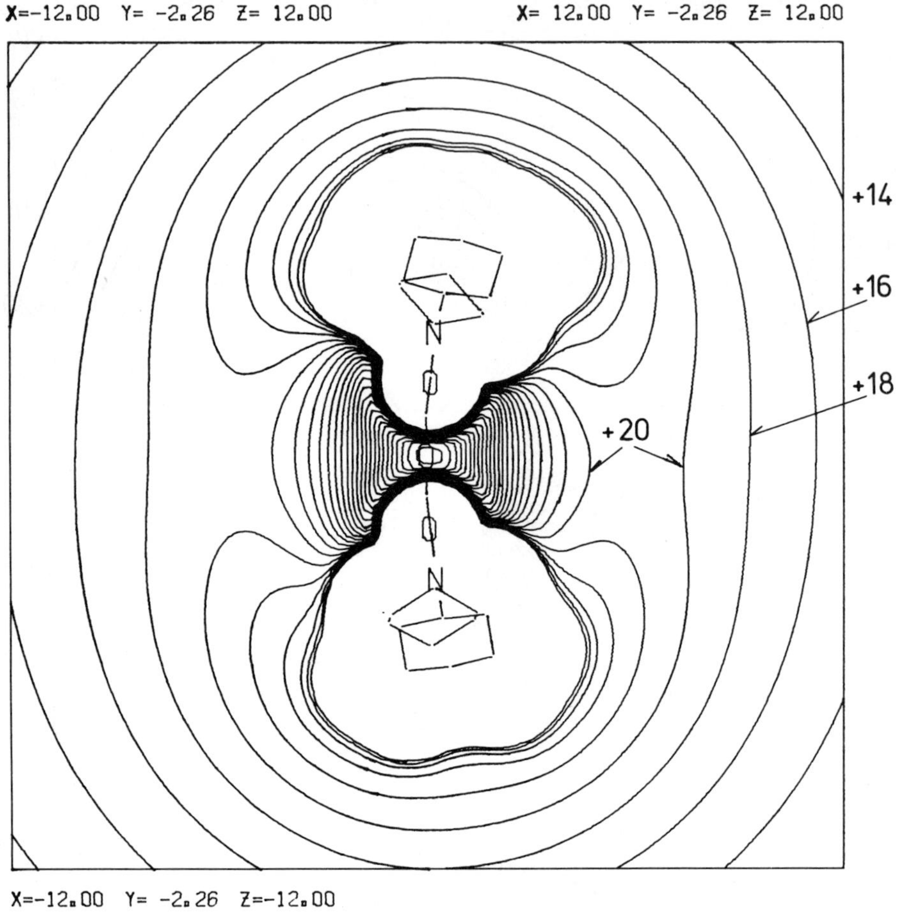

X=-12.00 Y= -2.26 Z= 12.00
X= 12.00 Y= -2.26 Z= 12.00
+14
+16
+18
+20
N
O
O
N
X=-12.00 Y= -2.26 Z=-12.00

Figure 22 Continued

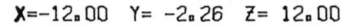

ACKNOWLEDGMENT

This work was partly supported by the Swiss National Science Foundation.

REFERENCES

1. Yu. A. Ovchinnikov, V. T. Ivanov, and A. M. Shkrob, *Membrane-Active Complexones* (BBA Library 12), Elsevier, Amsterdam, 1974.
2. C. Moore and B. C. Pressman, *Biochem. Biophys. Res. Commun.*, **15**, 562 (1964).
3. Z. Štefanac and W. Simon, *Chimia*, **20**, 436 (1966); *Microchem. J.*, **12**, 125 (1967).
4. C. J. Pedersen, *J. Am. Chem. Soc.*, **89**, 2495 (1967); **89**, 7017 (1967); **92**, 386 (1970); **92**, 391 (1970).
5. J. Petránek and O. Ryba, *Anal. Chim. Acta*, **72**, 375 (1974).
6. J. M. Lehn and J. P. Sauvage, *Chem. Commun.*, **1971**, 440.
7. B. Dietrich, J. M. Lehn, and J. P. Sauvage, *Tetrahedron Lett.*, **34**, 2885 (1969).
8. D. Ammann, R. Bissig, Z. Cimerman, U. Fiedler, M. Güggi, W. E. Morf, M. Oehme, H. Osswald, E. Pretsch, and W. Simon, in *Ion and Enzyme Electrodes in Biology and Medicine* (M. Kessler, L. C. Clark, Jr., D. W. Lübbers, I. A. Silver, and W. Simon, Eds.), Urban & Schwarzenberg, Munich, Berlin, Vienna, 1976, p. 22.
9. W. Simon, E. Pretsch, D. Ammann, W. E. Morf, M. Güggi, R. Bissig, and M. Kessler, *Pure Appl. Chem.*, **44**, 613 (1975).
10. D. Ammann, E. Pretsch, and W. Simon, *Tetrahedron Lett.*, **24**, 2473 (1972).
11. D. Ammann, R. Bissig, M. Güggi, E. Pretsch, W. Simon, I. J. Borowitz, and L. Weiss, *Helv. Chim. Acta*, **58**, 1535 (1975).
12. E. Weber and F. Vögtle, *Chem. Ber.*, **109**, 1803 (1976).
13. I. J. Borowitz, W. Lin, T. Wun, R. Bittman, L. Weiss, V. Diakiw, and G. B. Borowitz, *Tetrahedron*, **33**, 1697 (1977).
14. J. Schneider, E. Pretsch, and W. Simon, in preparation.
15. D. J. Cram, R. C. Helgeson, L. R. Sousa, J. M. Timko, M. Newcomb, P. Moreau, F. deJong, G. W. Gokel, D. H. Hoffman, L. A. Domeier, S. C. Peacock, K. Madan, and L. Kaplan, *Pure Appl. Chem.*, **43**, 327 (1975).
16. D. Bedeković, Dissertation ETHZ No. 5777, Juris, Zurich, 1976.
17. A. P. Thoma, Z. Cimerman, U. Fiedler, D. Bedekovic, M. Güggi, P. Jordan, K. May, E. Pretsch, V. Prelog, and W. Simon, *Chimia*, **29**, 344 (1975).
18. W. E. Morf and W. Simon, *Helv. Chim. Acta*, **54**, 794 (1971).
19. J. S. Muirhead-Gould and K. J. Laidler, *Trans Faraday Soc.*, **63**, 944 (1967).
20. H. Kistenmacher, H. Popkie, and E. Clementi, *J. Chem. Phys.*, **59**, 5842 (1973).
21. I. Džidić and P. Kebarle, *J. Phys. Chem.*, **74**, 1466 (1970).
22. *Handbook of Chemistry and Physics*, 44th ed., Chemical Rubber Publishing Co., Cleveland, Ohio, 1963.
23. L. Pauling, *The Nature of the Chemical Bond*, Cornell University Press, Ithaca, New York, 1960.
24. R. J. P. Williams, in *The Regulation of Intracellular Calcium* (C. J. Duncan, Ed.), Cambridge University Press, England, 1976, p. 1.

25. H. G. Hertz, *Angew. Chem.*, **82**, 91 (1970).
26. F. A. Cotton and G. Wilkinson, *Advanced Inorganic Chemistry*, Interscience, New York, 1966.
27. E. Clementi, in *Lecture Notes in Chemistry*, Vol. 2 (G. Berthier et al., Eds.), Springer-Verlag, Berlin, 1976.
28. J. M. Lehn and J. P. Sauvage, *J. Am. Chem. Soc.*, **97**, 6700 (1975).
29. B. Metz, D. Moras, and R. Weiss, *Chem. Commun.*, **1971**, 444.
30. W. E. Morf and W. Simon, *Helv. Chim. Acta*, **54**, 2683 (1971).
31. W. Simon, W. E. Morf, and P. Ch. Meier, in *Structure and Bonding*, Vol. 16 (J. D. Dunitz et al., Eds.), Springer-Verlag, Heidelberg, 1973, p. 113.
32. J.-M. Lehn, *Struct. Bonding*, **16**, 1 (1973).
33. A. E. Martell and M. Calvin, *The Chemistry of the Metal Chelate Compounds*, Prentice-Hall, New York, 1952, p. 149.
34. G. Schwarzenbach, *Helv. Chim. Acta*, **35**, 2344 (1952).
35. A. W. Adamson, *J. Am. Chem. Soc.*, **76**, 1578 (1954).
36. D. K. Cabbiness and D. W. Margerum, *J. Am. Chem. Soc.*, **91**, 6540 (1969).
37. A. B. Callear, I. Fleming, R. H. Ottewill, M. V. Twigg, S. G. Warren, and R. H. Prince, *Chem. Ind.*, 80 (1977).
38. R. D. Hancock and F. Marsicano, *J. Chem. Soc., Dalton Trans.*, 1096 (1976).
39. N. N. L. Kirsch and W. Simon, *Helv. Chim. Acta*, **59**, 357 (1976).
40. N. N. L. Kirsch, R. J. J. Funck, E. Pretsch, and W. Simon, *Helv. Chim. Acta*, **60**, 2326 (1977).
41. W. Simon, W. E. Morf, E. Pretsch, and P. Wuhrmann, in *Calcium Transport in Contraction and Secretion*, (E. Carafoli et al., Eds.), North-Holland, Amsterdam, Oxford, American Elsevier, New York, 1975, p. 15.
42. W. E. Morf, P. Wuhrmann, and W. Simon, *Anal. Chem.*, **48**, 1031 (1976).
43. W. Simon, Sixteenth Solvay Conference on Chemistry, Brussels, November 22-26, 1976.
44. G. Eisenman, Ed., *Membranes*, Vol. 2, Marcel-Dekker, New York, 1973.
45. G. Eisenman, in *Ion-Selective Electrodes* (R. A. Durst, Ed.), National Bureau of Standards Special Publication 314, Washington, 1969.
46. W. E. Morf and W. Simon, *Hung. Sci. Instrum.*, **41**, 1 (1977).
47. W. E. Morf and W. Simon, in *Ion-Selective Electrodes in Analytical Chemistry*, (H. Freiser, Ed.), Plenum, New York, 1978, in press.
48. W. E. Morf and W. Simon, in *Ion-Selective Electrodes*, (E. Pungor and I. Buzás, Eds.), Akadémiai Kiadó, Budapest, 1977, p. 25.
49. S. G. A. McLaughlin, G. Szabo, S. Ciani, and G. Eisenman, *J. Membr. Biol.*, **9**, 3 (1972).
50. G. Eisenman, S. Ciani, and G. Szabo, *Fed. Proc.*, **27**, 1289 (1968).
51. S. Ciani, G. Eisenman, and G. Szabo, *J. Membr. Biol.*, **1**, 1 (1969).
52. H.-R. Wuhrmann, W. E. Morf, and W. Simon, *Helv. Chim. Acta*, **56**, 1011 (1973).
53. W. E. Morf, D. Ammann, E. Pretsch, and W. Simon, *Pure Appl. Chem.*, **36**, 421 (1973).
54. G. A. Rechnitz and E. Eyal, *Anal. Chem.*, **44**, 370 (1972).
55. E. Eyal and G. A. Rechnitz, *Anal. Chem.*, **43**, 1090 (1971).

References

56. L. A. R. Pioda, V. Stankova, and W. Simon, *Anal. Lett.*, **2**(12), 665 (1969).
57. M. S. Frant and J. W. Ross, Jr., *Science*, **167**, 987 (1970).
58. A. A. Lev, V. V. Malev, and V. V. Osipov, in Ref. 44.
59. S. Lal and G. D. Christian, *Anal. Lett.*, **3**, 11 (1970).
60. R. Scholer, Dissertation ETH No. 4940, Zurich, 1972.
61. U. Fiedler, *Anal. Chim. Acta*, **89**, 111 (1977).
62. D. Ammann, M. Güggi, E. Pretsch, and W. Simon, *Anal. Lett.*, **8**(10), 709 (1975).
63. W. Simon, W. E. Morf, and D. Ammann, in *Calcium Binding Proteins and Calcium Function*, (R. H. Wasserman et al., Eds.), North-Holland, New York, Amsterdam, Oxford, 1977, p. 50.
64. R. Büchi and E. Pretsch, *Helv. Chim. Acta*, **58**, 1573 (1975).
65. K. Neupert-Laves and M. Dobler, *Helv. Chim. Acta*, **60**, 1861 (1977).
66. R. Büchi, E. Pretsch, and W. Simon, *Tetrahedron Lett.*, **20**, 1709 (1976).
67. R. Büchi, E. Pretsch, and W. Simon, *Helv. Chim. Acta*, **59**, 2327 (1976).
68. R. Büchi, E. Pretsch, W. E. Morf, and W. Simon, *Helv. Chim. Acta*, **59**, 2407 (1976).
69. A. Leo, C. Hansch, and D. Elkins, *Chem. Rev.*, **71**, 515 (1971).
70. E. Pretsch, R. Büchi, D. Ammann, and W. Simon, in *Analytical Chemistry, Essays in Memory of Anders Ringbom*, (E. Wänninen, Ed.), Pergamon, Oxford, New York, 1977, p. 321.
71. R. Bissig, U. Oesch, W. E. Morf, E. Pretsch, and W. Simon, *Helv. Chim. Acta*, **61**, 1531 (1978).
72. H. F. Osswald, D. Ammann, E. Pretsch, and W. Simon, *Helv. Chim. Acta*, in preparation.
73. K. Neupert-Laves and M. Dobler, *Helv. Chim. Acta*, in preparation.
74. P. R. Wells, *Linear Free Energy Relationships*, Academic, London, 1968.
75. M. S. Newman, *Steric Effects in Organic Chemistry*, Wiley, New York, 1956.
76. E. Pretsch, D. Ammann, and W. Simon, *Res. Dev.*, **25**(3), 20 (1974).
77. E. M. Arnett, in *Progress in Physical Organic Chemistry*, Vol. 1 (S. G. Cohen, A. Streitwieser, Jr., and R. W. Taft, Eds.), Interscience, New York, 1963, p. 223.
78. M. Güggi, E. Pretsch, and W. Simon, *Anal. Chim. Acta*, **91**, 107 (1977).
79. M. Güggi, U. Fiedler, E. Pretsch, and W. Simon, *Anal. Lett.*, **8**(12), 857 (1975).
80. M. Güggi, M. Oehme, E. Pretsch, and W. Simon, *Helv. Chim. Acta*, **59**, 2417 (1976).
81. R. Bissig, E. Pretsch, W. E. Morf, and W. Simon, *Helv. Chim. Acta*, **61**, 1520 (1978).
82. O. Ryba and J. Petránek, *J. Electroanal. Chem.*, **44**, 425 (1973).
83. E. W. Baumann, *Anal. Chem.*, **47**, 959 (1975).
84. R. J. Levins, *Anal. Chem.*, **43**, 1045 (1971); **44**, 1544 (1972).
85. H. K. Frensdorff, *J. Am. Chem. Soc.*, **93**, 600 (1971).
86. T. E. Jones, L. L. Zimmer, L. L. Diaddaria, D. B. Rorbacher, and L. A. Ochrymowycz, *J. Am. Chem. Soc.*, **97**, 7163 (1975).
87. B. Dietrich, Dissertation, Université Louis Pasteur, Strasbourg, 1973.
88. B. P. Nicolsky, *Zh. Fiz. Khim.*, **10**, 495 (1937).
89. G. Eisenman, Ed., *Glass Electrodes for Hydrogen and Other Cations*, Marcel-Dekker, New York, 1967.
90. R. P. Buck and J. R. Sandifer, *J. Phys. Chem.*, **77**, 2122 (1973).

91. R. P. Buck, *Anal. Chem.*, **48**, 23R (1976).
92. J. Koryta, *Ion-Selective Electrodes,* Cambridge University Press, Cambridge, 1975.
93. J. Koryta, *Anal. Chim. Acta,* **91**, 1 (1977).
94. P. C. Meier, D. Ammann, H. F. Osswald, and W. Simon, *Med. Prog. Technol.,* **5**, 1 (1977).
95. J. G. Schindler, R. Dennhardt, and W. Simon, *Chimia,* **31**, 404 (1977).
96. W. Simon, D. Ammann, H. F. Osswald, P. C. Meier, and R. E. Dohner, in *Advances in Automated Analysis, Technicon International Congress 1976,* Vol. 1 (E. C. Barton et al., Eds.), Mediad, Tarrytown, New York, 1977, p. 59.
97. P. C. Meier, D. Ammann, W. E. Morf, and W. Simon, in *Medical and Biological Applications of Electrochemical Devices,* (J. Koryta, Ed.), Wiley, in press.
98. H. A. Lardy, S. N. Graven, and S. Estrada-O., *Fed. Proc.,* **26**, 1355 (1967).
99. A. P. Thoma, A. Viviani-Nauer, S. Arvanitis, W. E. Morf, and W. Simon, *Anal. Chem.,* **49**, 1567 (1977).
100. W. E. Morf, *Anal. Chem.,* **49**, 810 (1977).
101. P. Wuhrmann, Dissertation ETHZ No. 5606, Juris, Zurich, 1976.
102. A. P. Thoma, Dissertation ETH No. 6062, Juris, Zurich, 1977.
103. W. E. Morf and W. Simon, in *Proceedings of the International Conference on Ion-Selective Electrodes,* Budapest, September 5-9, 1977 (E. Pungor, Ed.), in press.
104. H. T. Tien, *Bilayer Lipid Membranes (BLM), Theory and Practice,* Marcel-Dekker, New York, 1974.
105. P. Läuger and B. Neumcke, in Ref. 44.
106. S. M. Ciani, G. Eisenman, R. Laprade, and G. Szabo, in Ref. 44.
107. G. Szabo, G. Eisenman, R. Laprade, S. M. Ciani, and S. Krasne, in Ref. 44.
108. S. B. Hladky, *Biochim. Biophys. Acta,* **352**, 71 (1974).
109. S. M. Ciani, *J. Membr. Biol.,* **30**, 45 (1976).
110. G. Eisenman, S. Krasne, and S. Ciani, *Ann. N. Y. Acad. Sci.,* **264**, 34 (1975).
111. W. E. Morf, in preparation.
112. P. Läuger and G. Stark, *Biochim. Biophys. Acta,* **211**, 458 (1970).
113. P. Läuger, in *Physical Principles of Biological Membranes,* (F. Snell, Ed.), Gordon and Breach, New York, 1970, p. 230.
114. J. E. Hall, C. A. Mead, and G. Szabo, *J. Membr. Biol.,* **11**, 75 (1973).
115. P. Läuger, *Science,* **178**, 24 (1974).
116. P. Vuilleumier, Dissertation ETH, Zurich, in preparation.
117. D. A. Haydon, *Ann. N. Y. Acad. Sci.,* **264**, 2 (1975).
118. K.-H. Kuo and G. Eisenman, *Biophys. J.,* **17**, 212a (1977).
119. R. Margalit and G. Eisenman, *IUPS Satellite Symposium on Theory and Application of Ion-Selective Electrodes in Physiology and Medicine,* Arzneimittel-Forschung, **1977**, in press.
120. G. Dotsevi, Y. Sogah, and D. J. Cram, *J. Am. Chem. Soc.,* **97**, 1259 (1975).
121. E. Blasius, K.-P. Janzen, W. Adrian, G. Klautke, R. Lorscheider, P.-G. Maurer, V. B. Nguyen, T. Nguyen Tien, G. Scholten, and J. Stockemer, *Z. Anal. Chem.* **284**, 337 (1977).
122. P. Grossmann and W. Simon, *Anal. Lett.,* **10**(12), 949 (1977).

123. J. Chmielowiec and W. Simon, *Chromatographia,* **11,** 99 (1978).
124. B. E. Jepson and R. DeWitt, *J. Inorg. Nucl. Chem.,* 38, 1175 (1976).
125. A. Pullman, H. Berthod, and N. Gresh, *Int. J. Quantum Chem. Symp.,* **10,** 59 (1976).
126. P. Schuster, W. Jakubetz, and W. Marius, *Topics in Current Chemistry,* Vol. 60, Springer-Verlag, Berlin, 1975.
127. E. Clementi, F. Cavallone, and R. Scordamaglia, *J. Am. Chem. Soc.,* 99, 5531 (1977).

CHAPTER TWO
THE ROLE OF CROWN AND CRYPTAND COMPLEXATION OF CATIONS IN THE FORMATION OF METAL-AMINE AND METAL-ETHER SOLUTIONS

J. L. DYE

Department of Chemistry
Michigan State University
East Lansing, Michigan

1 Introduction	64
1.1 Overview of the Role of Crowns and Cryptands, 64	
1.2 Metal Solutions in Ammonia, 65	
1.3 Metal Solutions in Amines and Ethers, 67	
1.4 Solid Metal-Ammonia Compounds, 75	
2 Metal Solubility Enhancement by Crowns and Cryptands	76
2.1 Shifts of Equilibria, 76	
2.2 Solvent Effects, 77	
3 Effect of Crowns and Cryptands on the Species Present in Solution	79
3.1 Effect on Ion Pairs, 79	
3.2 Control of Metal Solution Composition, 81	
4 Solvent Free Systems	82
4.1 Crystallization of a Salt of the Sodium Anion, 83	
4.2 Films and Powdered Samples Prepared by Solvent Evaporation, 86	
5 Thermodynamics and Kinetics of Cation Complexation as Studied by Alkali Metal NMR Spectroscopy	89
5.1 Effect of the Solvent and Complexing Agent on Chemical Shifts, 90	
5.2 Linewidth Effects, 91	
5.3 Determination of Complexation Constants from Alkali Metal Chemical Shifts and/or Linewidths, 92	
5.4 Measurement of Exchange Rates by Alkali Metal NMR, 93	
5.5 Intepretation of Complexation Constants and Exchange Rates, 96	
5.6 Sandwich Complexes of Cs^+ with Crown Ethers, 99	
5.7 Evidence for Exclusive [2]-Cryptate Complexes, 101	
6 Alkali Metal NMR Studies of Metal Solutions	101
6.1 Static NMR Spectra, 102	
6.2 Dynamic Effects on NMR Spectra of Metal Solutions, 105	
Acknowledgments	107
References	107

1 INTRODUCTION

The ability of crowns and cryptands to complex alkali metal cations, and the resistance of these cyclic polyethers to chemical reduction have had profound effects on alkali metal solution chemistry. In the past, with few exceptions, we have had to rely upon a single solvent, ammonia, for studies of metal solutions at moderate and high concentrations (1-4). The tremendous solubility enhancement provided by crowns and cryptands has changed this (5-8). We can now prepare alkali metal solutions in the millimolar to molar concentration range in a variety of amine and ether solvents. This opens the door to the synthesis of new compounds via reactions in solution and to the study of such properties as alkali metal NMR (9-17). In this chapter we examine the nature of alkali metal solutions, the effect of crown and cryptand complexing agents on these solutions, and the new compounds that have been prepared by using them.

1.1 Overview of the Role of Crowns and Cryptands

Although the most interesting species in alkali metal solutions are anionic (solvated electrons and alkali metal anions), it is complexation of the alkali *cations* by crown ethers or cryptands which enhances the solubility. Because of this we can carry over from studies of simple salt solutions (18-37) the effect of cavity size and conformation on the complexation constant and selectivity of a particular crown or cryptand. Similarly the variation of the complexation constant with solvent and temperature can be examined by studying solutions of alkali metal salts. Finally, the rate of release of the alkali cation from the complex can also be determined by studies of salt solutions (22, 38-57), and the results may be used to predict the behavior of alkali metal solutions. Essentially the effect of the complexing agent is to break up contact ion pairs, shift ionic equilibria, and enhance the solubility of metals. It apparently plays only a secondary role in affecting the properties of solvated electrons and alkali metal anions.

Fortunately the crown ethers and cryptands are rather inert chemically. Even in the presence of powerful reducing agents and strong bases, the carbon-oxygen bonds of the ether remain intact at low temperatures as do the carbon-nitrogen bonds in the tertiary amine groups of cryptands. Recent work in our laboratory indicates that crown ethers are more resistant to decomposition by metal solutions than are cryptands.

It has been known for some time that the alkali metals are slightly soluble in polyethers such as dimethoxyethane and diglyme (58-63). The cation solvation energy provided by the ether oxygens combines with the polydentate character to overcome the lattice energy of the metal. It is now known that the anion produced by dissolving the metal in polyethers is primarily the alkali metal anion rather than the solvated electron. The latter species is present at relatively smaller concentrations. A crown ether or cryptand molecule provides an effective first solvation layer for the cation. Since anions are only weakly solvated, the effect of the complexing agent on species such as M^- and e^-_{solv} is minor.

Because of the strong interaction between alkali metal cations and the cyclic polyether complexing agents it is often possible to prepare crystalline salts that contain the complexed cation as an integral part of the crystal. The crystal structures of a number of such salts have been determined (64-76), and they provide useful information about the structure of the cationic complex and its interaction with the anion and with solvent of crystallization. It is this ability to isolate the cation from its counterion which provides a really new dimension to metal solution chemistry. On removal of solvent, or simply on crystallization from a saturated solution, new salts may be synthesized (9, 11, 13, 17) which contain the complexed cation together with either the alkali metal anion or a trapped electron; the complexing agent usually prevents precipitation of the metal.

The crown ethers and cryptands have thus provided three advantages in the study of metal solutions. First, the concentrations of the solutions can be greatly increased and new solvents can also be used. Second, by controlling the ratio of complexing agent to total metal, we can often control the stoichiometry of the solution, such that the anion can be either the solvated electron or the alkali metal anion. Finally, new solid compounds can be prepared which contain the complexed cation and either the alkali metal anion or the trapped electron.

1.2 Metal Solutions in Ammonia

Solutions of the alkali and alkaline earth metals in ammonia have been studied for well over a century, and an extensive literature exists in this field (1-4). It is not our purpose here to describe such solutions in detail but rather only to provide a basis for comparison with metal-amine and metal-ether solutions.

Liquid ammonia is a good solvent. Its relatively high dielectric constant ($D = 23$ at the boiling point of $-33.5°C$), small molecular size, and large dipole moment provide enough cation and electron solvation energy to yield very concentrated metal solutions. Indeed, at its melting point cesium is apparently completely miscible with ammonia. At the boiling point of ammonia saturated solutions of Na, K, Ca, Sr, Ba, and Yb contain about 15 mole % metal, Li is soluble up to 20 mole % metal, and cesium to about 65 mole % (1).

The solution properties range from electrolytic in dilute solutions to metallic in concentrated solutions with a nonmetal to metal transition occurring between 2 and 9 mole % metal. For all metals except cesium a two-phase liquid region exists in which a bronze, metallic phase is in equilibrium with a more dilute blue phase. Consolute temperatures range from a few degrees above the freezing point of ammonia in the case of rubidium to just below room temperatures for calcium-ammonia solutions. The nature of the species present in metal-ammonia solutions is still the subject of much speculation and controversy (77). There is general agreement that at very low concentrations ($< 10^{-3}$ M) the solutions contain solvated cations and solvated electrons, although the structure of the latter species is open to question. Based on theoretical arguments (78) and the considerable volume expansion that occurs when these solutions form, the most commonly accepted structure for $e^-_{NH_3}$ is that of an electron trapped in a cavity from which one or more ammonia molecules have been excluded (79-83). The electron is stabilized in the cavity by orientation of nearest-neighbor dipoles as well as by long-range polarization effects. Regardless of whether this picture is correct, or whether the electron is attached to extended solvent orbitals as a solvent anion (84), the solution properties become independent of the cation in dilute solutions. As the concentration is increased, cation-electron and electron-electron interactions become important. However, the variations with concentration of such properties as conductivity, optical spectrum, magnetic susceptibility, density, and activity coefficient are nearly independent of the cation, strongly suggesting that any new species that are produced (at concentrations below the onset of the nonmetal to metal transition) retain the essential characteristics of the solvated cation and probably also the solvated electron (85). The "ionic aggregation" scheme described below is consistent with the available data. However, alternative explanations, such as the formation of dielectrons (two electrons trapped in one cavity) (79, 86), alkali metal monomers (a solvated cation with an electron in an expanded orbital about it) (87), dimers (two monomers held together by "exchange forces") (87), and/or alkali metal anions (either as described later in this chapter or with two electrons in an expanded orbital about a solvated cation), cannot be ruled out. If such species exist, they do not give *specific* evidence of metal-dependent character.

The properties of dilute and moderately concentrated (up to about 0.5 M) metal-ammonia solutions may be at least qualitatively described by an *ionic aggregation model* (77, 85). According to this model the general features of the solvated cation and the solvated electron are retained. Normal ion pairing occurs between M^+ and e^-_{solv}, giving a nonconducting ion pair that leads to a relatively weak magnetic interaction of the unpaired electron spin with the metal nucleus, as evidenced by the paramagnetic (Knight) shifts seen in the alkali metal NMR spectra (88-93). Static susceptibilities (94, 95) and ESR spin susceptibilities (96-98) show clearly that the electron spins couple to form diamagnetic states that become more prominent as the concentration is increased. For example, at 0.1 M at least 90% of the electrons are spin paired at 23°C. At lower temperatures spin pairing is even more pronounced. The ionic aggregation model explains spin pairing via the formation of triple ions, $e^- \cdot M^+ e^-$, in which electron-electron interactions are strong enough to give a singlet gound state. It

1 Introduction

is necessary to assume an extra stabilization energy of about 2-3 kcal mol^{-1} compared to that expected for normal triple ions in ammonia. Presumably quadruple ions could also form by interaction of M$^+$ with the triple ion. Thus this model provides for the occurrence of species of stoichiometry M$^+$, e^-, M, M$^-$, and M$_2$ by relatively weak interactions between solvated cations and solvated electrons. An important unanswered question is whether an electron-paired species, e_2^{2-}, can form without a cation in the vicinity (78, 99-103). It is difficult to answer this question experimentally because, at the concentrations at which spin pairing is prevalent, one would not expect to find significant concentrations of doubly charged species. Even if such a species were energetically favorable, attraction of M$^+$ to form an ion pair, M$^+\cdot e_2^{2-}$, would be expected (77). It would be of considerable interest to determine the effect of a crown or cryptand complexing agent on the extent of spin pairing in metal-ammonia solutions to better assess the role of the cation in this process.

1.3 Metal Solutions in Amines and Ethers

1.3.1 Solubilities

Since the alkali and alkaline earth metals dissolve readily in liquid ammonia, it is not surprising that they also dissolve in primary amines, RNH$_2$. Lithium dissolves in methylamine (MeNH$_2$) to nearly the same extent as in ammonia (~15 mole %) (104) and in ethylenediamine (EDA) to a much lesser extent (1.9 mole %) (105). The solubility of the other alkali metals in these solvents is much lower. Table 1 gives the few solubilities that have been measured in amine and ether solutions, and some crude estimates for other cases. Solubilities lower than about 10^{-4} M are very difficult to measure reliably because (a) solution decomposition competes with solubilization, (b) the rate of dissolution can be very slow, and (c) impurities such as ammonia can greatly enhance the solubility.

Sodium is apparently insoluble in ethylamine, in the higher aliphatic amines, and in all ethers tested. Because of the intensity of the optical absorption band of Na$^-$, one should be able to detect by eye concentrations as low as 2 × 10^{-6} M. Therefore "insoluble" means that the solubility is at least this low. Lithium is soluble to an unknown extent in ethylamine and appears to be insoluble in ethers. It is difficult to prepare stable lithium solutions, presumably because of impurities (lithium metal cannot be distilled in glass or quartz). The heavier alkali metals, K, Rb, and Cs, are more soluble than sodium and are slightly soluble in tetrahydrofuran (THF) and polyethers (see Table 1 for some examples). None of the metals appears to be soluble in secondary or tertiary amines. Presumably the metals would dissolve in many polar solvents such as water and alcohols were it not for the reactivity of these solvents.

The order of solubility in ethylenediamine, Li > Cs > Rb > K > Na (105),

Table 1 Solubilities of Alkali Metals in Amines and Ethers in the Absence and Presence of Complexing Agents

Metal	Solvent[a]	Complexing[b] Agent	Solubility[c] (M)	Temp (°C)	Comments and Refs.[d]
Li	HMPA	—	>0.8	8.5	106
	EDA	—	0.29	25	105
	MA	—	$X_{Li} = 0.15$	—	104
Na	HMPA	—	~0.6	8.5	106
	EDA	—	2.4×10^{-3}	25	105
	MA	—	$>2 \times 10^{-4}$	−50	125
	MA	18C6	0.37	0	
	MA	C222	>0.06	−78	[e]
	{EA, THF, Et$_2$O, RNH$_2$}	—	$<10^{-6}$	—	Based on absence of color
	EA	C222	0.08	0	Na$^+$C222 Na$^-$ precipitates from solution
	EA	C222	~0.2	25	
	THF	C222	>0.09	25	[e]
	Et$_2$O	C222	$<10^{-4}$	0	Estimated from color
K	HMPA	—	>0.6	8.5	106
	EDA	—	1.04×10^{-2}	25	105
	1,2-DME	—	4.7×10^{-4}	−49.7	59
		—	6×10^{-4}	−61.9	59
	Polyether	—	~10^{-2}	−40	59
	THF	—	5×10^{-6}	24	118

	Solvent	Complexing agent	Concentration	Temp	Ref
	THF	29% Polyethylene oxide, M.W. 10,000–50,000	≤0.2	—	63
	MA	C222	>0.14	—	e
	MA	18C6	>0.06	—	e
	THF	C222	>0.12	—	e
Rb	EDA	—	1.31×10^{-2}	25	105
	MA	C222	>0.06	—	e
	EA	C222	>0.02	—	e
	Et$_2$O	C222	>0.04	—	e
	MA	18C6	>0.03	—	e
Cs	EDA	—	0.054	25	105
	MA	18C6	>2 × 10^{-3}	−50	125
	MA	C222	>0.07	—	e
	EA	C222	>0.1	—	e
	MA	C222	>0.05	−70	e
	THF	C222	>0.1	—	e

[a] Solvent abbreviations: HMPA = hexamethyl phosphoric triamide, EDA = ethylenediamine, THF = tetrahydrofuran, Et$_2$O = diethyl ether, RNH$_2$ = Primary amines with more than two carbons, 1,2-DME = 1,2-dimethoxyethane, Polyether = bis[2-(2 methoxyethoxy)ethyl] ether.
[b] 18C6 is the abbreviation for 18-crown-6 (see Chapter 3, Figure 1); C222 is the abbreviation for cryptand[2.2.2] (see Chapter 3, Figure 3).
[c] Moles of metal dissolved. When the solution composition is M$^+$C·M$^-$, the molarity of M$^-$ is half that given.
[d] When no references are listed, the observations refer to our unpublished observations.
[e] *Minimum* solubility based on the amount of complexing agent used in one or more experiments. Solubilities would probably be larger if more complexing agent were used.

may be easily explained (7) on the basis of the lattice energies of the metals, their ionization potentials, and the solvation energies of the gaseous cations. Thus, in spite of the fact that lithium has the most unfavorable lattice energy and ionization potential, it is most soluble in such "good" solvents as methylamine and ethylenediamine because of the large negative solvation energy of the cation. Sodium is the least soluble because its solvation energy, though more negative than that of potassium, for example, is not enough to overcome its relatively larger lattice energy and ionization potential (see Ref. 7 for details).

An interesting solvent whose solvation ability for the metals is exceeded only by ammonia is hexamethyl phosphoric triamide (HMPA), $[(CH_3)_2N]_3PO$. With a dielectric constant of about 30 at room temperature and a strong tendency to solvate cations, HMPA readily dissolves all the alkali metals. The solubility of sodium in HMPA is about 0.6 M (106); the solubilities of the other metals have not been measured, but are presumably even higher than this.

1.3.2 Cation-Dependent Properties

In contrast to metal-ammonia solutions, in which most properties are very nearly independent of the metal used, solutions in amines and ethers (and to a lesser extent in HMPA) show striking variations of properties with metal. The solubilities described above provide one example. With the exception of lithium solutions, the optical spectra show a new metal-dependent absorption band of M^- in addition to the IR band of the solvated electron. The conductivities of solutions that show only this metal-dependent band are drastically different from those that also show the IR band of e^-_{solv}. The chemical reactivity of Na^- is very different from that of e^-_{solv} as determined by both stopped-flow (107) and pulse-radiolysis studies (108). Finally, the ESR spectra show, in many cases, the presence of a species of stoichiometry M which is easily detected by the presence of hyperfine splitting (hfs) by the metal nucleus (109-122).

1.3.3 The Solvated Electron

The assignment of the metal-independent infrared absorption band to the solvated electron is relatively certain, since it may be corroborated by pulse-radiolysis studies (123). The correlation is reasonably good, as shown in Figure 1 (8). Many of the data shown in Figure 1 for metal solutions were obtained with the aid of crown and/or cryptand complexing agents. The correspondence of peak position and especially of peak shape with those obtained by pulse radiolysis of the pure solvent is not exact. This has led some workers to question whether the solvated electron interacts strongly with the complexing agent. However, this does not seem to be the case. The absorption bands of e^-_{solv} obtained by pulse

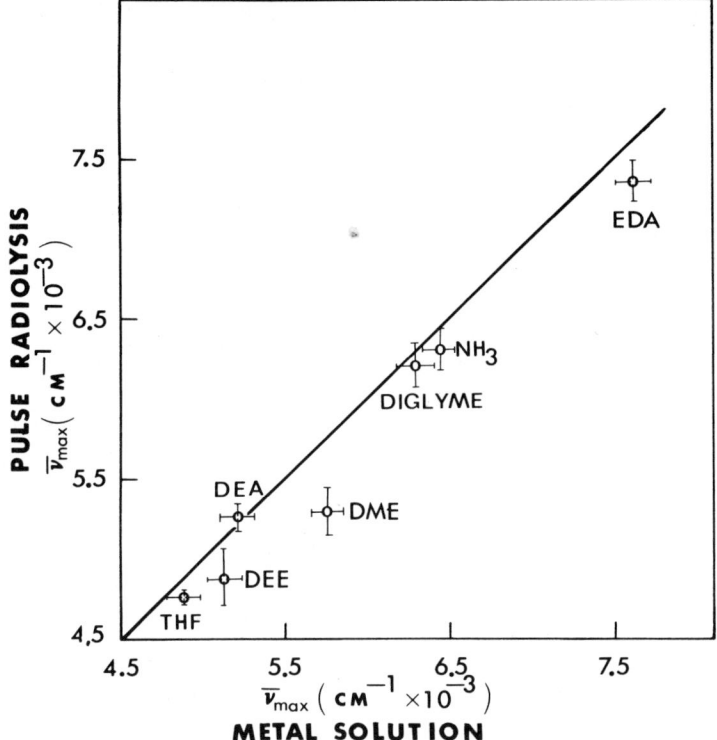

Figure 1 Relation between the peak position of e^-_{solv} obtained by dissolving metals and by pulse radiolysis. Taken from Ref. 8.

radiolysis are the same regardless of whether the complexing agent is present (124). It must be remembered that the metal solutions are several orders of magnitude more concentrated than those studied by pulse radiolysis. Therefore, changes in the absorption band caused by ion-pair formation and electron pairing are not surprising.

The conductances of solutions of cesium in methylamine (125) and in ethylenediamine (105) closely parallel the behavior of metal-ammonia solutions. Dilute solutions of cesium in these solvents show only the IR band of e^-_{solv}. When viscosity corrections are made, the limiting conductances agree with that of cesium in ammonia. The decrease of conductance with increasing concentration probably results from ion pairing. As expected from the lower dielectric constants, the ion-pair-formation constants in $MeNH_2$ and EDA are larger than in ammonia. These results show that the solvated electron behaves in these amine solvents in much the same way as it does in ammonia. It would be of great interest to have information on the magnetic susceptibilities of these solutions to find out whether electron-spin pairing can occur without the formation of M^-.

1.3.4 The "Monomer" Species M

Although a species of stoichiometry M is a minor component in comparison with M^- (and usually also compared with e^-_{solv}), it was easily identified by ESR studies of metal-amine solutions because of the "fingerprint" provided by hfs by the metal nucleus (109-122). Although the origin of the striking temperature, metal, and solvent dependence of the hfs as exemplified by Figures 2 and 3 has been the subject of considerable controversy over the years, recent correlations (14, 122, 126-128) of the ESR and optical spectra of M (the latter obtained by pulse radiolysis techniques) have provided a self-consistent picture of the nature of the "monomer" species. Because of the apparent effect of crowns and cryptands on the structure of this species, this model is described in more detail.

As indicated previously, the monomeric species in ammonia is probably best represented as an ion pair between the solvated cation and the solvated electron. Under such circumstances the unpaired electron concentration at the metal nucleus is low, electron exchange is fast, and only a single narrow ESR line is ob-

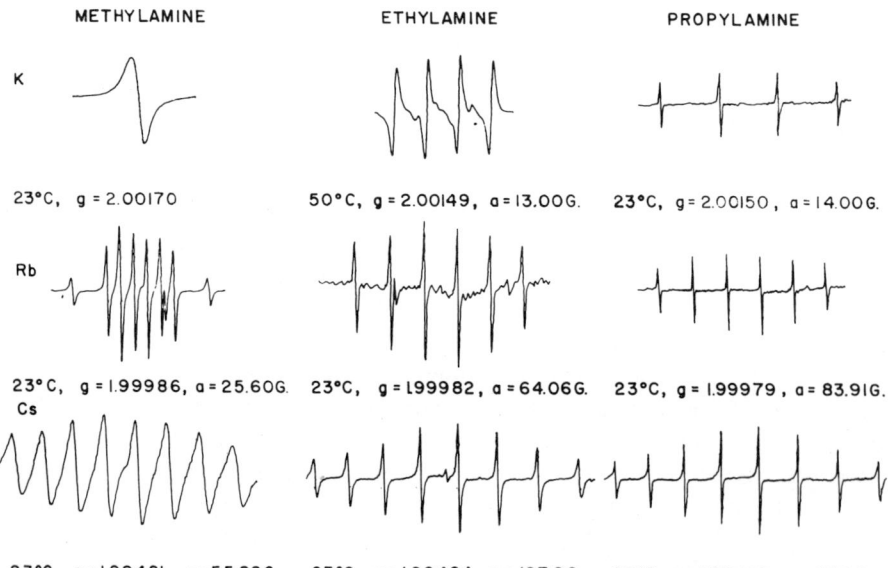

Figure 2 Variation of the monomer ESR signal with metal and solvent at 25°C. Taken from Ref. 115.

1 Introduction

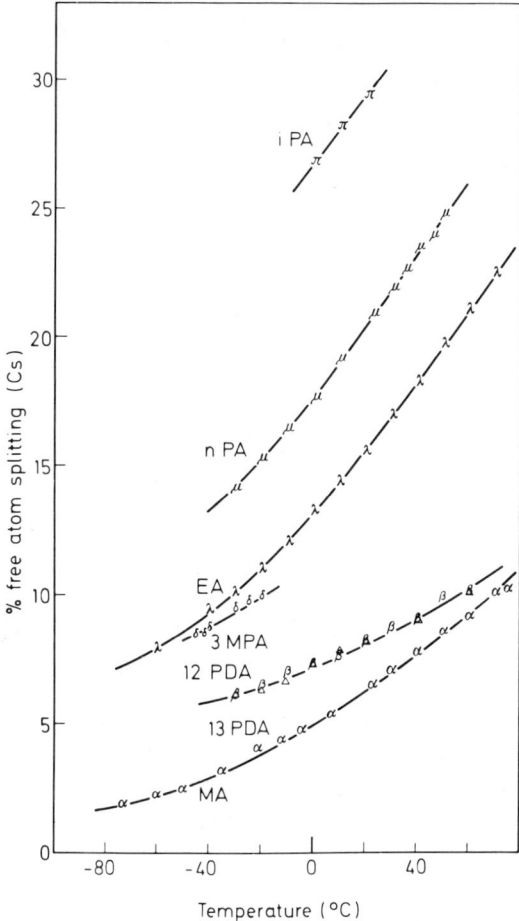

Figure 3 Dependence of the hfs of the Cs monomer on solvent and temperature. Solvents are iPA = isopropylamine; nPA = n-propylamine; EA = ethylamine; 3MPA = 3-methoxypropylamine; 12PDA and 13PDA = 1,2- and 1,3-propanediamine; MA = methylamine. Taken from Ref. 7.

served. In other solvents such as HMPA, EDA, and MeNH$_2$, which solvate the cation strongly, especially with Li$^+$ and Na$^+$, the structure of M is presumably also the solvent-separated or solvent-shared ion pair. The presumed structure of such a species is shown in Figure 4a. Strong support for such a species is provided by the ESR spectrum of solutions of lithium in ethylamine (112, 118). With these solutions no Li hfs is seen, but rather a nine-line pattern caused by the eight R-NH$_2$ protons of the Li(H$_2$NEt)$_4^+$ species. Thus the solvated electron is in close

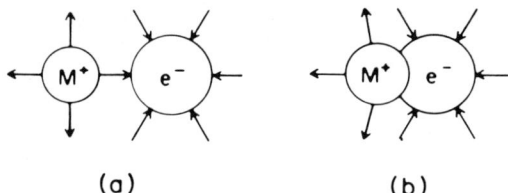

Figure 4 Schematic representation of (*a*) solvent-separated ion pair, and (*b*) contact ion pair.

contact with the rotating solvated cation, and the exchange of solvent from the cation is slow on the ESR time scale. An alternative explanation, which considers a spherically symmetric monomer with the unpaired electron in an expanded *s*-type orbital about the solvated lithium cation, is less attractive since one would expect substantial hyperfine splitting by the lithium nucleus in this case.

In "poor" solvents normal salts form contact ion pairs in which no solvent exists between the cation and the anion. When the solvated electron is the anion, we have the unique situation of an anion with no fixed "size"; that is, there are no core electrons to provide the strong repulsions that lead to a fixed cation-anion distance in normal ion pairs (14). Therefore the distance from the center of the negative charge of the solvated electron to the center of the cation is variable, depending on solvent and temperature. A decrease in solvent polarity or an increase in temperature leads to greater unpaired electron density at the metal nucleus. This in turn gives a larger hyperfine splitting and a blue shift of the optical absorption. The presumed structure of this contact ion pair is shown in Figure 4*b*.

1.3.5 Alkali Metal Anions

Explanations of the metal-dependent absorption bands in metal-amine and metal-ether solutions reflected a state of confusion for many years, and the literature prior to 1968 must be viewed with suspicion. In that year Hurley, Tuttle, and Golden (129) showed that the conflicting results were caused by extremely facile exchange of sodium ions from glass with other alkali metal ions in the solution. When dilute solutions were used for optical studies, contamination from sodium ions yielded an absorption resulting from the presence of Na^- even when no sodium metal had been added to the solution. By using apparatus made from fused silica, these contamination problems can be eliminated.

It was suggested by Golden, Guttman, and Tuttle in 1965 (84) that the diamagnetic species in metal-ammonia solutions is the alkali metal anion with two electrons in the outer *s*-orbital. As we have seen, there is no *specific* evidence for such a species in metal-ammonia solutions. However, it became a prime candidate for the species in metal solutions in amines and ethers giving metal-dependent optical bands. By comparing the solvent, metal, and temperature dependence of the optical band with the charge-transfer-to-solvent (ctts) bands of halide salts, Matalon, Golden, and Ottolenghi proposed in 1969 (130) that the diamagnetic species in metal solutions are indeed "genuine" alkali metal anions and not just clusters of solvated electrons and cations. Recently this identification has been confirmed by the isolation of a crystalline salt of Na^- (9, 11) and by alkali metal NMR solution studies (10, 12). Both these studies are discussed in detail later in this chapter.

Verification that the *stoichiometry* of the diamagnetic species responsible for the metal-dependent optical band is M^- came from a number of experiments. An oscillator strength of 1.9 for the absorption band of sodium in ethylenediamine (131) corresponds to two electrons for each sodium. The reaction of Na^+ with e^-_{solv} is indicated by pulse radiolysis (132) to be strictly second order in the e^-_{solv} concentration. The limiting equivalent conductances of sodium in methylamine (125) and in ethylenediamine (105) have values consistent with the presence of a salt Na^+, Na^-. The Faraday effect (133) shows that the absorption band arises from a diamagnetic species. In short, the evidence for stoichiometry M^- was overwhelming even before the isolation of a salt of Na^-.

1.4 Solid Metal-Ammonia Compounds

On the basis of vapor pressure measurements Kraus concluded in 1908 (134) that stable compounds of the type $Ca(NH_3)_6$ should exist. Since that time many studies have been made on the solid phases formed when metal-ammonia solutions are frozen. In some cases, for example, Na, K, Rb, and probably Cs, metal precipitates during the freezing process, and the solid consists of ammonia and precipitated metal. When saturated solutions of lithium, the alkaline earth metals Ca, Sr, and Ba, and the divalent lanthanides Eu and Yb are frozen, however, gold-colored solid compounds are formed (135).

Although it has not been possible to obtain single crystals of $Li(NH_3)_4$, many of the properties of this metallic solid have been studied. Powder X-ray studies combined with differential thermal analysis measurements show a cubic phase just below the eutectic temperature of 88.8°K and the transition to a hexagonal phase at 82.4°K. The stoichiometry is somewhat variable, and the compound is stable with either a slight excess or a slight deficiency of ammonia.

Lithium tetrammine is metallic in both the cubic and the hexagonal forms. A discussion of the electrical and magnetic properties is beyond the scope of this review, but it should be noted that the compound may be viewed as an "expanded metal" consisting of a lattice of $Li(NH_3)_4^+$ ions and delocalized electrons. Indeed, calculations indicate that it is the delocalization energy that gives the compound its stability (1). The term "expanded metal" refers to the rather large interionic distances in the solid. For the face-centered-cubic phase of $Li(NH_3)_4$ this distance is 6.75 Å.

The alkaline earth and lanthanide hexammines have body-centered-cubic structures with interionic distances ranging from 7.9 Å for $Ca(NH_3)_6$ to 8.6 Å for $Ba(NH_3)_6$. Like $Li(NH_3)_4$, these compounds appear to be expanded metals, but have larger interionic distances and two valence electrons contributing to the conduction band for each metal atom.

2 METAL SOLUBILITY ENHANCEMENT BY CROWNS AND CRYPTANDS

2.1 Shifts of Equilibria

The equilibria in alkali metal amine and ether solutions may be conveniently summarized (7) by the following equations:

$$2M(s) \rightleftharpoons M^+ + M^- \qquad (1)$$

$$M^- \rightleftharpoons M^+ + 2e^- \qquad (2)$$

$$M^+ + e^- \rightleftharpoons M \qquad (3)$$

In "good" solvents such as ammonia and HMPA the solvation energy of the cation predominates, and the major species in solution are M^+ and e^- (and the ionic aggregates that these species form). In methylamine and ethylenediamine, sodium tends to dissolve primarily as Na^+ and Na^-, whereas cesium forms Cs^+ and e^-_{solv}, at least in dilute solutions. In other amines and in ethers, M^+ and M^- predominate, with relatively lower concentrations of e^-_{solv} and M. As indicated earlier, the overall solubility of the alkali metals in amine and ether solvents is generally very low. It should also be noted that there is to date no strong evidence for Li^- although its existence in ethylamine has been proposed on the basis of optical spectra (136).

The effect of the addition of cation complexing agents such as crowns or cryptates (collectively represented by the symbol "C") is to add a fourth equilibrium:

$$M^+ + C \rightleftharpoons M^+C \qquad (4)$$

This markedly enhances the solubilities of the alkali metals and shifts reaction 2 to the right and reaction 3 to the left. *In the absence of excess metal*, therefore, the addition of the complexing agent produces M^+ and e^-_{solv}, mimicking the behavior of a good solvent. Solvent effects remain, however, influencing the extent to which this ionization can occur.

2.2 Solvent Effects

The influence of the solvent on reaction 2 is twofold. The predominant effect appears to be the solvation energy of the cation, but the free energy of solvation of e^-_{solv} must also be considered. There is no way to determine this directly, but the theories of electron solvation indicate that the position of the optical absorbance maximum should correlate with electron solvation energy (78, 80-83). This is borne out by the very low work function of metal-HMPA solutions, which even exhibit thermionic emission (137, 138). The absorbance maximum for e^-_{solv} in HMPA is at 2100 nm (139), compared to 1500 nm in NH_3 and 1300 nm in EDA. This may explain why spectral peaks corresponding to M^- are observed even in a solvent as good as HMPA (139-141). Solvation of the cation, but not the electron, is favored. The absorbance maxima for ethers are also in the vicinity of 2000 nm, indicating weaker electron solvation than in ammonia and EDA. Finally, in moving from ammonia to methylamine to ethylamine a gradual red shift of the absorption maximum appears, with room temperature values of 1600 nm, 1700 nm, and > 1800 nm, respectively (124). As the solvation energy of the electron decreases, it becomes more difficult to shift reaction 2 to the right even in the presence of a cation complexing agent.

The ability of a crown or cryptand to enhance metal solubility depends on the net free energy change in reaction 4. In those solvents that lead to the presence of primarily M^- rather than e^-_{solv} in the saturated solution,

$$\frac{S}{S_0} = (K_4[C] + 1)^{1/2} \approx (K_4[C])^{1/2} \qquad (5)$$

in which S is the total metal solubility, S_0 is the solubility in the absence of complexing agent, and $[C]$ is the concentration of free crown or cryptand in the solution. When the solubility enhancement is large ($K_4[C] \gg 1$), we obtain

$$S \approx \frac{K_4 S_0^2}{4} \left[\left(1 + \frac{16 C_0}{K_4 S_0^2}\right)^{1/2} - 1 \right] \qquad (6)$$

in terms of the *total* concentration of complexing agent C_0.*

The value of K_4 depends, of course, on the metal ion, complexing agent, and solvent. The same factors that cause S_0 to decrease as the solvating ability of the

---

*Equations 5 and 6 do not include activity coefficients or the effects of ion pairing.

solvent decreases will cause K_4 to increase. That is, the decrease in solvation energy of M^+ as one moves through the solvent sequence NH_3, HMPA, EDA, $MeNH_2$, polyethers, $EtNH_2$, THF, diethyl ether, diamines causes a drastic reduction in metal solubility S_0, but an increase in the complexation constant K_4. In general the effect on S_0 is dominant, and the solubility, even in the presence of the cation complexing agent, decreases as one moves along this solvent sequence. This analysis is valid when the only solid phase is the alkali metal. The overall solubility is then limited by precipitation of C or of some metal-containing species such as $M^+C \cdot M^-$.

Unfortunately there have been very few measurements of solubility in either the presence or absence of C. Table 1 gives the few data that are available. In general the results agree with the trends described above. In some solvents the metal dissolves without assistance, leading to an enhancement of solubility; in a number of others the metal will only dissolve when a cation complexing agent is used. The second group of solvents includes monoethers such as diethyl ether, diisopropyl ether and di-n-propyl ether, and secondary amines such as diethylamine and di-n-propylamine (8). It is interesting that the alkali metals will "dissolve" even in such nonpolar solvents as benzene and toluene in the presence of crowns and cryptands (142). In these solvents, however, the optical spectrum shows that there is electron transfer to the solvent to form the aromatic radical anion. When 18-crown-6, 18C6 (see Chapter 3, Figure 1, for structure), was used, long-lived ion pairs (on the ESR time scale) formed between the complexed cation and the radical anion. When the complexing agent was the cryptand, C222 (see Chapter 3, Figure 3, for structure), rapid exchange of the electron from one solvent molecule to another occurred. This effect dramatizes one of the more important differences between crowns and cryptands. In crowns one or two sites on the metal ion remain available for solvation or ion-pair formation. When the ion is completely enclosed in the cavity of a cryptand molecule, however, the ion is shielded from direct interaction with solvent molecules or counterions. Only when exclusive complexes form or when the solvent or counterion is small enough to penetrate into the cavity, do specific short-range interactions become important.

Although few quantitative data are available, the effect of solubility on metal solutions is striking. We can routinely prepare at least 0.1 M solutions of Na, K, Cs, and Rb in methylamine and ethylamine by using C222. In THF the solubilities are somewhat lower, especially for $Na^+C222 \cdot Na^-$. It is the lower solubility of the sodium salt which permits crystallization of this compound from ethylamine and THF through the cooling of saturated solutions. Because of the much greater tendency for solutions of the other metals to decompose and the slow rate of dissolution of all the metals at low temperatures, it is difficult to determine solubilities quantitatively.

3 EFFECT OF CROWNS AND CRYPTANDS ON THE SPECIES PRESENT IN SOLUTION

Since there is no specific evidence that either solvated electrons or alkali metal anions interact strongly with cyclic polyethers, the effect of crowns and cryptands on solution properties can be predicted by their effect on the alkali cations and the resultant shifts of equilibria.

3.1 Effect on Ion Pairs

It is well known that polyethers can interact with alkali cations strongly enough to prevent formation of contact ion pairs in nonaqueous solvents. Indeed, even the addition of relatively small concentrations of a polyether such as dimethoxyethane or dimethylglyoxime or of a good solvating agent such as HMPA can break up contact ion pairs between alkali cations and aromatic radical anions in a solvent such as THF (143). Crowns and cryptands in equimolar amounts can serve the same function. For example, the potassium hyperfine structure observed in ESR spectra of potassium anthracenide in THF is completely eliminated (6) by the addition of C222. This occurs because the cryptand prevents direct contact of the potassium cation with the anthracenide radical anion. Similarly the rate of protonation of the anthracenide radical anion by ethanol in THF is drastically reduced by the addition of C222 or dicyclohexano-18-crown-6 (DCC) to the solution in equimolar amounts (144). This has been attributed to the removal of contact ion pairs from the solution. Numerous other investigations have led to the same conclusion: the coordination of the cation by a good complexing agent provides a local environment that prevents contact between the bare ions.

An interesting question remains. Can crown-type complexes, which still have the axial position open, form contact ion pairs? Clearly the answer is "yes" in the case of the metal "solutions" in benzene and toluene referred to earlier (142).

3.1.1 Conductivity Studies

The replacement of contact ion pairs by crown- or cryptand-separated ion pairs can have a pronounced effect on the conductivity of salt solutions in nonaqueous solvents. The usual equilibrium in such solutions may be represented by

$$M^+ + X^- \xrightleftharpoons{K_1} M^+//X^- \xrightleftharpoons{K_2} M^+,X^- \qquad (7)$$

in which $M^+//X^-$ represents a solvent-shared or solvent-separated ion pair, and M^+,X^- represents a contact ion pair. The dependence of K_1 on the dielectric constant D and the cation-anion separation distance a is satisfactorily described by the Fuoss equation:

$$K_1 = \frac{4\pi a^3 N_{Av}}{3000} \cdot \exp\left(\frac{e^2}{aDkT}\right) \qquad (8)$$

where N_{Av} is Avogadro's number and e is the charge on the electron. If the interionic distance in the crown- or cryptand-separated ion pair is not very different from that of the solvent-shared ion pair, then in the corresponding equilibrium

$$M^+C + X^- \xrightleftharpoons{K_1'} M^+C \cdot X^- \qquad (9)$$

we can expect that $K_1' \approx K_1$ (145). Therefore in a solvent in which contact ion pairs *do not* normally form, either an increase or a decrease in equivalent conductance can result because of differences in the relative mobilities of the solvated cation and the complexed cation M^+C as well as from changes in the ion-pair-formation constant. However, in solvents that show extensive contact-ion-pair formation, the overwhelming effect is the dissociation of these ion pairs and the resulting large increase in the conductance (146). The effect of the complexation equilibrium on conductance forms the basis for one method of determining the complexation constant (145, 147-152). Since the conductance change is sensitive to changes in mobility as well as changes in ion pairing, this method must be applied with caution. It is most reliable in solvents in which ion pairing does not occur, that is, good donor solvents of high dielectric constant. The change in conductance with added complexing agent then represents differences in mobilities of M^+ and M^+C. If the complexation constant is not too large, the curvature in the specific conductance versus the concentration of added complexing agent is large enough to permit evaluation of the complexation constant.

3.1.2 Monomers in Metal Solutions

The formation of ion pairs or monomers in metal solutions by the following reaction:

$$M^+ + e^- \rightleftharpoons M \qquad (3)$$

is affected by complexation of M^+ with crowns and cryptands. When metal is not present in excess, reaction 3 is shifted to the left by the addition of C. This can cause the hfs by the alkali metal nucleus to vanish (6). However, in the presence of an excess of metal the concentration of the monomer can be expected to remain approximately constant because of the equilibrium

$$M(s) \rightleftharpoons M(soln) \qquad (10)$$

It has been reported that for solutions of potassium in THF the addition of the crown ether 18C6 causes an *increase* in the monomer ESR signal intensity and that the hfs remains (153). This increase in intensity could result simply because the more concentrated solutions do not decompose as readily as dilute solutions. Alternatively the crown-complexed cation M^+C might be able to form a contact ion pair with e^-_{solv}.

The optical spectra of alkali metal monomers in various solvents have been studied by pulse-radiolysis techniques (124). Except in ammonia and ethylenediamine a blue shift accompanies formation of the monomer compared with the spectrum of e^-_{solv}. The magnitude of this blue shift is larger in "poorer" solvents and correlates reasonably well with the "percent atomic character" of the monomer as determined by ESR hfs (122, 127, 128). The radiolysis of solutions that contain crowns or the cryptand C222 results in elimination of this blue shift (154). Again the complexing agent seems to prevent formation of contact ion pairs between M^+ and e^-_{solv}.

3.2 Control of Metal Solution Composition

In the presence of excess metal the ratio $[M^-]/[e^-_{solv}]$ should be independent of the presence of a cation complexing agent except for effects of unequal activity coefficients. Therefore Na^- can be expected to be the predominant anionic species when sodium metal is present in excess in all solvents except NH_3 and HMPA. On the other hand, even in the presence of excess cesium metal a mixture of Cs^- and e^-_{solv} is expected in $MeNH_2$, EDA, and $EtNH_2$, whereas Cs^- predominates in other amines and in ethers. One would expect solutions of K and Rb to be intermediate in their behavior.

With an appropriate choice of metal and solvent it is therefore possible, by permitting contact with an excess of metal, to prepare solutions that contain predominantly M^+C and M^-. In addition to these major species small concentrations of the monomer or ion pair M, and the solvated electron are present in all solvents. If the ion pair $M^+C \cdot e^-_{solv}$ is stabilized, appreciable concentrations of this species may also be present.

When a deficiency of metal exists, the ratio $[M^-]/[e^-_{solv}]$ decreases because of a shift of the reaction

$$M^- + C \rightleftharpoons M^+C + 2e^- \tag{11}$$

to the right. Since reaction 11 is the sum of reactions 2 and 4, $K_{11} = K_2 K_4$. The sodium anion Na^- is so stable that even in as good a solvent as $EtNH_2$ reaction 11 is not shifted far to the right in the presence of excess C222. In THF only Na^- can be detected optically in sodium solutions, even when a large excess of cryptand is used. On the other hand the equilibrium of solutions of potassium in ethylamine in the presence of an equimolar amount of cryptand C222 is shifted almost completely to the right so that such solutions contain mainly K^+C and e^-_{solv}. Even in THF reaction 11 lies far to the right with K and Rb in the presence of an excess of C222, and with Cs when a larger cryptand such as C322 is used. Whenever solutions of lithium can be prepared, e^-_{solv} seems to form in preference to Li^-. Because of the general instability of lithium solutions, however, this has not been verified in many solvents.

In summary the introduction of crown and cryptand complexing agents for alkali cations permits at least partial control of the stoichiometry of metal-amine and metal-ether solutions. By controlling the ratio of complexing agent to total metal, one can prepare solutions that range in composition from M^+C,M^- to M^+C,e^-_{solv}.

4 SOLVENT FREE SYSTEMS

The geometry of the complexes of cations with crowns and cryptands is most directly obtained by X-ray crystallography of solid salts of the type M^+CX^-. Although the structure of the complex in the crystalline state may not be the same as in solution, it is often assumed that they are the same. Thus X-ray-diffraction studies of salts of both Na^+C222 and K^+C222 show that the cation is located in the center of the cavity in both cases. With the smaller sodium cation the strands of the cryptand are twisted into an antiprismatic arrangement of the six ether oxygen atoms (70), while the potassium cation gives nearly a prismatic arrangement (69). It can be assumed that similar configurations exist in the solution. Some structures show that the solvent or counterion can interact with the cation, even though the latter is inside the cavity (67). Presumably these interactions through "holes" in the cryptand can also exist in solution.

It is more difficult to relate the crystal structures of crown-type complexes (64, 65, 72-76) to their structures in solution because of the tendency to form sandwich-type complexes when the ion size is large compared to the hole size of the crown ring, and also because of the open axial positions which can interact strongly with the counterion or the solvent. For example, 1:1 stoichiometry could exist in a solid, yet each cation could interact with two crown rings. In solution such a sandwich complex would require 2:1 stoichiometry.

4 Solvent Free Systems

When it became possible to prepare concentrated solutions of M^+C,M^- and M^+C,e^-_{solv}, it seemed likely that crystalline solids of these stoichiometries might be prepared either by crystallization from a saturated solution or by evaporation of the solvent. The stabilization of the cation by the complexing agent might be great enough to prevent formation of metal by one of the following reactions:

$$M^+C + M^- \rightleftharpoons 2M(s) + C \qquad (12)$$

$$M^+C + e^- \rightleftharpoons M(s) + C \qquad (13)$$

Alternatively the rate of loss of M^+ from the complex might be slow enough to permit formation of solid salts of the type M^+C,M^- and M^+C,e^- before either reaction 12 or 13 could occur. Whether permitted by thermodynamic or by kinetic stability of M^+C, solids have been prepared both by crystallization from solution and by rapid solvent evaporation. The preparation and characterization of these solids are now described.

4.1 Crystallization of a Salt of the Sodium Anion

It is possible to prepare solutions of sodium in ethylamine in the presence of cryptand C222 whose sodium concentration at 0°C is about 0.08 M. If the solution is permitted to remain in contact with an excess of sodium, the solution stoichiometry approaches Na^+C,Na^- with only small concentrations of excess C222 and e^-_{solv} remaining. When such solutions are rapidly cooled to -78°C, a gold-colored precipitate of composition Na^+C,Na^- forms (9, 11, 13, 16). This microcrystalline solid can be washed with diethyl ether and stored indefinitely in vacuo in glass tubes at -10°C. Some samples, which have been kept in the freezer for over two years, now appear to have formed a small amount of free sodium metal, but whether this is the result of release of sodium via reaction 12 or of decomposition of the cryptand moiety is not known. In any event the freshly prepared precipitate can be redissolved in ethylamine and crystals grown by slow cooling from +10°C to -20°C over a period of about a day. The crystals grow either as thin hexagonal plates or as thicker crystals with faces having three-fold symmetry.

The structure of $Na^+C222 \cdot Na^-$, determined by x-ray diffraction (11), is best described as hexagonal closest packing of the ions Na^+C222 and Na^- similar to that of the crystalline salt $Na^+C222 \cdot I^-$ (70). The structure of the complexed cation shows the same antiprismatic arrangements of the ether oxygen atoms as found in the iodide salt. A comparison of some of the distances is found in Table 2, and a drawing showing the cryptated cation and the locations of the Na^- anions is given in Figure 5. The elemental analysis of the crystalline solid is in accord with the proposed stoichiometry. The absence of prior oxidation has been con-

Table 2 Comparison of Distances in
$Na^+C222 \cdot Na^{-a}$ and $Na^+C222 \cdot I^{-b}$

	$Na^+C \cdot Na^-$	$Na^+C \cdot I^-$
Bonded Distances (Å)[c]		
Na^+-N	2.72 (1)	2.75 (2)
Na^+-O	2.57 (1)	2.58 (2)
N-C1	1.48 (2)	1.47 (2)
C1-C2	1.29 (2)	1.44 (3)
C2-O	1.41 (1)	1.44 (2)
O-C3	1.40 (2)	1.40 (5)
C3-C3	1.31 (3)	1.40 (4)
Nonbonded Distances		
Na^+...C1	3.45 (2)	3.53 (2)
Na^+...C2	3.46 (2)	3.46 (2)
Na^+...C3	3.34 (3)	3.39 (2)
N...N	5.44 (2)	4.40 (2)
	$A^- = Na^-$	$A^- = I^-$
Na^+...A^-	7.06 (1)	7.40 (1)
A^-...N	5.54 (2)	5.39 (2)
A^-...O	5.72 (2)	5.22 (2)

[a]From Ref. 11.
[b]From Ref. 69.
[c]The number in parentheses after the distances refers to the estimated standard deviation of the last significant digit.

firmed by measuring the amount of hydrogen produced when a sample of $Na^+C222 \cdot Na^-$ is decomposed with water. Similarly the proton NMR spectrum of decomposed samples shows that the integrity of the cryptand structure is maintained. Rapid heating of a powdered sample of $Na^+C222 \cdot Na^-$ yields a melting point of 83°C compared to 68°C for the pure cryptand and 97.5°C for sodium metal. Melting is accompanied by the rapid formation of sodium metal and the free molten cryptand along with some decomposition of the cryptand moiety. Thus $Na^+C222 \cdot Na^-$ is not thermodynamically stable in the liquid state. Solid $Na^+C222 \cdot Na^-$ can be precipitated from solutions in ethylamine and in THF at

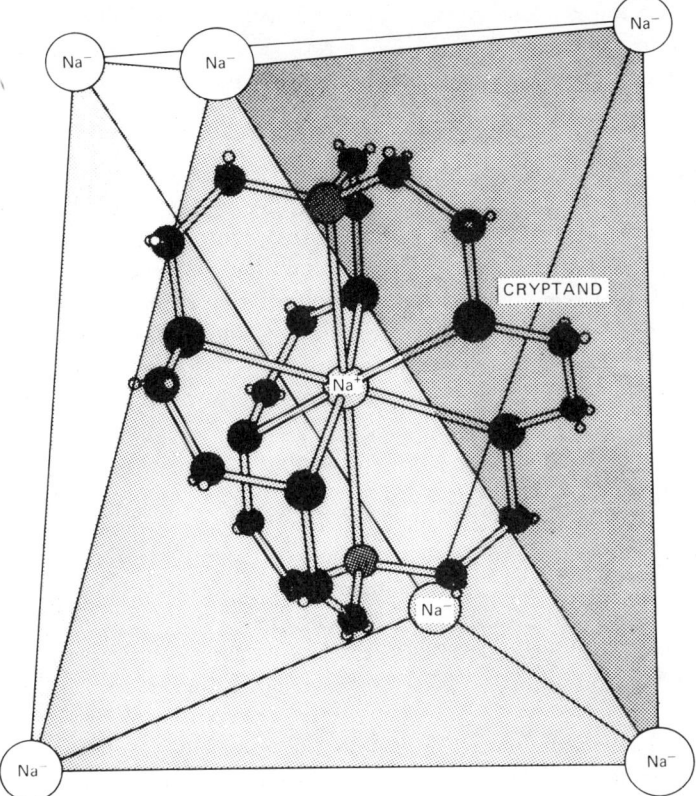

Figure 5 Structure of the central Na^+C222 ion and the placement of six surrounding Na^+ ions in the $Na^+C222 \cdot Na^-$ crystal. Taken from I. J. Tehan, B. L. Barnett, and J. L. Dye, *J. Am. Chem. Soc.*, **96** 7203 (1974), and from J. L. Dye "Anions of the Alkali Metals," copyright © July 1977 by Scientific American, Inc., all rights reserved. Used by permission.

temperatures of 0°C and below in the presence of sodium metal, strongly suggesting that the solid is thermodynamically stable at these temperatures. The rate constant for the loss of Na^+ from C222 in THF at 0°C, as estimated from the exchange rate at higher temperatures, is about 1 sec^{-1}, which should be fast enough to permit precipitation of sodium by reaction 12 if $Na^+C222 \cdot Na^-$ were not thermodynamically stable. However, it is possible that the relative kinetics of crystallization of $Na(s)$ and of $Na^+C222 \cdot Na^-(s)$ are sufficiently different to permit formation of the latter compound even if it is not thermodynamically stable.

Attempts to grow crystals from solutions containing other metals have been hampered by the greater tendency toward decomposition shown by solutions of lithium, potassium, rubidium, and cesium compared with solutions of sodium.

To date, precipitates have been observed to form spontaneously from solutions that contain $Na^+C222 \cdot Na^-$, $K^+C222 \cdot Na^-$, $K^+C222 \cdot K^-$, $Rb^+C222 \cdot Rb^-$, and $Cs^+C222 \cdot Cs^-$. Only the first of these has been characterized.

Corbett and co-workers (155-160) have utilized alkali cation stabilization by cryptand C222 to prepare crystalline salts of the "Zintl anions" (161-163). These are polyanions of the elements Pb, Sb, Te, and Sn which form when alloys of sodium with the corresponding metal are dissolved in ammonia or amines. By complexing Na^+ with C222, Corbett has succeeded in preparing crystalline salts that contain polyanions such as Sb_7^{3-}, Sn_5^{2-}, Pb_5^{2-}, Sn_9^{4-}, and Te_3^{2-}. In the absence of a complexing agent removal of the solvent simply yields amorphous solids that revert to the alloy of sodium and the corresponding metal.

4.2 Films and Powdered Samples Prepared by Solvent Evaporation

By using a volatile solvent such as methylamine to prepare solutions of stoichiometry $M^+C \cdot M^-$ and $M^+C \cdot e^-_{solv}$, followed by low-temperature evaporation of methylamine, thin solid films and/or finely divided powdered samples can be prepared which are free of solvent. Because of the high stability of sodium solutions and of $Na^+C222 \cdot Na^-(s)$, films made from solutions of $Na^+C222 \cdot Na^-$ can be readily prepared and studied. These occur as microcrystalline solids with the same gold-bronze metallic appearance obtained by precipitation of $Na^+C222 \cdot Na^-(s)$. The appearance, stoichiometry, and transmission spectra of such films strongly suggest that they contain the compound $Na^+C222 \cdot Na^-$ (17). Films can be repeatedly redissolved in methylamine and re-formed without change in appearance or spectra. We therefore assume that films formed by evaporation of solutions of $M^+C222 \cdot M^-$ are similar to the crystalline salts of the same stoichiometry.

The absorption spectra of films of $Na^+C222 \cdot Na^-$ always show three distinct features, as seen in Figure 6. The major peak at 650 nm correlates well with the peak of Na^- in solution (164), which has been attributed to a ctts band. A similar carryover of the ctts absorption bands from solution to crystals in observed with halide salts. This band of Na^- shows a slight temperature shift of the peak position (-1.6 cm^{-1} deg^{-1}) and half-width ($+1.1$ cm^{-1} deg^{-1}). The origin of the pronounced shoulder at 530 nm and the small peak at 410 nm is unknown. These two features are not present in a film produced by evaporation of solvent from a solution that contains $K^+C222 \cdot Na^-$.

Films prepared from solutions of the alkali metals K, Rb, and Cs in methylamine in which excess metal is present to ensure the formation of M^- also show the characteristic anion bands, as indicated in Figure 7. In all cases the films are metallic bronze in appearance by reflected light and blue by transmitted light. It thus appears possible to prepare solid, solventfree salts of Na^-, K^-, Rb^-, and Cs^- although only the salt of Na^- has been fully characterized.

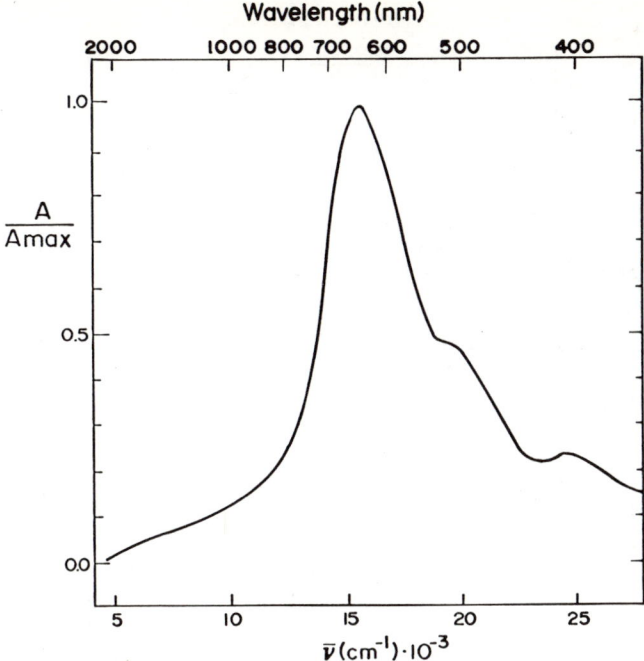

Figure 6 Absorption spectrum of a thin film of $Na^+C222 \circ Na^-$. Taken from Ref. 17.

Films prepared from solutions that contain sodium and 18-crown-6 exhibit interesting behavior (13). The characteristic absorption band of Na^- and the gold-bronze color of the films are observed only when methylamine vapor is present. When the methylamine is removed by pumping or by condensation in a side arm with liquid nitrogen, the films change color from gold to purple or gray. The spectrum of the purple films suggests the precipitation of finely divided sodium metal. The purple and gray films can be redissolved in methylamine, and gold films can again be formed by solvent evaporation. This process can be repeated a number of times. The most likely explanation of this sequence of events is that $Na^+ \cdot 18C6 \cdot Na^-$ films are only stable when the axial positions on $Na^+ \cdot 18C6$ are occupied by solvent molecules. Therefore removal of solvent probably results in the formation of metal and free 18-crown-6. The presence of free metal in the purple films has not been proven, so they could be a new solventfree phase.

Of potentially great interest are the films formed by solvent evaporation from solutions of stoichiometry $M^+C \cdot e^-_{solv}$. By using cryptand C222 with potassium in methylamine, deep blue, solventfree films can be obtained which show either an

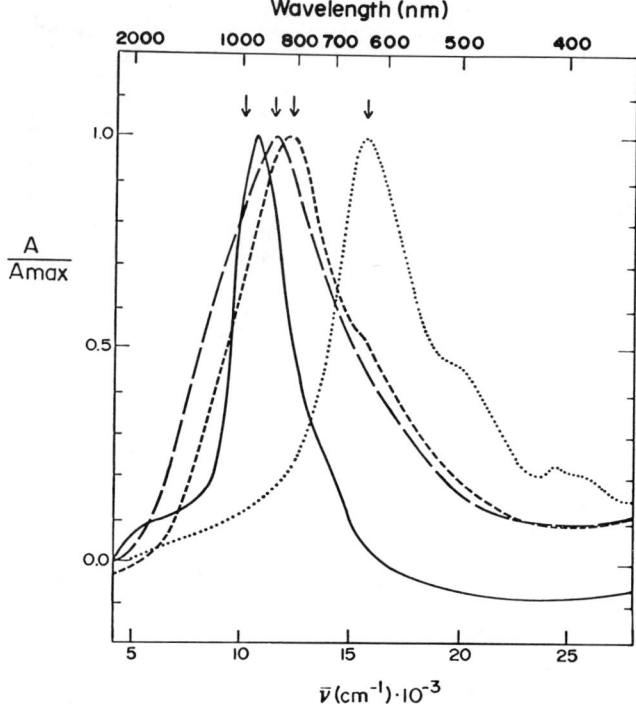

Figure 7 Absorption spectra of solventfree thin films produced by rapid evaporation of solutions of M^+C, M^- in methylamine. From left to right (peak positions) M = Cs, Rb, K, and Na, respectively. The arrows indicate the position of the absorption maxima for the corresponding anion in ethylenediamine. Taken from Ref. 17.

absorption band in the IR region at 1350 nm or the band of K^- at 840 nm or both, as shown in Figure 8. When equimolar amounts of potassium and cryptand are used, only the IR band is observed. Since the spectra of solutions prepared in this way show only the peak of e^-_{solv}, the IR band in the films can be considered characteristic of trapped electrons. Because potassium and the cryptand are present in equimolar amounts, the stoichiometry of the films is consistent with the formulation $K^+C222 \cdot e^-$, and these solventfree films may represent a new class of compounds called "electrides." Similar spectra are observed with other combinations of metal and complexing agents as well.

When excess potassium metal is present, the solution stoichiometry is best represented by $K^+C \cdot K^-$. The spectra of films prepared by solvent evaporation from such solutions show only the band of K^- (also shown in Figure 8). Although much more needs to be done to fully characterize the films formed by solvent evaporation, the results to date suggest that under favorable circum-

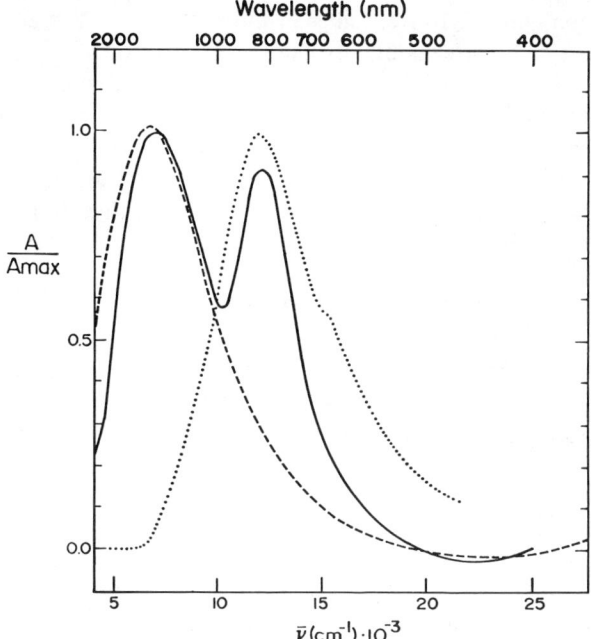

Figure 8 Absorption spectra of solventfree thin films produced by evaporation of methylamine from solutions containing C222 and various relative amounts of potassium. Dashed line (electride); lowest relative potassium content; solid line, intermediate concentrations of e^- and K^-; dotted line, $K^+C \cdot K^-$ film. Taken from Ref. 17.

stances the composition of the films closely parallels that in solution. This is not always the case, however, as is shown by our attempts to prepare films of $Rb^+C222 \cdot Rb^-$ by evaporation of a solution of rubidium in diethyl ether in the presence of cryptand C222. Only a heterogeneous gray mixture, presumably containing Rb(s) and C222, resulted. This mixture could be redissolved to give a dark blue solution. Evidently the reaction

$$Rb^+C222 + Rb^- \rightarrow 2Rb(s) + C222 \qquad (14)$$

is both thermodynamically favored and rapid in this case.

5 THERMODYNAMICS AND KINETICS OF CATION COMPLEXATION AS STUDIED BY ALKALI METAL NMR SPECTROSCOPY

Proton, ^{13}C, and alkali metal NMR studies have provided much detailed information about the nature of cation complexation by crowns and cryptands. The power of proton and ^{13}C NMR studies in the elucidation of structures in solution is demonstrated, for example, by the recent work of Live and Chan (165), who were able to deduce the solution structures of complexes of dibenzo-18-crown-6, benzo-18-crown-6, and dibenzo-30-crown-10 as well as the conformations of the free complexes. Both proton and alkali metal NMR (22, 38-44, 51, 54, 55, 57, 166-170) have been used to determine the thermodynamics and kinetics of exchange, leading to the general conclusion that exchange proceeds via a dissociative mechanism

$$M^+C \rightleftharpoons M^+ + C \qquad (15)$$

In this section we describe the use of alkali metal NMR spectroscopy to study the thermodynamics and kinetics of exchange. This sets the stage for a description of alkali metal NMR studies of metal solutions in the final section of this chapter.

5.1 Effect of the Solvent and Complexing Agent on Chemical Shifts

For all alkali cations except Li^+ a solvent-induced paramagnetic (downfield) shift relative to the gaseous cation dominates over the weaker diamagnetic shift. For example, the paramagnetic shifts* from the corresponding gaseous atoms are -60.5, -212, and -344 ppm for aqueous Na^+, Rb^+, and Cs^+, respectively (12, 172). Although the gaseous cations are paramagnetically shifted from the atoms, the effect is expected to be small. For example, a paramagnetic shift of -5.1 ppm is calculated for $Na^+(g)$ relative to $Na(g)$ (171). The large solvent paramagnetic shifts originate from electron-pair donation from the solvent to the outer p-orbitals of the cations (173-177). Calculated shifts based on the general formulation of Ramsey (178) are in reasonable qualitative agreement with the observed shifts, although in general the observed shifts are greater than those calculated. A direct indication of the role of electron-pair donation by the solvent is given by the good correlation between the $^{23}Na^+$ chemical shifts and the Guttmann donor number of the solvent (179-183). The correlation is not as good for $^{39}K^+$ (184) and $^{133}Cs^+$ (185) shifts, although the same trends are observed.

On the basis of the correlation with solvent donicity one is tempted to equate the stability of the cation-solvent complex with the extent of paramagnetic shift. To a degree this is a valid comparison, but since the shift depends on repulsive overlap, steric effects are expected to be important as well.

*The sign of the chemical shift used in this chapter is the same as that of the shielding constant; that is, a negative chemical shift is paramagnetic. The opposite convention is now commonly accepted.

When the coordinating solvent molecules are replaced by a crown or cryptand, we have a new local "solvent," and the lone-pair donors become the ether oxygens and, in the case of cryptands, the tertiary amine nitrogens as well. When the solvent is effectively excluded from the first coordination shell, as with Li$^+$C211 (166), Na$^+$C222 (51), Cs$^+$C222 (inclusive) (55), and Cs$^+$(18C6)$_2$ (sandwich complex) (57), the chemical shift becomes insensitive to the solvent and counterion, as expected. When, however, the solvent or counterion can still interact with the complexed cation, as with Cs$^+\cdot$18C6 (57) and Cs$^+$C222 (exclusive) (55), the chemical shift remains dependent on the nature of the solvent. It appears that this sensitivity (or lack of it) to the solvent might prove to be a useful diagnostic tool to assess the degree of isolation of the cation from the solvent.

Another striking effect of the interaction of the cation with the complexing agent is the correlation of the magnitude of the chemical shift with the "tightness" of fit of the ion in the complex. On theoretical grounds the extent of the paramagnetic shift is expected to depend on the short-range repulsive overlap of the electron donor with the cation. This seems to be borne out by experiment. For example, the inclusive complex Cs$^+$C222 shows a solvent-independent shift of -245 ppm (paramagnetic) relative to Cs$^+$ (aq) (55). This is larger by a factor of nearly 2 than the entire range of shifts of Cs$^+$ with solvent from the poorest donor, nitromethane (+60 ppm), to the "best" solvent, DMSO (-68 ppm) (185). The most reasonable explanation for this large paramagnetic shift of Cs$^+$ *inside* the cryptand cavity is that in this environment the repulsions between the ether oxygen and the cesium cation are very strong. The increased overlap of the lone-pair electrons with the outer p-orbitals of Cs$^+$ causes a very large paramagnetic shift. By contrast the solvent-independent shift of +48 ppm in Cs$^+$(18C6)$_2$ (57) is about that expected for an ether-type solvent, indicating that the Cs$^+$-O interactions are essentially relaxed in this complex. A similar trend is observed with Na$^+$C222 (51), Na$^+$C221 (184, 186), and Na$^+$C211 (184, 186), in which chemical shifts are about +12, +4.5, and -11 relative to Na$^+$(aq). Again the increased paramagnetic shift correlates with the tightness of fit. However, tightness of fit is not synonymous with stability of the complex. Indeed, in the case of sodium the Na$^+$C221 complex is thermodynamically most stable.

5.2 Linewidth Effects

Except for ^{7}Li and ^{133}Cs, linewidths in alkali metal NMR spectra are dominated by quadrupolar broadening. The magnitude of this broadening is dependent on the nuclear spin I, the nuclear quadrupole moment Q, the asymmetry parameter η ($\eta = 0$ for a symmetric field gradient), the z component of the electric field gradient at the nucleus ($\partial^2 V/\partial z^2$), caused by fluctuations of the environment of the ion, and τ_c, the correlation time that characterizes these fluctuations. The appropriate equation for this quadrupolar relaxation is (187)

$$\frac{1}{T_1} = \frac{1}{T_2} = \frac{3(2I+3)}{40I^2(2I-1)}(1+\frac{\eta^2}{3})[\frac{eQ}{h}(\frac{\partial^2 V}{\partial z^2})]^2 \tau_c \qquad (16)$$

When the field is not spherically symmetric, as is the case for alkali cations trapped by crown ethers, the lines can be very broad because of modulation of the z-component of the field gradient as the complex rotates in solution. However, even for systems that show time-average spherical symmetry such as the hydrated alkali cations, time fluctuations of the short-range repulsions can lead to marked quadrupolar broadening. Hertz (188), who treated the problem of aqueous alkali cation linewidths at infinite dilution, found good agreement between calculated and observed linewidths. Although the treatment is somewhat parameterized, it certainly accounts well for the changes that are observed from one alkali cation to another. According to Hertz the stronger the first-layer coordination, the narrower will be the line, since relaxation effects must then arise from molecules outside the first coordination sphere. Weakening of the first-layer solvent binding will tend to broaden the lines (after correction for viscosity and size differences) in solvents that do not strongly coordinate the cation. Of course, deviations from a spherically symmetric electric field caused by steric factors also tend to broaden the lines and can be the dominant factors in many nonaqueous solvents. The effect of deviations from spherical symmetry is dramatically demonstrated by the linewidth of Na^+, Na^+C222, and $Na^+ \cdot 18C6$ in methylamine. The corresponding linewidths (full widths at half-height) are 8, 31, and ~200 Hz, respectively (12). At least part of this broadening represents an increase in the rotational correlation time. However, the change from the cryptate complex to the crown complex is undoubtedly mainly the result of the decreased symmetry of the field at the nucleus in the latter case.

5.3 Determination of Complexation Constants from Alkali Metal Chemical Shifts and/or Linewidths

When the exchange rate of M^+ with M^+C is fast on the NMR time scale, only a single exchange-averaged line results. If the free and the complexed cation have different chemical shifts, the variation of the chemical shift with $[C]/[M^+]$ can be used to evaluate the complexation constant. The observed chemical shift δ_{obs} is related to those of M^+ and M^+C by the equation

$$\delta_{obs} = X_{M^+}\delta_{M^+} + X_{M^+C}\delta_{M^+C} \qquad (17)$$

in which $X_{M^+} = 1 - X_{M^+C}$ is the relative fraction of uncomplexed ions. Of course, if M^+ exists in appreciable concentrations as ion pairs $M^+ \cdot X^-$ or $M^+C \cdot X^-$, then the contributions of these species must be included in the calculation of δ_{obs}. To evaluate the complexation constant at a given temperature (for the case of formation of only a 1:1 complex) three sets of measurements are required:

1. Chemical shift versus concentration for a salt M^+, X^-. If δ is independent of concentration, effects of ion pairing are absent. If δ depends on concentration, the data may be used to obtain the ion-pair-formation constant.
2. Chemical shift versus concentration for the complex M^+C,X^-. If δ is independent of concentration, effects of ion pairs between M^+C and X^- are absent.
3. Chemical shift versus molar ratio $[C]/[M^+]$ for a fixed total concentration of M^+. If this gives two straight lines with an intersection at 1:1 molar ratio, the association constant is too large to permit its evaluation by this method. On the other hand, if the complexation constant is very small, the determination of δ_{M^+C}, even by computer fitting of Eq. 17 to the data, becomes difficult. Depending on the precision of the measurements and the lowest concentration that can be studied, the range of reliable values of the association constant lies between 10^1 and 10^5. Numerical methods are available for evaluating the association constant and limiting chemical shift by a nonlinear least-squares procedure (169). Provision can be made for the inclusion of calculated activity coefficients by using the Debye-Hückel equation if desired. A similar variation of linewidth and/or relaxation time could also be used to evaluate complexation constants. This method is particularly useful when one of the two sites has a much larger linewidth than the other.

5.4 Measurement of Exchange Rates by Alkali Metal NMR

Lehn and co-workers (22) used proton NMR to show that the rate of release of alkali cations from cryptands can be slow on the NMR time scale. The large difference in linewidth between the cation complexed by a crown ether (dibenzo-18-crown-6) and the solvated cation was used by Shchori and co-workers (38, 40) and by Shporer and Luz (44) to study the rate of release of the cation from the crown ether complex. Similar line broadening has been used to study exchange rates of Na^+ with a series of biologically important complexing agents (189-191). In 1973 Ceraso and Dye (39) reported the first measurement of exchange rates in which clearly distinguishable separate alkali metal NMR signals could be observed (see Figure 9). The solvated and cryptated sodium ions in EDA gave separate ^{23}Na NMR peaks at room temperature which coalesced into a single peak at 50°C. Complete NMR lineshape analysis was used to evaluate rate constants and activation enthalpies and entropies. This method has

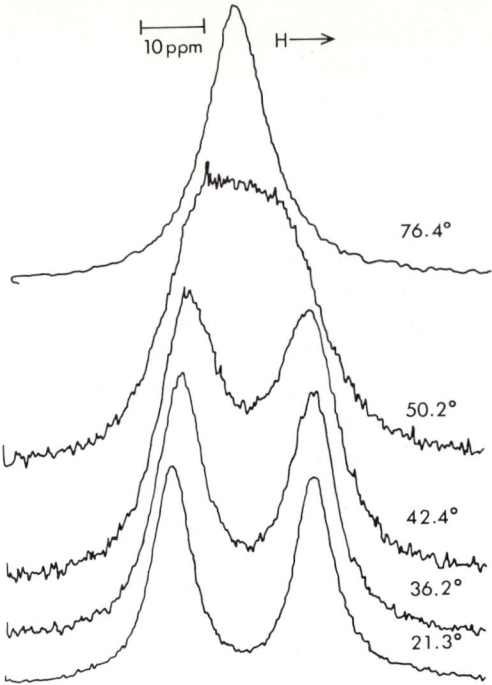

Figure 9 ^{23}Na NMR spectra at various temperatures in ethylenediamine. Solution contains 0.3 M C222 and 0.6 M NaBr. Taken from Ref. 51.

since been used for various crown and cryptate complexes containing Na$^+$, Li$^+$, and Cs$^+$. A detailed analysis of experimental conditions and the fit of the Bloch equations to the experimental lineshapes has been given for the two-site exchange case by Ceraso and co-workers (51). The use of such a generalized nonlinear least-squares procedure is preferred over the common method of lineshape simulation, since the former provides statistical information about the uncertainties in the fitting parameters and is not subject to arbitrary decisions about the "best" lineshape to use. For complex NMR patterns, however, it is necessary to simulate the NMR spectra by the density-matrix technique, and a least-squares fit of the equations to the data can be prohibitively expensive.

The determination of exchange rates and activation parameters by NMR studies of complexes is particularly favorable because the linewidths and resonance frequencies in the absence of exchange can be accurately determined *as functions of temperature.* Therefore the only adjustable parameter other than those used to determine instrumental response is the exchange time τ, which is related to the rate constants that describe the mechanism of exchange.

5 Thermodynamics and Kinetics of Cation Complexation

For all cases studied, the exchange mechanism obtained from the relaxation time τ as a function of the concentrations of M^+ and C is the dissociative mechanism

$$M^+ + C \underset{k_{-1}}{\overset{k_1}{\rightleftharpoons}} M^+C \qquad (18)$$

This yields

$$\frac{1}{\tau} = \frac{k_{-1}}{p_B} \qquad (19)$$

in which p_B is the fractional concentration of M^+ present in the complex. That is,

$$p_B = \frac{[M^+C]}{[M^+] + [M^+C]} \qquad (20)$$

An alternative mechanism

$$*M^+ + M^+C \underset{k_2}{\overset{k_2}{\rightleftharpoons}} *M^+C + M^+ \qquad (21)$$

predicts the exchange-time relation

$$\frac{1}{\tau} = 2k_2 [M^+]_t \qquad (22)$$

in which $[M^+]_t$ is the total concentration of cation in its solvated and complexed form. Since the exchange time actually depends inversely on p_B and is independent of the total concentration $[M^+]_t$ (for those cases for which this has been tested), the dissociative mechanism is indicated. This mechanism is also indicated by proton NMR studies (22) and by stopped-flow measurements (45, 56).

A limitation imposed by this mechanism is that only k_{-1} can be determined from lineshape analysis of either alkali metal NMR (in which case $[M^+]_t$ is greater than $[C]_t$) or by proton NMR (in which case $[C]_t$ is greater than $[M^+]_t$). To obtain k_1 it is necessary to know the equilibrium constant. In principle the lineshape could be used to adjust both τ and p_B to obtain both k_{-1} and k_1. However, this is only feasible when the complexation constant is sufficiently small that appreciable concentrations of both free complexing agent and uncomplexed cations are present in solution. When this is the case, the exchange is usually rapid and only a single population weighted-average signal is obtained.

5.5 Intepretation of Complexation Constants and Exchange Rates

A number of compilations have been made of $\Delta G°$, $\Delta H°$, and $\Delta S°$ for the complexation of alkali cations by various macrocyclic complexing agents in a variety of solvents (20, 30, 33). An excellent and comprehensive interpretation of the enthalpic and entropic contributions for complexation in aqueous solutions has been given by Kauffmann, Lehn, and Sauvage (33), and will not be repeated here. Data for exchange rates are also available from a number of sources. In this section we limit our discussion to the general trends that can be extracted from these data, particularly the effect of solvent and complexing agent on the thermodynamics and kinetics of complexation.

The thermodynamics of complexation of M^+ by macrocyclic complexing agents shows a remarkable range of enthalpic and entropic contributions. In order for M^+C to be detected in reasonably dilute solutions, it is necessary that $\Delta G°$ be negative (hypothetical ideal unit molalities or molarities are used as standard states). However, only the range $-1 > \Delta G° \gtrsim -7$ kcal mole^{-1} is accessible from NMR measurements, since a more positive $\Delta G°$ than -1 kcal mole^{-1} gives no detectable complexation ($K < 10$) whereas a more negative $\Delta G°$ than -7 kcal mole^{-1} gives such a large complexation constant ($K > 10^5$) that it cannot be measured directly by a "titration" technique. When potentiometric measurements are possible (usually only in aqueous solutions or in mixed solvents containing water), this range can be extended to about $\Delta G° = -15$ kcal mole^{-1}. Of course $\Delta H°$ can be measured calorimetrically regardless of the strength of the complexation.

The contributions to $\Delta H°$ and $\Delta S°$, discussed in detail by Kauffmann and co-workers (33), are just summarized here. The value of $\Delta H°$ is influenced by (a) enthalpy of bond replacement (ligand-to-ion interaction replaces the first solvation shell interaction), (b) the Born term (arises from changes in ion-solvent interaction beyond the first solvation or ligation shell), (c) changes in interbinding site repulsions (these are "locked in" by the framework of the macrocyclic ligand, but for the solvated ion lead to destabilization of high coordination number solvation), (d) changes in ligand solvation enthalpy, (e) steric deformations of the ligand caused by inclusion of the cation.

Changing the solvent from water to a poorer donor solvent, especially if the dielectric constant also decreases, is expected to yield negative contributions to $\Delta H°$ from effects (a), (b), and (d). We expect effect (e) to be relatively solvent independent, but effect (c) is difficult to assess; certainly it depends on the size of the solvent molecules and on the degree of localization of the solvent dipole moment.

Again, following the description of Kauffmann and co-workers, the entropy change accompanying complex formation is determined by the following: (a) entropy increase caused by release of the solvation shell about the ion, (b) entropy increase caused by release of solvent bound to the free ligand, (c) translational entropy loss on formation of a single complex from two species (for the usual standard states of 1 mole l^{-1} this can amount to -15 to -25 e.u. depending on the motional freedom remaining in the complex); (d) decrease in solvent entropy caused by solvent structure formation about the large cryptated cation (presumably this is only important in highly structured solvents such as water), (e) changes in ligand internal entropy caused by orientation, rigidification, and conformational changes.

It would be very useful to be able to describe the solvent dependence of $\Delta H°$ and $\Delta S°$ of complexation, but few data are available. Results for water and for water-ethanol mixtures suggest that the major effect is a more negative value of $\Delta H°$ as water is replaced by methanol, although this is usually accompanied by a negative shift of $\Delta S°$ as well. The result is to make $\Delta G°$ markedly more negative, that is, to increase the stability of the complexes. Since the selectivity is largely dictated by $\Delta H°$, the effect of a more negative ionic solvation enthalpy is to increase the selectivity in methanol-water mixtures compared to that in water alone.

The complexation of cesium cations by various crowns and cryptands in a number of solvents is weak enough to permit determination of the complexation constants by ^{133}Cs NMR (54, 55, 57, 167, 169). The results are given in Table 3. Except for the anomalous position of pyridine in the solvent list, a decrease in the donor number of the solvent leads to an increase in the stability of the complex. This is the expected effect when enthalpy dominates.

The equilibrium constant for the complexation of Cs$^+$ by C222 has been studied as a function of temperature in acetone, propylene carbonate (PC), and N,N-dimethylformamide (DMF) (55). Although this case is complicated by the formation of both exclusive and inclusive complexes, the overall values of $\Delta H°$ and $\Delta S°$ for formation of the inclusive complex show that the complex is enthalpy stabilized and entropy destabilized. The data for acetone, PC, and DMF yield $\Delta S°$ values of -32, -21, and -19 cal mole^{-1} deg^{-1}, and $\Delta H°$ values of -15.4, -11.5, and -8.3 kcal mole^{-1}, respectively. The data for acetone are very limited and should be interpreted with caution. However, for PC and DMF the results should be valid to within at least ± 3 cal mole^{-1} deg^{-1} for $\Delta S°$ and ± 0.8 kcal mole^{-1} for $\Delta H°$ (twice the standard deviation estimates). The values of $\Delta S°$ are consistent with the value of -23.7 cal mole^{-1} deg^{-1} found by Kauffmann and co-workers (33) for C222 complexation of Cs$^+$ in 95 wt. % methanol-water mixtures. An interesting question arises in connection with $\Delta S°$: what is the origin of the large negative value of $\Delta S°$? Its appearance in five solvents (the value of Cs$^+$ complexation by C222 in H$_2$O is unknown, but Rb$^+$ yields $\Delta S° = -19.8$ cal mole^{-1} deg^{-1}) seems to rule out solvent structure making by the large cryptated cation. Values of $\Delta S°$ for crown ether complexation of Cs$^+$ are limited to water and 70 wt. % methanol-water mixtures. For 18C6 the value of $\Delta S°$ changes from

Table 3 Formation constants of 1:1 and 2:1 (C/Cs$^+$) Complexes of Cs$^+$ with 18C6, Dibenzo-18-crown-6 (DBC), and Dicyclohexano-18-crown-6 (DCC) in Various Solvents at 25°C (from Ref. 57)

Solvent	18C6	DBC	DCC (mixture)
Pyridine	$K_1 > 5 \times 10^5$ $K_2 = 74 \pm 2^c$	$(7 \pm 2) \times 10^3$ 230 ± 20	$> 10^5$ a
Propylene carbonate	$K_1 = (1.5 \pm 0.6) \times 10^4$ $K_2 = 10.9 \pm 4$	$\sim 10^3$ a	$\sim 10^4$ a
Acetone	$K_1 > 2 \times 10^5$ $K_2 = 34.0 \pm 0.5$	$> 10^3$ a	$> 10^4$ a
Dimethylformamide	$K_1 = (9 \pm 3) \times 10^3$ $K_2 = 2.44 \pm 0.05$	$K_1 = 30 \pm 3^b$	$K_1 = (2.8 \pm 0.9) \times 10^3$
Dimethyl sulfoxide	$K_1 = (1.1 \pm 0.1) \times 10^3$ $K_2 = 1.0 \pm 0.4$	$K_1 = 22 \pm 3^b$	$K_1 = (1.6 \pm 0.1) \times 10^2$
Acetonitrile	$K_1 > 10^4$ $K_2 = 3.7 \pm 0.6$	$K_1 = 35 \pm 2^b$	$K_1 > 10^4$ a

^{a}The variation of the chemical shift suggests formation of the 2:1 (C/Cs$^+$) complex, but K_2 cannot be determined.
^{b}It is possible that both 1:1 and 2:1 complexes form with DBC in these solvents. However, a good fit of the data is obtained by assuming formation of only the 1:1 complex.
^{c}Uncertainties are estimated standard deviation.

−8.1 cal mole^{-1} deg^{-1} in water to −14.1 cal mole^{-1} deg^{-1} in 70% methanol. It would be of great interest to have $\Delta S°$ values for crown ether complexation of Cs$^+$ in other solvents, especially in view of the anomalous values of $\Delta S^{\ddagger}$ described below.

In addition to negative entropy contributions from structure-making in the solvent, changes in configurational entropy of the ligand may well be very pronounced. Particularly with a large ion such as cesium, the ring or cavity is strained so that freedom of the "backbone" to take on many configurations is severely limited. The configurational entropy of the free ligand should be large, since at least the —CH$_2$— groups should be free to rotate. This effect is counterbalanced by solvation of the free ligand which is lost when the complex forms. If Δn represents the average loss in the number of conformations available to the ligand on complex formation, then the conformational contribution to the entropy of complexation is given by

$$\Delta S°_{conf} = - R \ln(\Delta n) \qquad (23)$$

Since this must be large enough to overcome the entropy *increase* associated with desolvation of the ion and the free ligand, we might expect $\Delta S°_{conf}$ to be at least as negative as −30 cal mole^{-1} deg^{-1}. This would require that Δn be $\approx$ 3 ×10^6. This corresponds to an average "conformation-locking" of about five conformers for each —CH$_2$-CH$_2$— unit in cryptand C222, assuming independence of these units.

Comparison of the values of $\Delta H^{\ddagger}$ and $\Delta S^{\ddagger}$ for the release of Li$^+$ (43), Na$^+$ (51), and Cs$^+$ (55, 57) from [2]-cryptates with the corresponding overall values of $\Delta H°$ and $\Delta S°$ leads one to conclude that the transition state resembles the final state of the solvated ion and the free cryptand. For example, $\Delta S°$ for the *release* of Cs$^+$ from an inclusive complex with C222 in DMF is +19 cal mole^{-1} deg^{-1}, whereas $\Delta S^{\ddagger}$ is +17 cal mole^{-1} deg^{-1} (55). The corresponding value of $\Delta H°$ is +8.3 kcal mole^{-1}, whereas $\Delta H^{\ddagger}$ is 12.9 kcal mole^{-1}. A puzzling contrast can be made with the behavior of crown ethers. $\Delta S^{\ddagger} = -14$ cal mole^{-1} deg^{-1} for the release of Cs$^+$ from DCC(A) in PC and from 18C6 in pyridine (57). The value of $\Delta S°$ is not known, but the results in water and in water-methanol mixtures suggest that it is probably positive. This may mean that the transition state is appreciably different from the final state of the free crown ether and solvated cation. Perhaps the conformations accessible to the crown ether molecule in the transition state are limited. More thermodynamic data in a variety of nonaqueous solvents are definitely needed to answer some of these questions.

5.6 Sandwich Complexes of Cs$^+$ with Crown Ethers

The formation of cation complexes with crown ethers with stoichiometry other than 1:1 was reported early in the studies of complexation by both Pedersen

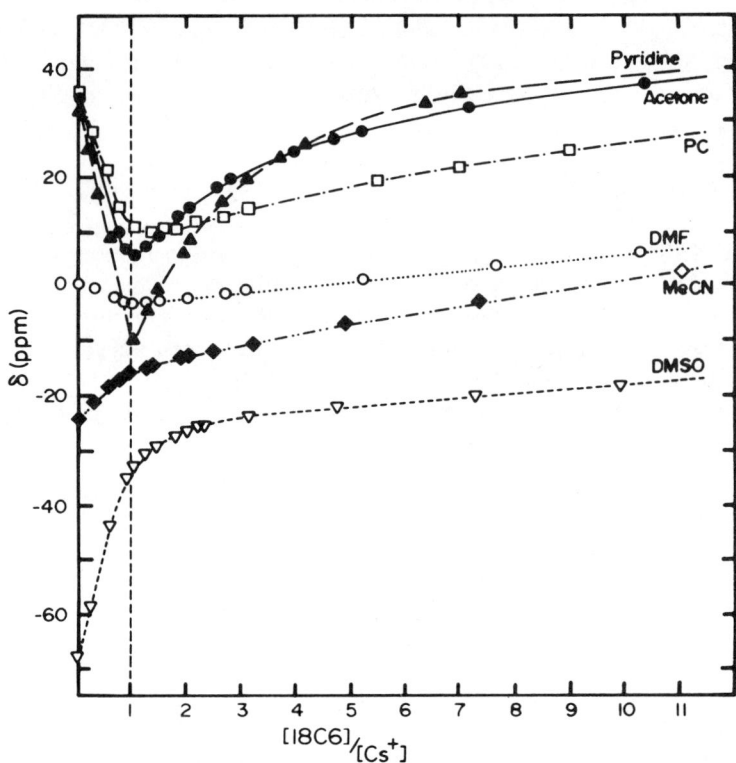

Figure 10 ^{133}Cs NMR chemical shifts vs. 18C6/Cs$^+$ molar ratio in various solvents. The solutions contain 0.01 M cesium tetraphenylborate. Taken from Ref. 57.

and Frensdorff (18, 19, 23, 28). Apparently, when the cation radius is larger than the hole size, sandwich complexes, in which the cation is located between two molecules of crown ether, can form in solution. The formation of such complexes is readily detected by alkali metal NMR techniques, especially when the 1:1 complex is strong. The variation of the ^{133}Cs chemical shift with molar ratio, $[18C6]/[Cs^+]_t$ in a number of solvents is shown in Figure 10 (57). In two solvents, DMF and dimethylsulfoxide (DMSO), it was possible to obtain both K_1 and K_2, referring to the first and second complexation steps, respectively,

$$Cs^+ + 18C6 \underset{}{\overset{K_1}{\rightleftharpoons}} Cs^+ \cdot 18C6 \qquad (24)$$

$$Cs^+ \cdot 18C6 + 18C6 \underset{}{\overset{K_2}{\rightleftharpoons}} Cs^+ \cdot (18C6)_2 \qquad (25)$$

by a fit of all of the data to the appropriate equations. For the other solvents studied K_2 could be obtained, but K_1 was so large that only a lower limit could be evaluated. Studies of the ^{133}Cs NMR spectra in the presence of dibenzo-18-crown-6 (DBC) and dicyclohexano-18-crown-6 (DCC) also gave evidence for 2:1 complexes, but in most cases the variation of chemical shift with $[C]/[Cs^+]_t$ after the 1:1 stoichiometry was reached was too slight to permit evaluation of K_2. The [2]-cryptands C222 and monobenzo C222 showed no evidence of complexes of shoichiometry higher than 1:1. Relaxation effects have also been used to detect the formation of sandwich complexes (170).

5.7 Evidence for Exclusive [2]-Cryptate Complexes

Inclusive complexes with [2]-cyptands have chemical shifts that are relatively independent of solvent and temperature. The same is true for the 2:1 sandwich complexes of 18C6 with Cs^+. However, the 1:1 crown complexes and exclusive [2]-cryptate complexes have chemical shifts that depend on solvent and temperature. This variation of the limiting chemical shift, and the fact that Cs^+ complexation by cryptands C221 and even C211 occurs even though these ligands clearly have cavities that are too small for the bulky cesium cation, led us to postulate (55) the simultaneous existence of two types of complex, exclusive and inclusive, between Cs^+ and cryptand C222.

The limiting chemical shift for the 1:1 complex $Cs^+\cdot C222$ is strongly dependent on solvent and temperature, but extrapolates to a solvent-independent shift of -245 ± 5 ppm [paramagnetic from Cs^+(aq)] at low temperatures in acetone, PC, and DMF. Both the temperature dependence of the limiting shift and the lineshape analysis at temperatures low enough to give separate signals from Cs^+ and $Cs^+\cdot C222$ are consistent with the intermediate formation of an exclusive complex. We visualize such a complex as a "wraparound" complex in which all the ether oxygens and the amine nitrogens can interact with Cs^+ but which has one face of the cryptand open so that interaction with the solvent and/or counterion can still occur. Down to the freezing point of the solvents used, the in-out exchange process is rapid on the NMR time scale, even well after the exchange with free Cs^+ has been "frozen out." However, some broadening of the $Cs^+\cdot C222$ peak occurs, indicating that the in-out exchange may be slowing down. The inclusive complex is only slightly favored over the exclusive complex, with $\Delta G°$ ranging from -0.8 kcal mole^{-1} at 300°K to -1.5 kcal at 200°K.

6 ALKALI METAL NMR STUDIES OF METAL SOLUTIONS

The alkali metal NMR studies described in the preceding section were initiated as a preview to the study of alkali metal NMR spectra in metal solutions. Although the NMR peaks of monomers or any species that is paramagnetic or exchanges rapidly with a paramagnetic species would be too broad to detect, we felt that the diamagnetic alkali metal anions might be detectable. This has, indeed, proven to be the case, and the resulting spectra tell us much about the nature of M^-.

6.1 Static NMR Spectra

The ability to detect M^- by its NMR spectrum depends on three factors: (a) The concentration of M^- must be high enough for detection. Because of the narrow line, this requires only approximately 0.02, 0.05, and 0.005 M solutions for Na^-, Rb^-, and Cs^-, respectively. Because the signal of M^+C is broad, however, its detection requires higher concentrations. (b) The exchange of M^+C with M^- must be slow on the NMR time scale. Since the rate of release of M^+ from complexes with [2]-crypands is slow, this does not prove to be a problem. However, with crown ethers exchange can be rapid (c) Paramagnetic broadening by solvated electrons and/or monomers must be minimal.

6.1.1 Detection of the ^{23}Na NMR Signal from Na^-

Because sodium solutions in the presence of cryptand C222 contain almost exclusively $Na^+ \cdot C222$ and Na^-, and because the loss of Na^+ from the former is slow on the NMR time scale, it is easy to detect ^{23}Na NMR signals (10) from both Na^+C222 and Na^-, as shown in Figure 11. Both signals have chemical shifts that are independent of solvent within ±1 ppm. Because Na^+ in the cryptate complex is shielded from the solvent, solvent independence for the cation is expected. The linewidth of Na^+C in the presence of Na^- in THF is comparable to the (extrapolated) linewidth of Na^+C in the presence of tetraphenylborate (51).

The chemical shift of Na^- from its value in the gas phase [calculated from the experimentally known shift of Na(g) with the aid of analytic Hartree-Fock wavefunctions (171)] is -0.3, -1.0, and -1.2 ppm (downfield) in THF, ethylamine, and methylamine, respectively. These small shifts probably represent essentially zero shift within the accuracy of the calculation, although the slight paramagnetic shift and the distinct broadening of the Na^- signal (3, 6-9, 11 Hz, respectively) as one moves from THF to methylamine are real. The absence of a paramagnetic shift of Na^- is the strongest evidence we have that this species is a "genuine" anion in solution with two electrons in the outer s-orbital. If it consisted of a *solvated* cation interacting with a pair of electrons, there would be a paramagnetic shift at the sodium nucleus because of solvent lone-pair donation to Na^+. The narrowness of the line is also indicative of a spherically symmetric species with weak solvent interactions.

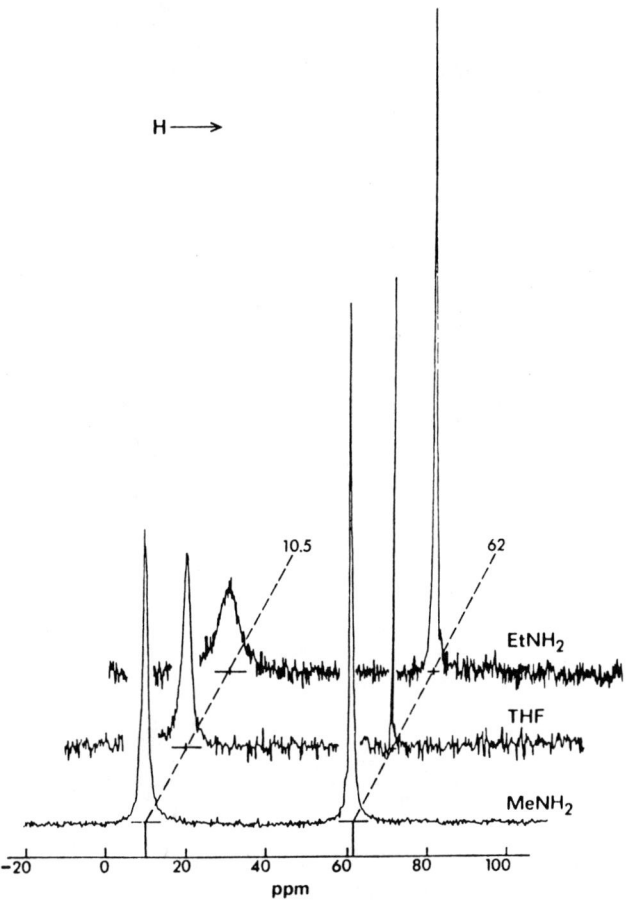

Figure 11 ^{23}Na NMR spectra of Na$^+$C222,Na$^-$ solutions ($\sim$0.1 M) in three solvents. All chemical shifts are referenced to aqueous Na$^+$ at infinite dilution. Negative chemical shifts are paramagnetic. Taken from Ref. 12.

6.1.2 NMR Spectra of Rb$^-$ and Cs$^-$

Solutions of rubidium in ethylamine and in THF in the presence of cryptand C222 show a single peak at −26.2 and −14.4 ppm, respectively, relative to the gaseous atom (12). The corresponding linewidths are 220 and 15 Hz in the two solvents. Since ^{87}Rb has a large quadrupole moment, one would not expect Rb$^+$C222 to be detectable at the concentrations used. The increase in width of the Rb$^-$ signal and its downfield shift on changing from THF to ethylamine probably

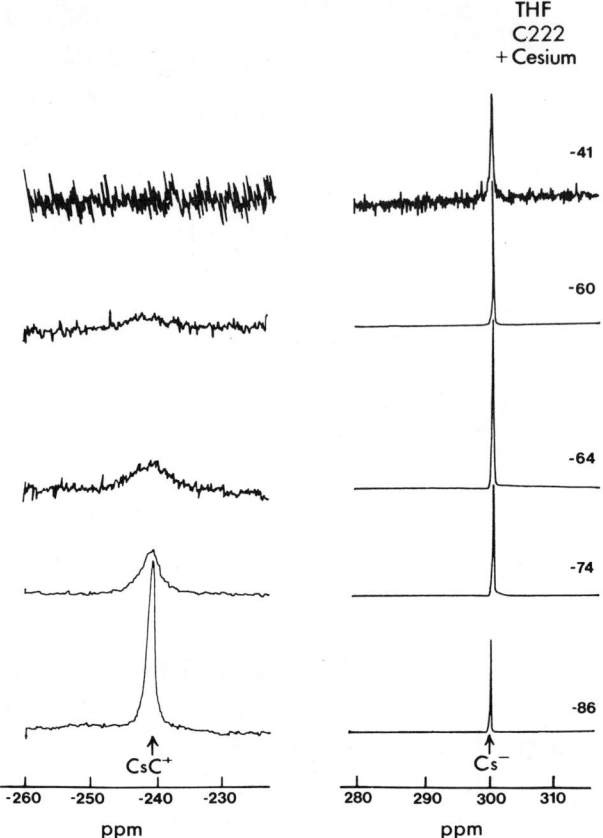

Figure 12 Effect of temperature on the ^{133}Cs NMR signal for a solution of Cs$^+$C222,Cs$^-$ in tetrahydrofuran (192). The amplitude of the Cs$^+$C222 signal on the left has been multiplied by 15 for clarity of presentation. However, the areas under the peaks are equal within experimental error.

represent the effect of interaction of Rb$^-$ with paramagnetic species such as the monomer or ion pair and/or the solvated electron. Any of the following exchange processes would tend to broaden the line of M$^-$ and shift it paramagnetically:

$$*M + M^- \rightleftharpoons *M^- + M \qquad (26)$$

$$M^- \rightleftharpoons M + e^-_{solv} \qquad (27)$$

$$M^- \rightleftharpoons M^+ + 2e^-_{solv} \qquad (28)$$

When the NMR signal of Cs⁻ was first detected in THF [−52.3 ppm from Cs(g), linewidth = 10 Hz at 202°K], we were puzzled by the absence of a signal for Cs⁺C222 (12). The linewidths of ^{133}Cs⁺ resonances are generally narrow, and this should have been easily detected. It has since become clear (192) that some process other than exchange of Cs⁺C222 and Cs⁻ is broadening the cation resonance. Figure 12 shows the effect of temperature on the spectra in THF. At low temperatures signals of both Cs⁺·C222 and Cs⁻ are apparent, but the peak from Cs⁺·C222 rapidly broadens and disappears into the baseline without a corresponding broadening of the Cs⁻ signal. This and other dynamic phenomena are discussed in Section 6.2.

6.1.3 Conclusions from Static Spectra

The NMR spectra confirm the existence of spherically symmetric alkali metal anions with paired electrons in the outer s-type orbital. The signals from Na⁻ and Rb⁻ in THF probably represent the narrowest known signals from the corresponding nuclei because of high spherical symmetry and effective shielding of the p-electrons. Even the residual linewidth appears to result from exchange processes rather than from natural linebroadening. The sensitivity of the peak of M⁻ to paramagnetic and exchange effects is considered in the next section.

6.2 Dynamic Effects on NMR Spectra of Metal Solutions

When solutions of sodium in methylamine in the presence of either 18C6 or 15C5 are examined by ^{23}Na NMR techniques, the effects of cation-anion exchange readily become apparent. This is shown in Figure 13 for the 18C6 case (12, 192). A classic coalescence pattern is observed which can be well represented by a two-site-exchange process. The net effect is the exchange of the sodium nucleus between the complex with 18C6 and the sodium anion. However, the variation of τ with concentration makes it clear that the direct exchange process

$$*Na^- + Na^+ \cdot 18C6 \rightleftharpoons *Na^+ \cdot 18C6 + Na^- \tag{29}$$

is not responsible. Even the simple dissociative mechanism

$$Na^+ \cdot 18C6 \rightleftharpoons Na^+ + 18C6 \tag{30}$$

$$*Na^+ + Na^- \rightleftharpoons *Na^- + Na^+ \tag{31}$$

is not compatible with the observed variation of the exchange time with concentration and with the concentration of 18C6. This indicates that processes such as

$$Na^- \rightleftharpoons Na + e^-_{solv} \tag{32}$$

and/or

$$Na^- \rightleftharpoons Na^+ + 2e^-_{solv} \qquad (33)$$

may be important. The complete mechanism has not yet been determined.

The disappearance of the signal of $Cs^+ \cdot C222$ before that of Cs^- for solutions in THF as shown in Figure 12 cannot be explained by an overall two-site-exchange mechanism, since this would require that both peaks broaden by the same amount. Evidently a third site is involved. The chemical shift of the Cs^+C222 signal in THF [−250 ppm, downfield from $Cs^+(aq)$] indicates the presence of the inclusive complex. Failure to detect any other species probably means that a paramagnetic species such as the cesium monomer Cs is involved in the

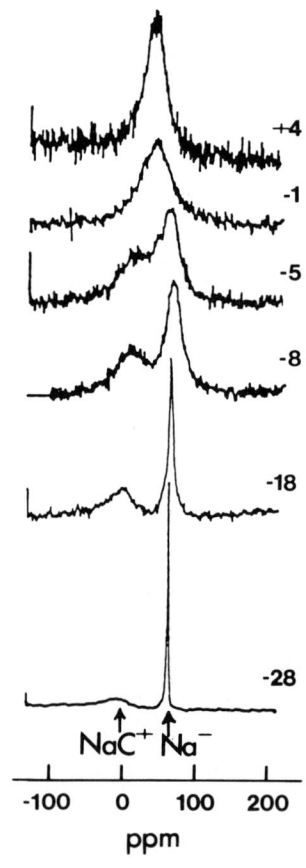

Figure 13 ^{23}Na NMR spectra at various temperatures for solutions of $Na^+ \cdot 18C6\ Na^-$ in methylamine.

exchange. This would require that an overall exchange process such as

$$*Cs + Cs^+ \cdot C222 \rightleftharpoons *Cs^+ \cdot C222 + Cs \qquad (34)$$

be faster than

$$*Cs + Cs^- \rightleftharpoons *Cs^- + Cs \qquad (35)$$

The situation is even more puzzling for solutions of cesium in ethylamine in the presence of C222. The optical spectra clearly indicate the presence of high concentrations of Cs^-. However, the only species detected by ^{133}Cs NMR is the complexed cation $Cs^+ \cdot C222$. As the temperature is increased above about $-40°C$, this signal also broadens and disappears. The simplest explanation is that an exchange such as that represented by Eq. 35 is rapid compared with that represented by (34), although the reason for the reversal of relative rates from THF to ethylamine is certainly not clear. Obviously more work is needed to understand these complex dynamic effects. Nevertheless, because of the wealth of information available from chemical shifts, linewidths, and variation of the lineshape with exchange, alkali metal NMR will surely prove to be a valuable probe of the structure of metal solutions in amines and ethers.

ACKNOWLEDGMENTS

Support of the research described in this chapter by the U.S. Energy Research and Development Adminstration under Contract No. EY-76-S-02-0958, and by the National Science Foundation under Grant No. DMR77-22975 is gratefully acknowledged. Thanks are due to M. DaGue, P. Smith, and M. Yemen for help with the references and for useful comments about the manuscript.

REFERENCES

1. For a detailed discussion of the properties of metal-ammonia solutions see: J. C. Thompson, *Electrons in Liquid Ammona,* Oxford University Press, Oxford, 1976.
2. Review articles on metal-ammonia solutions include the following: (a) C. A. Kraus, *J. Chem. Educ.,* **30**, 83 (1953); (b) W. L. Jolly, *Prog. Inorg. Chem.,* **1**, 235 (1959); (c) M. C. R. Symons, *Q. Rev.,* **13**, 99 (1959); (d) U. Schindewolf, *Angew. Chem.,* **80**, 165 (1968); *Angew. Chem., Int. Ed. Engl.,* **7**, 190 (1968); (e) T. P. Das, *Adv. Chem. Phys.,* **4**, 303 (1962); (f) J. L. Dye, *Sci. Am.,* **216**, 77 (February 1967); (g) M. H. Cohen and J. C. Thompson, *Adv. Phys.,* **17**, 857 (1968).
3. A collection of "landmark papers" in the metal-ammonia field has been given by W. L. Jolly, *Metal-Ammonia Solutions,* Dowden, Hutchinson, Ross, Stroudsberg, Pennsylvania, 1972.

4. Papers resulting from conferences in the field of metal-ammonia solutions and the broader general area of electrons in fluids include the following: (a) G. Lepoutre and M. J. Sienko Ed., *Metal-Ammonia Solutions, Colloque Weyl I,* W. A. Benjamin, New York, 1964; (b) R. F. Gould (Ed.) *Advances in Chemistry Series,* No. 50, American Chemical Society Publications, Washington, D. C., 1965; (c) J. J. Lagowski and M. J. Sienko (Ed.), *Metal-Ammonia Solutions, Colloque Weyl II,* IUPAC, Butterworths, London, 1970; (d) U. Schindewolf (Ed.), *Ber. Bunsenges. Phys. Chem.,* **75** (1971); (e) J. Jortner and N. R. Kestner (Ed.), *Electrons in Fluids, Colloque Weyl III,* Springer-Verlag, Berlin, 1973; (f) "Colloque Weyl IV. Electrons in Fluids—The Nature of Metal-Ammonia Solutions," *J. Phys. Chem.,* 79(26) (1975); (g) *Can. J. Chem.* **55**(11) 1796-2277 (1977), *International Conference on Electrons in Fluids,* Banff, Alberta, Canada, September 5-11, 1976.
5. J. L. Dye, M. G. DeBacker, V. A. Nicely, *J. Am. Chem. Soc.,* **92**, 5226 (1970).
6. J. L. Dye, M. T. Lok, F. J. Tehan, R. B. Coolen, N. Papadakis, J. M. Ceraso, and M. DeBacker, Ref. 4d, p. 659.
7. J. L. Dye, in Ref. 4e, p. 77.
8. M. T. Lok, F. J. Tehan, and J. L. Dye, *J. Phys. Chem.,* **76**, 2975 (1972).
9. J. L. Dye, J. Am. Ceraso, M. T. Lok, B. L. Barnett, and F. J. Tehan, *J. Am. Chem. Soc.,* **96**, 608 (1974).
10. J. M. Ceraso and J. L. Dye, *J. Chem. Phys.,* **61**, 1585 (1974).
11. F. J. Tehan, B. L. Barnett, and J. L. Dye, *J. Am. Chem. Soc.,* **96**, 7203 (1974).
12. J. L. Dye, C. W. Andrews, and J. M. Ceraso, *J. Phys. Chem.,* **79**, 3076 (1975).
13. J. L. Dye, C. W. Andrews, and S. E. Mathews, *J. Phys. Chem.,* **79**, 3065 (1975).
14. J. L. Dye, *Pure Appl. Chem.,* **49**, 3 (1977).
15. J. L. Dye, *J. Chem. Educ.,* **54**, 332 (1977).
16. J. L. Dye, *Sci. Am.,* **237**, 92 (July 1977).
17. J. L. Dye, M. R. Yemen, M. G. DaGue, and J.-M. Lehn, *J. Chem. Phys.,* **68**, 1665 (1978).
18. C. J. Pedersen, *J. Am. Chem. Soc.,* **89**, 7017 (1967); **92**, 386 (1970).
19. C. J. Pedersen, *Fed. Proc.,* **27**, 1305 (1968).
20. R. M. Izatt, J. H. Rytting, D. P. Nelson, B. C. Haymore, and J. J. Christensen, *Science* **164**, 443 (1969).
21. B. Dietrich, J.-M. Lehn, and J. P. Sauvage, *Tetrahedron Lett.,* 2885, 2889 (1969).
22. J.-M. Lehn, J. P. Sauvage, and B. Dietrich, *J. Am. Chem. Soc.,* **92**, 2916 (1970).
23. H. K. Frensdorff, *J. Am. Chem. Soc.,* **93**, 600 (1971).
24. R. M. Izatt, D. P. Nelson, J. H. Rytting, B. L. Haymore, and J. J. Christensen, *J. Am. Chem. Soc.,* **93**, 1619 (1971).
25. J.-M. Lehn and J. P. Sauvage, *Chem. Commun.,* 440 (1971).
26. J. J. Christensen, J. O. Hill, and R. M. Izatt, *Science,* **174**, 459 (1971).
27. W. E. Morf and W. Simon, *Helv. Chim. Acta,* **54**, 2683 (1971).
28. C. J. Pedersen and H. K. Frensdorff, *Angew. Chem., Int. Ed. Engl.,* **11**, 16 (1972).
29. J.-M. Lehn, *Struct. Bonding,* **16**, 1 (1973).
30. J. J. Christensen, D. J. Eatough, and R. M. Izatt, *Chem. Rev.,* **74**, 351 (1974).
31. J.-M. Lehn and J. P. Sauvage, *J. Am. Chem. Soc.,* **97**, 6700 (1975).
32. G. A. Anderegg, *Helv. Chim. Acta,* **58**, 1218 (1975).

References

33. E. Kauffmann, J.-M. Lehn, and J. P. Sauvage, *Helv. Chim. Acta,* **59**, 1099 (1976).
34. A. C. Knipe, *J. Chem. Educ.,* **53**, 618 (1976).
35. R. M. Izatt, R. E. Terry, B. L. Haymore, L. D. Hansen, N. K. Dalley, A. G. Avondet, and J. J. Christensen, *J. Am. Chem. Soc.,* **98**, 7620 (1976).
36. R. M. Izatt, R. E. Terry, D. P. Nelson, Y. Chan, D. J. Eatough, J. S. Bradshaw, L. D. Hansen, and J. J. Christensen, *J. Am. Chem. Soc.,* **98**, 7626 (1976).
37. J.-M. Lehn, *Acc. Chem. Res.,* **11**, 49 (1978).
38. E. Shchori, J. Jagur-Grodzinski, Z. Luz, and M. Shporer, *J. Am. Chem. Soc.,* **93**, 7133 (1971).
39. J. M. Ceraso and J. L. Dye, *J. Am. Chem. Soc.,* **95**, 4432 (1973).
40. E. Shchori, J. Jagur-Grodzinski, and M. Shporer, *J. Am. Chem. Soc.,* **95**, 3842 (1973).
41. J. P. Kintzinger and J.-M. Lehn, *J. Am. Chem. Soc.,* **96**, 3313 (1974).
42. J.-M. Lehn and M. E. Stubbs, *J. Am. Chem. Soc.,* **96**, 4011 (1974).
43. Y. M. Cahen, J. L. Dye, and A. I. Popov, *J. Phys. Chem.,* **79**, 1292 (1975).
44. M. Shporer and Z. Luz, *J. Am. Chem. Soc.,* **97**, 665 (1975).
45. V. M. Loyola, R. G. Wilkins, and R. Pizer, *J. Am. Chem. Soc.,* **97**, 7382 (1975).
46. G. W. Liesegang, M. M. Farrow, N. Purdie, and E. M. Eyring, *J. Am. Chem. Soc.,* **98**, 6905 (1976).
47. L. J. Rodriguez, G. W. Liesegang, R. D. White, M. M. Farrow, N. Purdie, and E. M. Eyring, *J. Phys. Chem.,* **81**, 2118 (1977).
48. G. W. Liesegang, M. M. Farrow, F. A. Vazquez, N. Purdie, and E. M. Eyring, *J. Am. Chem. Soc.,* **99**, 3240 (1977).
49. B. G. Cox and H. Schneider, *J. Am. Chem. Soc.,* **99**, 2809 (1977).
50. J. M. Ceraso and J. L. Dye, *J. Am. Chem. Soc.,* **95**, 4432 (1973).
51. J. M. Ceraso, P. B. Smith, J. S. Landers, and J. L. Dye, *J. Phys. Chem.,* **81**, 760 (1977).
52. K. Henco, B. Tümmler, and G. Maass, *Angew Chem., Int. Ed. Engl.,* **16**, 538 (1977).
53. B. Tümmler, G. Maass, E. Weber, W. Wehner, and F. Vögtle, *J. Am. Chem. Soc.,* **99**, 4683 (1977).
54. E. Mei, J. L. Dye, and A. I. Popov, *J. Am. Chem. Soc.,* **99**, 5308 (1977).
55. E. Mei, A. I. Popov, and J. L. Dye, *J. Am. Chem. Soc.,* **99**, 6532 (1977).
56. V. M. Loyola, R. Pizer, and R. G. Wilkins, *J. Am. Chem. Soc.,* **99**, 7185 (1977).
57. E. Mei, A. I. Popov, and J. L. Dye, *J. Phys. Chem.,* **81**, 1677 (1977).
58. J. L. Down, J. Lewis, B. Moore, and G. W. Chinson, *Proc. Chem. Soc. (Lond.),* 209 (1957).
59. F. Cafasso and B. R. Sundheim, *J. Chem. Phys.,* **31**, 809 (1959).
60. F. S. Dainton, D. M. Wiles, and A. N. Wright, *J. Chem. Soc. (Lond.),* 4283 (1960); *J. Polym. Sci.,* **45**, 111 (1960).
61. T. R. Tuttle, Jr., and S. I. Weissman, *J. Am. Chem. Soc.,* **80**, 5342 (1960).
62. E. S. Petrov, M. I. Belousova, and A. I. Shatenshtein, *J. Gen. Chem. USSR (Engl.),* **34**, 2477 (1964).
63. I. M. Panayotov, Ch. B. Tsvetanov, I. V. Berlinova, and R. S. Velichkova, *Makromol. Chem.,* **134**, 313 (1970).
64. D. Bright and M. R. Truter, *Nature,* **225**, 176 (1970); *J. Chem. Soc., B,* 1544 (1970).

65. M. A. Bush and M. R. Truter, *Chem. Commun.*, 1439 (1970).
66. R. Weiss, B. Metz, and D. Moras, *Proc. Int. Conf. Coord. Chem., 13th*, **2**, 85 (1970).
67. B. Metz, D. Moras, and R. Weiss, *J. Am. Chem. Soc.*, **93**, 1806 (1971); *Chem. Commun.*, 444 (1971).
68. D. Moras, B. Metz, B. Herceg, and R. Weiss, *Bull. Soc. Chim. Fr.*, 551 (1972).
69. D. Moras, B. Metz, and R. Weiss, *Acta Crystallogr., Sect. B*, **29**, 383, 388 (1973).
70. D. Moras and R. Weiss, *Acta. Crystallogr., Sect. B*, **29**, 400 (1973).
71. B. Metz, D. Moras, and R. Weiss, *Acta Crystallogr., Sect. B*, **29**, 1377, 1382, 1388 (1973).
72. J. D. Dunitz, M. Dobler, P. Seiler, and R. P. Phizackerley, *Acta Crystallogr., Sect. B*, **30**, 2733 (1974).
73. J. D. Dunitz and P. Seiler, *Acta Crystallogr., Sect. B.*, **30**, 2739 (1974).
74. M. Dobler, J. D. Dunitz, and P. Seiler, *Acta Crystallogr., Sect. B*, **30**, 2741, 2744 (1974).
75. M. Dobler and R. P. Phizackerley, *Acta. Crystallogr., Sect. B*, **30**, 2746; 2748; 2750 (1974).
76. M. R. Truter, *Struct. Bonding*, **16**, 71 (1973).
77. J. L. Dye, in Ref. 4c, p. 1.
78. For references to theoretical papers see M. D. Newton, *J. Phys. Chem.*, **79**, 2795 (1975).
79. R. A. Ogg, Jr., *J. Am. Chem. Soc.*, **68**, 155 (1946); *J. Chem. Phys.*, **14**, 114, 295 (1946); *Phys. Rev.*, **69**, 243, 668 (1946).
80. J. Jortner, *J. Chem. Phys.*, **30**, 839 (1959).
81. D. A. Copeland, N. R. Kestner, and J. Jortner, *J. Chem. Phys.*, **53**, 1189 (1970).
82. N. R. Kestner, Ref. 4e, p. 1.
83. N. R. Kestner and J. Jortner, *J. Chem. Phys.*, **77**, 1040 (1973).
84. S. Golden, C. Guttman, and T. R. Tuttle, Jr., *J. Am. Chem. Soc.*, **87**, 135 (1965); *J. Chem. Phys.*, **44**, 3791 (1966).
85. M. Gold, W. L. Jolly, and K. S. Pitzer, *J. Am. Chem. Soc.*, **84**, 2264 (1962).
86. J. Kaplan and C. Kittel, *J. Chem. Phys.*, **21**, 1429 (1953).
87. E. Becker, R. H. Lindquist, and B. J. Alder, *J. Chem. Phys.*, **25**, 971 (1956).
88. D. E. O'Reilly, *J. Chem. Phys.*, **41**, 3729 (1964).
89. J. V. Acrivos and K. S. Pitzer, *J. Phys. Chem.*, **66**, 1693 (1962).
90. E. Duval, P. Rigny, and G. Lepoutre, *Chem. Phys. Lett.*, **2**, 237 (1968).
91. J. P. Lelieur, in Ref. 4e, p. 305.
92. J. P. Lelieur and P. Rigny, *J. Chem. Phys.*, **59**, 1148 (1973).
93. J. P. Lelieur, P. Damay, and G. Lepoutre, *J. Phys. Chem.*, **79**, 2879 (1975).
94. E. Huster, *Ann. Phys.*, **33**, 477 (1938).
95. S. Freed and N. Sugarman, *J. Chem. Phys.*, **11**, 354 (1943).
96. C. A. Hutchison, Jr., and R. C. Pastor, *Rev. Mod. Phys.*, **25**, 285 (1953); *J. Chem. Phys.*, **21**, 7959 (1953).
97. A. Demortier and G. Lepoutre, *C. R. Acad. Sci. Paris*, **268**, 453 (1969).
98. A. Demortier, M. DeBacker, and G. Lepoutre, *J. Chim. Phys.*, **69**, 380 (1972).
99. R. H. Land and D. E. O'Reilly, *J. Chem. Phys.*, **46**, 4496 (1967).

References

100. K. Fueki, *J. Chem. Phys.*, **50**, 5381 (1969).
101. D.-F. Feng, K. Fueki, and L. Kevan, *J. Chem. Phys.*, **58**, 3281 (1973).
102. D. A. Copeland and N. R. Kestner, *J. Chem. Phys.*, **58**, 3500 (1973).
103. N. R. Kestner and J. Logan, *J. Phys. Chem.*, **79**, 2815 (1975).
104. Y. Nakamura, Y. Horie, and M. Shimoji, *Trans. Faraday Soc. I*, **70**, 1376 (1974).
105. R. R. Dewald and J. L. Dye, *J. Phys. Chem.*, **68**, 128 (1964).
106. N. Gremmo and J. E. B. Randles, *Trans. Faraday Soc. I*, **70**, 1480 (1974).
107. J. L. Dye, *Acc. Chem. Res.*, **1**, 306 (1968).
108. J. L. Dye, M. T. Lok, F. J. Tehan, and L. M. Dorfman, unpublished observations; see also Ref. 132.
109. K. D. Vos and J. L. Dye, *J. Chem. Phys.*, **38**, 2033 (1963).
110. K. Bar-Eli and T. R. Tuttle, Jr., *Bull. Am. Phys. Soc.*, **8**, 352 (1963).
111. M. Ottolenghi, K. Bar-Eli, H. Linschitz, and T. R. Tuttle, Jr., *J. Chem. Phys.*, **40**, 3729 (1964).
112. K. Bar-Eli and T. R. Tuttle, Jr., *J. Chem. Phys.*, **40**, 2508 (1964).
113. L. R. Dalton, J. D. Rynbrandt, E. M. Hansen, and J. L. Dye, *J. Chem. Phys.*, **44**, 3969 (1966).
114. R. Catterall, M. C. R. Symons, and J. W. Tipping, *J. Chem. Soc. A*, 1529 (1966); 1234 (1967).
115. J. L. Dye and L. R. Dalton, *J. Phys. Chem.*, **71**, 184 (1967).
116. R. Catterall, J. Slater, and M. C. R. Symons, *J. Chem. Phys.*, **52**, 1003 (1970).
117. V. A. Nicely and J. L. Dye, *J. Chem. Phys.*, **53**, 119 (1970).
118. R. Catterall, M. C. R. Symons, and J. W. Tipping, in Ref. *4c*, p. 317 (1970).
119. R. Catterall, J. Slater, and M. C. R. Symons, in Ref. *4c*, p. 329 (1970).
120. R. Catterall, I. Hurley, and M. C. R. Symons, *J. Chem. Soc., Dalton Trans.*, 139 (1972).
121. R. Catterall and P. P. Edwards, *J. Phys. Chem.*, **79**, 3010 (1975).
122. R. Catterall, J. Slater, W. A. Seddon, and J. W. Fletcher, *Can. J. Chem.*, **54**, 3110 (1976).
123. L. M. Dorfman, F. Y. Jou, and R. Wagemen, *Ber. Bunsenges. Phys. Chem.*, **75**, 681 (1971).
124. For a recent view of the role of pulse radiolysis in the identification of e^-_{solv}, M, and M$^-$, see J. W. Fletcher and W. A. Seddon, *J. Phys. Chem.*, **79**, 3055 (1975).
125. R. R. Dewald and K. W. Browall, *J. Phys. Chem.*, **74**, 129 (1970).
126. R. Catterall, in discussion portion of Ref. 124.
127. W. A. Seddon, J. W. Fletcher, F. C. Sopchyshyn, and R. Catterall, *Can. J. Chem.*, **55**, 3356 (1977).
128. W. A. Seddon, J. W. Fletcher, and R. Catterall, *Can. J. Chem.*, **55**, 2017 (1977).
129. I. Hurley, T. R. Tuttle, Jr., and S. Golden, *J. Chem. Phys.*, **48**, 2818 (1968).
130. S. Matalon, S. Golden, and M. Ottolenghi, *J. Phys. Chem.*, **73**, 3098 (1969).
131. M. G. DeBacker and J. L. Dye, *J. Phys. Chem.*, **75**, 3092 (1971).
132. J. L. Dye, M. G. DeBacker, J. A. Eyre, and L. M. Dorfman, *J. Phys. Chem.*, **76**, 839 (1972).
133. T. R. Tuttle, Jr., *Chem. Phys. Lett.*, **20**, 371 (1973).

134. C. A. Kraus, *J. Am. Chem. Soc.*, **30**, 653 (1908).
135. For a detailed discussion of the properties of metal-ammonia compounds and the pertinent references, see Ref. 1, Chapter 7.
136. K. Bar-Eli and G. Gabor, *J. Phys. Chem.*, **77**, 323 (1973).
137. B. Baron, P. Delahay, and R. Lugo, *J. Chem. Phys.*, **53**, 1399 (1970).
138. N. Gremmo and J. E. B. Randles, *Trans. Faraday Soc. I*, **70**, 1488 (1974).
139. J. M. Brooks and R. R. Dewald, *J. Phys. Chem.*, **72**, 2655 (1968).
140. R. R. Dewald, in Ref. 4c, p. 497.
141. E. I. Mal'tsev, P. M. Alpatova, and A. V. Vannikov, *Élektrokhimiya*, **13**, 203 (1977), and references therein.
142. B. Kaempf, S. Raynal, A. Collet, F. Schué, S. Boileau, and J.-M. Lehn, *Angew. Chem., Int. Ed. Engl.*, **13**, 611 (1974).
143. M. Szwarc, *Ions and Ion-Pairs in Organic Reactions*, Vol. 1, Wiley-Interscience, New York, 1972, p. 115ff.
144. E. R. Minnich, L. D. Long, J. M. Ceraso, and J. L. Dye, *J. Am. Chem. Soc.*, **95**, 1061 (1973).
145. E. Shchori and J. Jagur-Grodzinski, *Isr. J. Chem.*, **11**, 243 (1973).
146. S. Boileau, P. Mernery, and J. C. Justice, *J. Solution Chem.*, **4**, 873 (1975).
147. J. J. Christensen, D. J. Eatough, and R. M. Izatt, *Chem. Rev.*, **74**, 351 (1974).
148. R. Ungaro, B. ElHaj, and J. Smid, *J. Am. Chem. Soc.*, **98**, 5198 (1976).
149. N. Matsuura, K. Umemoto, Y. Takeda, and A. Sasaki, *Bull. Chem. Soc. Jpn*, **49**, 1246 (1976).
150. J. Jagur-Grodzinski, *Bull. Chem. Soc. Jpn.*, **50**, 3077 (1977).
151. N. Matsuura, *Bull. Chem. Soc. Jpn.*, **50**, 3078 (1977).
152. N. Nae and J. Jagur-Grodzinski, *J. Am. Chem. Soc.*, **99**, 489 (1977); *J. Chem. Soc., Faraday I*, 1951 (1977).
153. A. Friedenberg and H. Levanon, *Chem. Phys. Lett.*, **41**, 84 (1976).
154. Discussion following the paper by B. Bockrath, J. F. Gavlas, and L. M. Dorfman, *J. Phys. Chem.*, **79**, 3064 (1975).
155. J. D. Corbett, D. G. Adolphson, D. J. Merryman, P. A. Edwards, and F. J. Armatis, *J. Am. Chem. Soc.*, **97**, 6267 (1975).
156. J. D. Corbett and P. A. Edwards, *J. Chem. Soc., Chem. Commun.*, 984 (1975).
157. D. G. Adolphson, J. D. Corbett, and D. J. Merryman, *J. Am. Chem. Soc.*, **98**, 7234 (1976).
158. A. Cisar and J. D. Corbett, *Inorg. Chem.*, **16**, 632 (1977).
159. J. D. Corbett and P. A. Edwards, *J. Am. Chem. Soc.*, **99**, 3313 (1977).
160. P. A. Edwards and J. D. Corbett, *Inorg. Chem.*, **16**, 903 (1977).
161. E. Zintl, J. Goudeau, and W. Dullenkopf, *Z. Phys. Chem., Abt. A*, **154**, 1 (1931).
162. E. Zintl and A. Harder, *Z. Phys. Chem., Abt. A*, **154**, 47 (1931).
163. E. Zintl and W. D. Dullenkopf, *Z. Phys. Chem., Abt. B*, **16**, 183 (1932).
164. R. R. Dewald and J. L. Dye, *J. Phys. Chem.*, **68**, 121 (1964).
165. D. Live and S. I. Chan, *J. Am. Chem. Soc.*, **98**, 3769 (1976).
166. Y. M. Cahen, J. L. Dye, and A. I. Popov, *Inorg. Nucl. Chem. Lett.*, **10**, 899 (1974); *J. Phys. Chem.*, **79**, 1289 (1975).

References

167. E. Mei, J. L. Dye, and A. I. Popov, *J. Am. Chem. Soc.*, **98**, 1619 (1976).
168. A. Hourdakis and A. I. Popov, *J. Solution Chem.*, **6**, 299 (1977).
169. E. Mei, L. Liu, J. L. Dye, and A. I. Popov, *J. Solution Chem.*, **6**, 771 (1977).
170. F. W. Wehrli, *J. Magn. Reson.*, **25**, 575 (1977).
171. G. Malli and S. Fraga, *Theor. Chim. Acta*, **5**, 275 (1966).
172. A. Beckman, K. D. Böklen, and D. Elke, *Z. Phys.*, **270**, 173 (1974).
173. C. Deverell and R. E. Richards, *Mol. Phys.*, **10**, 551 (1966).
174. A. Saika and C. P. Slichter, *J. Chem. Phys.*, **22**, 26 (1954).
175. C. Deverell, *Mol. Phys.*, **16**, 491 (1969).
176. D. Ikenberry and T. P. Das, *Phys. Rev.*, **138**, A822 (1965).
177. D. W. Hofmeister and W. H. Flygare, *J. Chem. Phys.*, **43**, 795 (1965); **44**, 3584 (1966).
178. N. F. Ramsey, *Phys. Rev.*, **77**, 567 (1950); **78**, 699 (1950); **83**, 540 (1951); **86**, 243 (1952).
179. E. G. Bloor and R. G. Kidd, *Can. J. Chem.*, **46**, 3425 (1968).
180. R. H. Erlich, E. Roach, and A. I. Popov, *J. Am. Chem. Soc.*, **92**, 4989 (1970).
181. R. H. Erlich and A. I. Popov, *J. Am. Chem. Soc.*, **93**, 5620 (1971).
182. M. Herlem and A. I. Popov, *J. Am. Chem. Soc.*, **94**, 1431 (1972).
183. M. S. Greenberg, R. L. Bodner, and A. I. Popov, *J. Phys. Chem.*, **77**, 2449 (1973).
184. J. S. Shih and A. I. Popov, to be published.
185. W. J. DeWitte, L. Liu, E. Mei, J. L. Dye, and A. I. Popov, *J. Solution Chem.*, **6**, 337 (1977).
186. J. P. Kintzinger and J.-M. Lehn, *J. Am. Chem. Soc.*, **96**, 3313 (1974).
187. A. Abragam, *The Principles of Nuclear Magnetism,* Oxford University Press, London, 1961.
188. H. G. Hertz, *Ber. Bunsenges. Phys. Chem.*, **77**, 531 (1973).
189. D. H. Haynes, B. C. Pressman, and A. Kowalsky, *Biochemistry*, **10**, 852 (1971).
190. D. H. Haynes, *FEBS Lett.*, **20**, 221 (1972).
191. P. B. Chock, *Proc. Natl. Acad. Sci. USA*, **69**, 1939 (1972).
192. P. B. Smith and J. L. Dye, to be published.

CHAPTER THREE
MULTIDENTATE MACROMOLECULES: PRINCIPLES OF COMPLEXATION WITH ALKALI AND ALKALINE EARTH CATIONS

N. S. POONIA
Department of Chemistry
University of Indore
Indore, India

1 Introduction	116
2 Crowns	120
2.1 Metal-Crown Complexes: Stoichiometry and Conformation, 123	
2.2 Stoichiometry of Metal-Crown Complexes: Ion-Cavity-Radius Concept, 125	
2.3 Metal-Crown Complexation: Role of the Anion, 126	
2.4 Metal-Crown Complexation: Charge Density of the Cation, 133	
2.5 Differentiation of Na/K and Ca/Mg with Crown Ethers, 136	
3 Sensors	137
4 Cryptands	137
4.1 Bicyclic Cryptands, 138	
4.2 Tricyclic Cryptands, 139	
5 Antibiotics	139
5.1 Metal-Antibiotic Complexes, 140	
6 Crowns and Antibiotics—Similarity in Principles of Complexation	143
6.1 Synthetic and Natural Multidentates—Further Resemblance in Principles of Complexation, 145	
7 Summary	147
References	149

1 INTRODUCTION

Extensive work is being carried out on alkali (M^+) and alkaline earth (M^{2+}) metal ions (general abbreviation M) with natural antibiotics and synthetic tri-, bi-, and monocyclic macromolecules (1-5). One aim of the work is to understand conditions that lead to the discrimination of seemingly alike pairs of cations—potassium and sodium, and magnesium and calcium—by the cell membrane during transport in living systems (6, 7); the other is to explore the analytical value (3) of the diverse macromolecules for alkali and alkaline earth cations.

To understand the biological implication of alkali and alkaline earth cations, work has been executed with the following three points in mind (6, 7):

1. The cell membrane is essentially a lipid bilayer studded with polar sites.
2. The membrane employs transport mediators to take up (transport) a cation from aqueous bathing medium; during transport the transport mediators separate the cation from its counteranion and solvating liquid—*anionic species.*
3. In most natural systems the membrane preferably takes up (8, 9) potassium and magnesium over sodium and calcium, respectively.

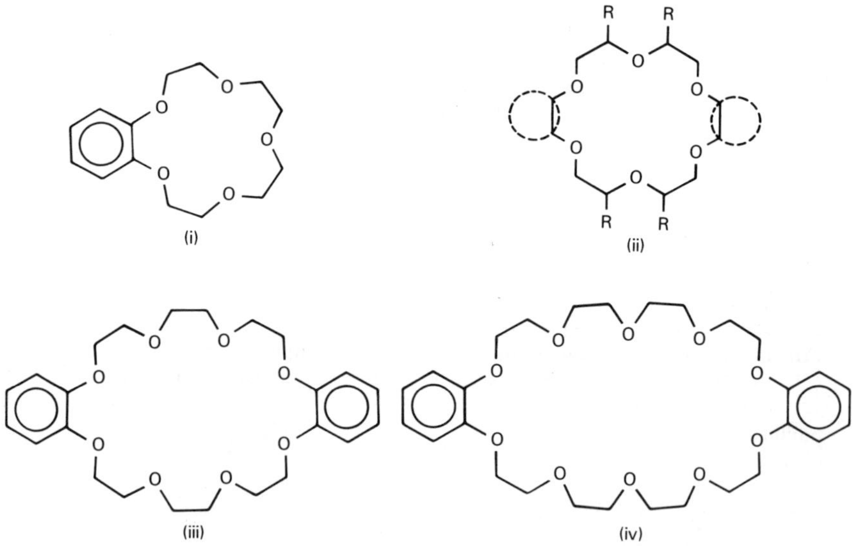

Figure 1 Crowns: (*i*) benzo-15-crown-5 (B15C5); (*ii*) R = H, broken circle representing no attached functional group, 18-crown-6 (18C6); R = H, broken circle representing benzene, dibenzo-18-crown-6 (DB18C6); R = H, broken circle representing cyclohexane, dicyclohexano-18-crown-6 (DC18C6); R = CH_3, broken circle representing benzene, TEMF; (*iii*) dibenzo-14-crown-8 (DB24C8); (*iv*) dibenzo-30-crown-18 (DB30C10).

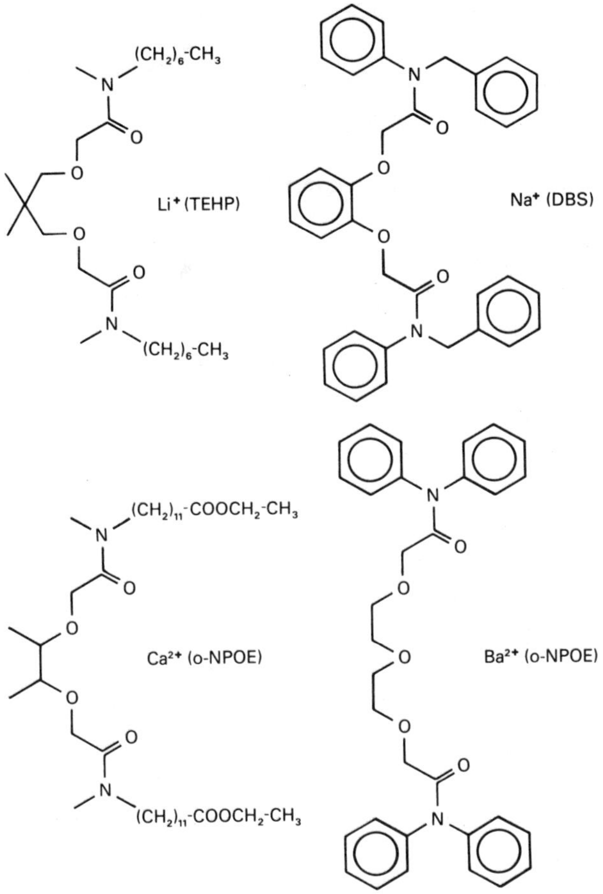

1 Introduction

Figure 2 Sensors.

Point (1) has led to the execution of in vitro work with the concerned cations using open-chain as well as cyclic macromolecules possessing hydrophobic exteriors and hydrophilic interiors. Such macromolecules include (*a*) crowns, neutral synthetic cyclic polyethers (Figure 1); (*b*) sensors, neutral synthetic open-chain multidentates (Figure 2); (*c*) cryptands, synthetic heteroatom bi- and tricyclic multidentates (Figure 3); and (*d*) antibiotics, shown in Figures 4 (ionizable open-chain antibiotics of the nigericin group) and 5 (cyclic antibiotics of the valinomycin group).

Point (2) has stimulated work involving (*a*) the use of monocyclic polyethers (crown ethers) as models for the transport mediators (13-15) in an attempt to

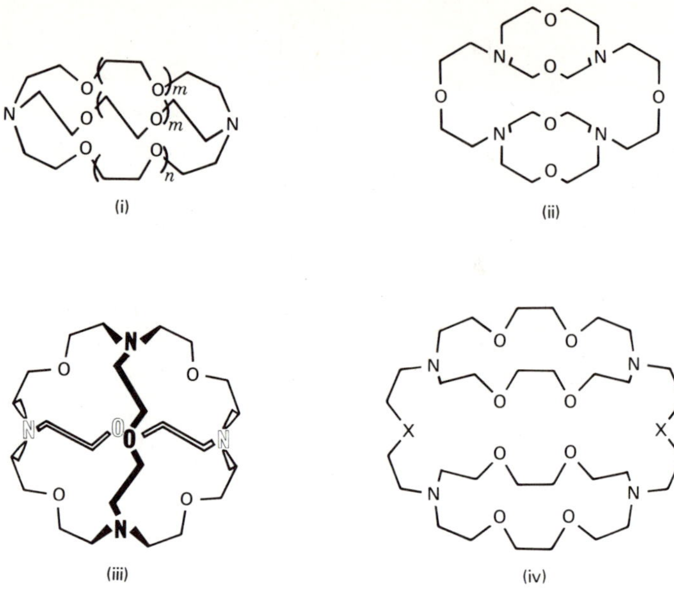

Figure 3 Cryptands: (*i*) a bicyclic molecule

$m = n = 0$	= [1.1.1]
$m = 0, n = 1$	= [2.1.1]
$m = 1, n = 0$	= [2.2.1]
$m = n = 1$	= [2.2.2]
$m = 1, n = 2$	= [3.2.2]
$m = 2, n = 1$	= [3.3.2]
$m = n = 2$	= [3.3.3]

(*ii*) a tricyclic molecule; (*iii*) a tricyclic spheroidal molecule; (*iv*) a tricyclic cylindrical molecule.

understand conditions that lead to a complete separation of M from anionic species and to the formation of ion-separated complete encapsulates of M in crown ethers; (*b*) the use of various macromolecular ligands— crown ethers and antibiotics—to promote extraction of the cations (16-20) from an aqueous to an organic phase; and (*c*) the use of macromolecules in the study of transport reactions of M across artificial or natural membranes (3, 12, 18, 19, 21-27).

Point (3) has stimulated in vitro work (3, 4, 28, 29) with M-macromolecule systems in the solution state to determine the trends of cation specificities shown by the macromolecule toward M.

In this chapter the M-macromolecule systems are discussed with special reference to the crown ethers. The aim of the discussion is to introduce some new principles that are involved during complexation of these macrocyclic molecules with M and to show that these principles are more common to those

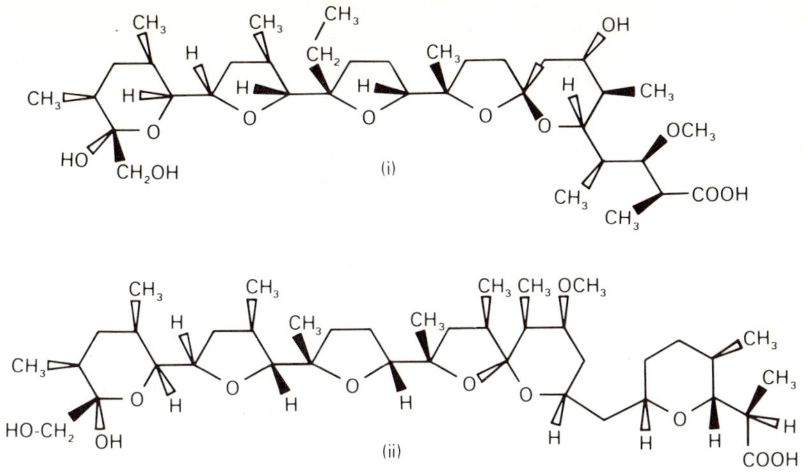

1 Introduction

Figure 4 Ionizable open-chain antibiotics of the nigericin group: (*i*) monensin; (*ii*) nigericin.

involved during complexation with cryptands and the cyclic antibiotics than has been thought so far (18, 29-35). Until now M-macrocyclic molecule systems have been discussed by considering the size of the cation and the cavity size of the ligand: the ion-cavity radius concept. Here these systems are better explained by considering the charge density of the cation and conformation energy aspect of the ligand. Experimental evidence is given to show that during the formation of an ion-separated complete encapsulate of a small-cavity crown ether with a low charge density cation the crown plays the dominant role on thermodynamic grounds (ligand encapsulation), whereas during the formation of an encapsulate of this crown with a high charge density cation, the cation itself plays a dominant role in managing its own encapsulation through electrostatic forces (self-encapsulation).

The effect of an anion on the interaction of the macromolecule with M is well known (1-3). In this chapter the role played by organic (L) and inorganic (X) anions is discussed in detail to show that (*a*) the thermodynamic solution stability of the complex is enhanced as the nucleophilicity (basicity) of the anion diminishes, and that (*b*) the possibility of crystallization of a complex and the stoichiometry of the latter in the solid state are both determined by the anion.

The M-crown systems have been examined in the solid state (13, 15, 36-45) for the purpose of understanding the stereochemistry and stoichiometry of the solid phases. Crown ethers (4, 46-48) and the diverse multidentate macromolecules (1, 3, 4, 17, 46-66) have also been studied in solution to understand their cation specificities under different conditions. This chapter shows that some common

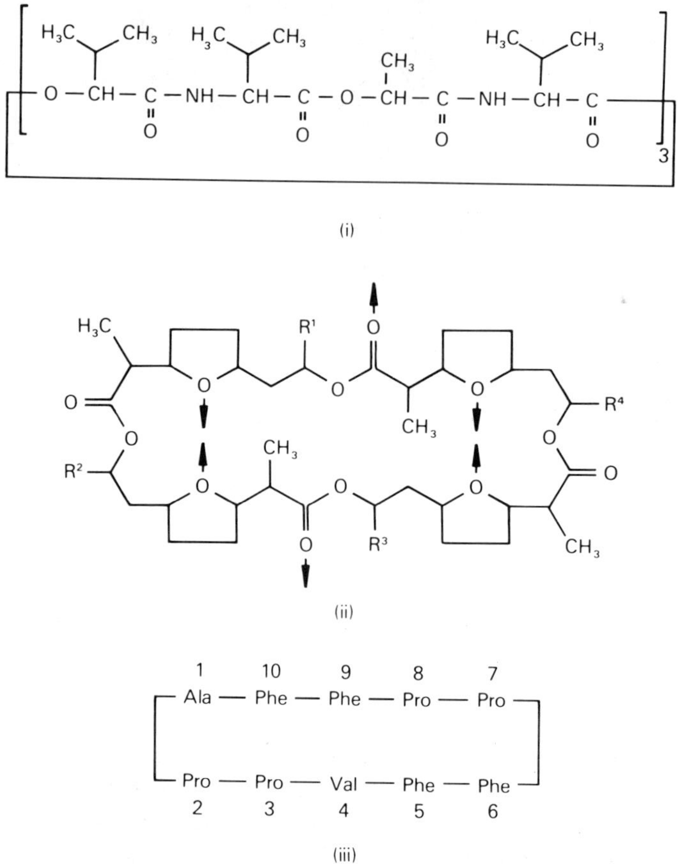

Figure 5 Cyclic antibiotics of the valinomycin group: (*i*) valinomycin; (*ii*) actins— nonactin (R^1-R^4 = CH_3), monactin (R^4 = C_2H_5; R^1, R^2, R^3 = CH_3), dinactin (R^2, R^4 = C_2H_5; R^1, R^3 = CH_3), trinactin (R^2, R^3, R^4 = C_2H_5; R^1 = CH_3), and tetranactin (R^1-R^4 = C_2H_5); (*iii*) antamanide.

principles are involved during interaction of macromolecules with M in both phases. In conclusion, complementary to the existing (1-5) quantitative data on such systems, information is being provided which helps to understand complexation and the stoichiometry of the resulting complexes.

2 CROWNS

A detailed description of the crown ethers can be found in papers published by

Pedersen (67, 71) and by Pedersen and Frensdorff (32). They are macrocyclic polyethers in which donor oxygens are arranged in a ring, ususally each separated from the other by two carbon atoms, $-CH_2-CH_2-$. The rings are of different sizes, containing from 4 to 20 oxygen atoms. In most crown ethers the ring carries one or more aromatic nuclei (Figure 1). A large number of crown ethers have been synthesized (67-70) and used (13-15, 44-48, 67, 68, 71-74) as ligands for M^+, and the subject has been covered in recent reviews (1-5, 32, 75-77). Complexes between the crowns and M^{2+} ions have also been claimed (67), but with the smaller cations, magnesium and calcium, convincing results have only recently been obtained (14, 78). Of all the crown ethers synthesized, benzo-15-crown-5 (B15C5), 18-crown-6 (18C6), dibenzo-18-crown-6 (DB18C6), dicyclohexano-18-crown-6 (DC18C6), dibenzo-24-crown-8 (DB24C8), and dibenzo-30-crown-10 (DB30C10), shown in Figure 1, have been found to be the best ligands for alkali and alkaline earth cations (32). These compounds illustrate the important cavity sizes and molecular flexibilities of the family members, Crown ethers 18C6 and DC18C6 are of the cavity size of DB18C6, but are more flexible (79).

Crown ether B15C5 is highly solubilized in a polar medium—even in 1:1 water-methanol, whereas DB18C6 is only moderately soluble even in acetone. As described below, B15C5 yields a variety of interesting results with the cations in question, whereas DB18C6 constantly produces 1:1 ion-paired encapsulates. Because of these considerations B15C5 is one of the most useful crown ligands; DB18C6, although most popularly known to those interested in macromolecular ligands, is least so.

In the solid phase all antibiotics (80-88) and polycyclic synthetic (89-94) macromolecules have been found to produce complete encapsulates with most cations. The complexes (Li-antamanide)$BrCH_3CN$ (95), (bicyclic)$Ba(NCS)_2 \cdot H_2O$ (94), and (tricyclic)$AgNO_3$ (96) [where (bicyclic) and (tricyclic) are bicyclic and tricyclic ligands, respectively; representative structures are shown in Figure 3] are among the exceptional cases that are ion paired (X-ray analysis). Obviously the results obtained with these ligands are monotonous for our purpose, and during reactions with these ligands it becomes difficult to single out the coordinative role played by the cation. Through variation of experimental conditions, however, it is possible to obtain ion-paired as well as ion-separated encapsulates of crown ethers. Consideration of the conditions leading to these differences in the nature of the products makes it possible to distinguish between the roles played by a cation under two different conditions or between two different cations under a given set of conditions. In natural systems differentiation between any two cations takes place under a given set of conditions, and crown ethers are capable of carrying out such a differentiation. This is a point of great merit for crown ethers and is one of the main reasons for interest in the discussion of these macromolecules.

Table 1 Stoichiometries of Complexes of Some Crown Ethers with Inorganic and Organic Salts of Some Alkali and Alkaline Earth Cations

Crown	K^+	Na^+	Li^+	Ca^{2+} (Mg^{2+})
B15C5	1:2 for X or L[a-c]	1:1 (aq) for halides 1:1 for NCS or L	1:1 (aq) for Br, I, or Pic 1:1 for NCS	1:1 for L 1:2 for Pic[d]
DB18C6	1:1 (aq) for X or L	1:1 (aq) for X or L	1:1 (aq) for Pic	1:1 (aq) for Pic
DB24C8	2:1 for NCS or L	1:1 (aq) for X 2:1 for NCS[e] or L	1:1 (aq) (?) for Pic	1:1 + 2:1 (?) for Pic
DB30C10	1:1 for X or Pic[c,f]	2:1 (aq) for I, NCS, or Pic	2:1 (aq) for Pic	1:1 (aq) for Pic

[a] General abbreviations for anions: X = Br, I, or NCS; L = 2-nitrophenolate (Onp), 2,4-dinitrophenolate (Dnp), 2-hydroxybenzoate (Sal), or 2,4,6-trinitrophenolate (Pic). Except for LiX with B15C5, the inorganic salts of Li, Ca, and Mg do not produce isolable complexes with any crown.
[b] Rb and Cs also produce 1:2 complexes.
[c] The anion is usually solvated.
[d] When the synthesis medium contains some proton donor - H$_2$O or PicH. However, it is as yet not certain that 1:2 complexes of calcium are sandwiches.
[e] When the synthesis medium (ethanol) is dehydrated.
[f] Rb and Cs also produce 1:1 complexes.

2 Crowns

2.1 Metal-Crown Complexes: Stoichiometry and Conformation

A generalized summary of the stoichiometries of representative metal-crown complexes is provided in Table 1. Regardless of the solvent molecule the complexes have 1:1 or 1:2 (sandwich) or 2:1 (bimetallic) stoichiometry with respect to salt:crown; bimetallic complexes are those having two cations complexed in the cavity of a crown. Ever since the appearance of the first comprehensive report by Pedersen (67), it has been believed that complexation of a cation with a crown ether and the stoichiometry of the resulting metal-crown complex are determined basically by the fit of M in the cavity of the ligand. For those having the best fit, 1:1 complexes are postulated to be formed in solution as well as in the solid state. In such a complex the ring oxygens of the ligand arrange equatorially around the cation, which remains exposed on the axial sides to the anionic species—the counteranion and/or solvent molecule—as shown in Figure 6(i). The cavity size of DB18C6 and its analogues (diameter ~3Å) is most suited for a 1:1 fit with potassium (ionic diameter, 2.66 Å) (13, 32, 44, 67, 71), but these crowns produce 1:1 complexes with cations of other sizes too. Thus Na(DB18C6) (39, 43), Na(18C6) (97), Na(DC18C6) (41), Rb(DB18C6) (39, 71), Ca(18C6) (78), and Ca (or Mg) (DB18C6) (14) have 1:1 stoichiometry in solid state, and the systems Na (or K) (DB18C6) and M^+(DC18C6) (M^+ = Na, K, or Rb) have 1:1 stoichiometry in solution (46). Cations smaller than potassium enter the cavity of the crown (40, 41), but the larger ones stay outside the plane of the donor ring (39); in the complex Na(18C6) (97), sodium distorts the cavity of the crown to suit its ionic size.

As the cavity size of the crown ether increases for a given size of M, 1:1 complexes are still obtained, but the ligand tends to be folded around the cation.

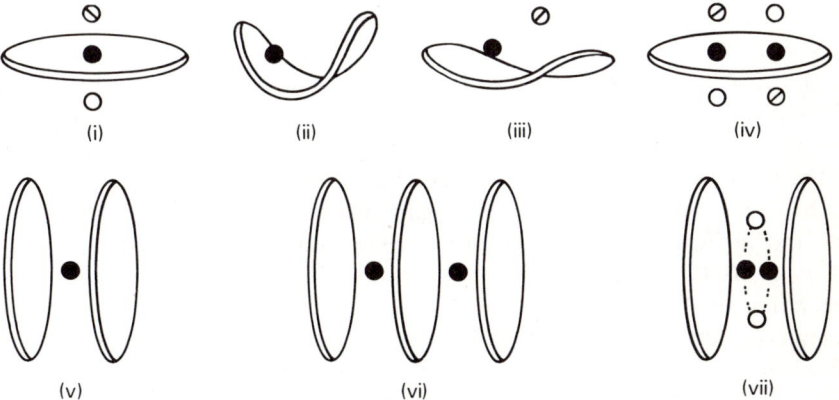

Figure 6 Schematic representation of different M-crown complexes. A loop stands for a crown ether, a solid dot for M, an open circle for an anion, and crossed circle for a solvent (water) molecule.

Folding of the ligand takes place for DB24C8 and DB30C10, for DB30C10 around potassium in the complex K(DB30C10)I (98) [Figure 6(*ii*)], and for DB24C8 around sodium, as expected in the monohydrated 1:1 complexes of NaX (44) [Figure 6(*iii*)]. Alternatively bimetallic complexes are produced [Figure 6(*iv*)] in which two cations are complexed in the unfolded cavity of the ligand, as found by X-ray analysis of the complexes $(KNCS)_2(DB24C8)$ (36) and (Na-*o*-nitrophenolate)$_2$(DB24C8) (99), and as expected for the bimetallic complexes of sodium (44) and lithium (14) with DB30C10.

When the size of M exceeds the cavity size of the crown ether, 1:2 or 2:3 complexes are formed. The 1:2 complexes have been postulated (71) to be sandwiches [Figure 6(*v*)], and 2:3 ones club sandwiches [Figure 6(*vi*)]. Sandwich complexes are formed by B15C5 with potassium, rubidium, and cesium in solution (46, 100) as well as in the solid (13, 44, 77) state. Sandwiches are also formed by DB18C6 with rubidium (71) or cesium (77) in the solid state as well as in solution (46), as well as by DC18C6 with cesium in solution (101). Club sandwiches are postulated for DB18C6 with rubidium (67) and for DB18C6, DC18C6, and benzo-18-crown-6 with cesium (71). These complexes, however, are usually undefined (71), and the complex $[Rb_2(DB18C6)_3]$ 2NCS (39) has been revealed to have actually an actual stoichiometry of 1:1, where the additional molecule of the crown ether is loose in the lattice. The formation of club sandwiches therefore appears doubtful. In addition the ligand in the middle of the system cannot be expected to make effective M-O contacts on two axial sides at a time.

Recently a new class of 1:1 metal-crown complexes has been recognized, but with the help of X-ray analysis these have been found to be dimeric (2:2) in the crystal lattice [Figure 6(*vii*)]. All such complexes (42, 102, 103) are those of the thiocyanates of larger M^+ ions (cesium and rubidium) with crowns having a cavity size of DB18C6, namely, CsNCS(18C6) (102), RbNCS(18C6) (103), and CsNCS(TEMF) (42), where TEMF is DB18C6 with four methyl groups attached to the ring methylenes. This stoichiometry in the solid state is apparently an alternative to the rather expected sandwich structure. The pronounced tendency of NCS for ion pairing (13, 104) and bridging (36) with a cation deters ligation of the second molecule of the crown ether with the cation.

The donor ring of the crown is roughly planar in complexes of the types (i) and (iv) in Figure 6. Except for a few complexes such as $[Na(B15C5)] I \cdot H_2O$ (37) and $[Na(DC18C6)] NCS \cdot H_2O$ (97), where the molecule of water causes ion separation of the system, most such complexes are ion paired. We term such complexes *ion-paired encapsulates*. The complex shown in Figure 6(*iii*) is obtained when a large cavity crown such as DB24C8 undergoes 1:1 complexation with a small cation such as sodium or lithium (14). The crown ether wraps around the cation, which continues to be solvated with a molecule of the solvent. Such a complex can be termed an *incomplete encapsulate*. The complex in Figure 6(*ii*)–designated "wraparound encapsulate"– is obtained when the cavity size of the crown far exceeds the ionic radius of M; the crown completely strips M of anionic species by wrapping around it. The complex shown in Figure 6(*v*)–designated "sandwich encapsulate"– is obtained when the cavity size of the crown ether falls short of the size of M (according to the existing conceptions); the crown ether completely strips M of anionic species by sandwiching. The sandwich encapsulate and the wraparound encapsulate may be termed *complete encapsulates*. These encapsulates are comparable to those obtained from cyclic antibiotics (80-82), and information derived from the formation of the former is used to understand the latter. The results in Table I show that the complete encapsulates are formed either with low charge density cations (K^+, Rb^+, and Cs^+) or with a high charge density cation (Mg^{2+}), and ordinarily not with medium charge density cations (Li^+ and Na^+), a point of great significance that is elaborated in Section 2.2.

2.2 Stoichiometry of Metal-Crown Complexes: The Ion-Cavity-Radius Concept

The ion-cavity-radius concept has been used widely to explain the stoichiometry of metal-crown systems (1-5, 32, 67, 71). Doubt arose about the validity of this concept (5), however, when it was discovered that silver (32) and thallous (44) fomed 1:1 complexes with B15C5 as compared with the similar-sized potassium (105), which formed (13, 44, 67, 71) 1:2 sandwiches with the same crown. Some other results that cannot be explained by this concept include the following:

1. B15C5 produces 1:2 encapsulates with larger cations, such as potassium, rubidium, and cesium (13, 44, 46, 67, 71, 77, 100), as well as with the smaller magnesium (14). However, with lithium, which is about the size of magnesium (105), this crown ether ordinarily forms 1:1 encapsulates (14).
2. DB30C10 forms 1:1 encapsulates with Mg(Pic)$_2$ and Ca(Pic)$_2$ (Pic = 2,4,6-trinitrophenolate). However, this crown ether yields 2:1 bimetallic encapsulates with Li(Pic) (14), Na(Pic) (45), and NaX (X = NCS or I) (44) even though size of these cations are similar to those of magnesium and calcium.
3. DB30C10 produces a 1:1 wraparound encapsulate with potassium, as revealed by X-ray analysis of K(DB30C10)I (98). However, a crown of the smaller cavity, namely, DB24C8, can form a bimetallic product with the same cation, as determined by X-ray analysis of (KNCS)$_2$ (DB24C8) (36).
4. The ionic size of rubidium and cesium suits the formation of 1:2 sandwiches with crowns of the cavity size of 18C6 (71). However, the complexes RbNCS(18C6) (103), CsNCS(18C6) (102), and CsNCS(TEMF) (42) have been found to be 1:1 complexes that dimerize in the crystal lattice.

Further, by changing the polarity of the synthesis medium and/or by changing the counteranion, it becomes possible to alter the stoichiometry of a given metal-crown system. Examples include the following:

5. From ethanol B15C5 yields 1:1 complexes with $M(Pic)_2$ (M = Mg or Ca) salts, irrespective of the proportion of the reactants in ethanol. However, 1:2 complexes are produced from ethanol to which water and/or PicH has been added.
6. For most anions, such as Br, I, NCS, and Pic, sodium forms 1:1 complexes with B15C5 in solution (46, 100) as well as in the solid state (13, 44, 67). However, when spherical bulky anions, such as ClO_4 and Ph_4B, are used as the counterions, this cation yields 1:2 sandwiches (106) with the same crown ether.
7. NaX (X = Br, I, or NCS) salts form 1:1 incomplete encapsulates with DB24C8. However, when NaNCS is used with this crown ether, an anhydrous 2:1 bimetallic product can also be isolated from carefully dehydrated ethanol.
8. NaX salts produce monohydrated 1:1 encapsulates with DB24C8 under ordinary conditions. NaL [L = 2-nitrophenolate (99) or 2-hydroxybenzoate] salts, however, produce (44) 2:1 bimetallic products with the same crown under all conditions.

The ion-cavity-radius concept, therefore, is inadequate to explain complexation of alkali and alkaline earth cations with crown ethers. From an examination of the foregoing results, on the contrary, it appears that the charge density of M, the nature (nucleophilicity) of the anion, and conformation energy aspects of the crown ether are important factors to be considered in the formation and stoichiometry of metal-crown complexes. In the text that follows the role of these parameters is discussed in detail.

2.3 Metal-Crown Complexation: Role of the Anion

Although it is known (1, 3) qualitatively that the counteranion influences the strength of metal-crown interaction in solution, no systematic work is reported which demonstrates that by regulating the nucleophilicity of the anion it is possible to control the interaction of a macromolecule with a cation. Some experimental observations are now discussed to establish this point and to show that the anion plays a more fundamental role in the chemistry of metal-crown systems than is known so far.

2.3.1 Effect of Anion on Synthesis of Complexes

The simplest observation is that NaX (1:1) and KX (1:2) salts (X = Br or I) produce isolable complexes with B15C5 (13, 32, 44, 67, 71), whereas MgX_2 and CaX_2 salts recrystallize uncomplexed from aqueous organic as well as organic media (15). Metal-crown complexes crystallize only when they exist in solution (67). This indicates that complexation of B15C5 with M^{2+} in either type of medium is not favored to the extent it is with M^+ ions, even though M^{2+} ions are stronger polarizers, because of the stronger involvement of M^{2+} ions with X, which is a result of their high charge density.

The following results with the organic salts of magnesium and calcium (ML_2) illustrate that metal-crown interaction and the crown-M ratio in the solid product are both favored if a less nucleophilic anion L is used or if the anion is stabilized by the use of strong proton donors in the reaction medium (14). Thus when B15C5 is reacted with $M(Onp)_2$ (Onp = 2-nitrophenolate) salts, no solid complex is formed in ethanol or 80% aqueous ethanol, and the reactants are only recrystallized. $M(Dnp)_2$ (where Dnp is 2,4-dinitrophenolate, which is comparatively more delocalized) salts produce 1:1 complexes in either medium. $M(Pic)_2$ (where Pic is 2,4,6-trinitrophenolate, the most self-stabilized of the three) salts produce 1:1 complexes from ethanol and 1:2 complexes from 80% aqueous ethanol. These results are independent of the proportion of the reactants; thus 1:2 complexes crystallize from 1:1 reaction mixtures, and 1:1 complexes crystallize even from 1:2 reaction mixtures. Obviously metal-crown interaction is favored as the anion is changed from the most highly ion-associating Onp to the most self-stabilized Pic, and as Pic is stabilized through heteroconjugation with water present in the reaction medium. For $M(Pic)_2$ systems 1:2 complexes can also be isolated from 1:1 reaction mixtures if picric acid instead of water is used in the reaction mixture, since Pic gets stabilized through the formation of a Pic.$(HPic)_n$ type of homocongugate.

If metal-crown interaction is carried out in the absence of an anion, 1:2 complexes of $M(Pic)_2$ with B15C5 can be isolated even without the use of water or the parent acid of the anion as the foreign proton donor in the reaction medium. Thus employing an MCl_2-PicH-B15C5 (1:1:1) reaction mixture in ethanol, it is possible (14) to crystallize complexes of the type $M(Pic)_2(B15C5)_2$, although in the reaction mixture there is neither water, nor picric acid in excess of the stoichiometric requirement, nor B15C5 sufficient for 1:2 complexation. In an "indirect procedure" of this type metal-crown interaction and Cl . . .HPic bonding take place simultaneously. The prospective charge neutralizer, Pic, is produced in situ by the decomposition of the heteroconjugate, Cl^-. . HPic, after metal-crown interaction has taken place. Since there is no competing anion throughout the process of metal-crown complexation, 1:2 complexes are isolated even in the absence of conditions favorable for their formation.

2.3.2 Effect of the Anion on Complexation in Solution

The effect of the anion on complexation in solution obviously cannot be the same as in solid complexes. Nevertheless it does exist and it can be demonstrated that metal-crown interaction is hampered as the nucleophilicity of the charge neutralizer is enhanced. Paper chromatographic migrations (107) of the ML (M = Li, Na, K, Mg, or Ca; L = Onp, Dnp, or Pic) salts are shown in Table 2. Though immobile in pure toluene, each salt migrates when toluene-ethanol (9:1) is used as a developer. Due to frontal analysis (108) of the developer on the paper, the mobile phase in this system should also be essentially toluene. Obviously ethanol solvates M to produce M (ethanol)$_n$-type complexes, which migrate through their solubilization in toluene; the migrations of M$^+$ ions, follow the order Li > Na > K, which is in agreement with their solvating power. However, the R_f value of each cation is decreased by the anion in the order Onp > Dnp > Pic, although hydrophobicity of the anions increases in the order Pic < Dnp < Onp and the respective salts should have been increasingly soluble in toluene. The same order of migrations is observed when B15C5 is present in the developer, indicating that complexation of a cation with ethanol or B15C5 is hindered as the nucleophilicity of L is enhanced. Further, when these developers are used, magnesium should migrate faster than calcium under all conditions due to the possibility of its stronger complexation with neutral ligands. However, with the help of the developer without B15C5 it moves faster than calcium only when it is present as a salt of Pic and Dnp; as a salt of Onp it migrates rather slowly. Obviously the effect of cation-anion involvement on metal-ligand interaction is attributable also to the high charge density of the cation, which invites a strong pairing with the counteranion.

Solubility measurements of salts in the presence of the crown ether indicate (32) that the solution stability of K(DC18C6)X in various organic solvents is decreased by X in the order Cl > Br > I. Also the stability of the 1:1 complex of sodium with B15C5 in methanol decreases (15) as the counteranion is changed from Pic (log K = 5.49) to Dnp (log K = 4.36) to Onp (log K = 3.52). Furthermore the stability of complexes of a given cation changes predictably as the "effective" nucleophilicity of the counteranion during complexation is regulated. Thus in acetone the stability of the K-ligand bond (15) is higher for Pic (log $K \sim$ 2) than for Onp (log $K \sim$ O) when dichlorotetraglycol is the ligand. However, the stability sequence gets reversed (Pic: log $K \sim$ 2; Onp: log K = 4.51) when the ligand is tetraglycol. These results are attributed to a high nucleophilicity difference between the two charge-neutralizing anions. In the

Table 2 Paper Chromatographic Mobility of ML and ML$_2$ Salts at about 22°C: Method of Development, Radial; Time of Development, 1.5-2 hr

Developer	Anion (L)	Metal (M)[a]				
		K	Na	Li	Ca	Mg
Toluene-ethanol (9:1)	Pic	63	66	80	69	74
	Dnp	87-BD(65)[b]	91-BD(25)	93-BD(75)	90-BD(25)	92-BD(45)
	Onp	25	50	75	SD[c]	0
Toluene-ethanol (9:1) containing 1% B15C5	Pic	100	100	100	100	100
	Dnp	100-BD(75)	100-BD(42)	100-BD(23)	100-BD(24)	100-BD(62)
	Onp	SD	SD	SD	SD	0-FD-(70)[d]

[a] For convenience all R_f values are multiplied by 100.
[b] The main spot is at R_f 87 and shows back diffusion (BD) up to R_f value 65; the lower the R_f value of BD, the weaker is the M-ethanol interaction.
[c] SD = slight diffusion.
[d] The spot is immobile, but shows forward diffusion (FD) up to R_f 70; the higher the value of FD, the stronger is the metal-crown interaction.

tetraglycol the ether oxygens stabilize only potassium ion, while -OH groups can also stabilize the anion through bonding. This is especially true of the highly nucleophilic Onp, as is evident from the decolorization of the yellow color of this anion in the presence of this ligand. The K(tetraglycol)$_n$Onp complex is therefore ion separated and fairly stable, unlike K(tetraglycol)Pic. In the dichloroether there is no such possibility for Onp to be stabilized, and it continues to lower the polarizing ability of potassium ion more effectively than Pic and weakens the K-ligand bond.

Complexation of a cation with a multidentate ligand and bridging of the counteranion with the polar proton of the latter are simultaneous processes (77), and the strength of these interactions can be correlated with the acidity (charge density) of the cation and the basicity (nucleophilicity) of the anion. This explains why an ionizable open-chain crown ether (Figure 7) can be used in the

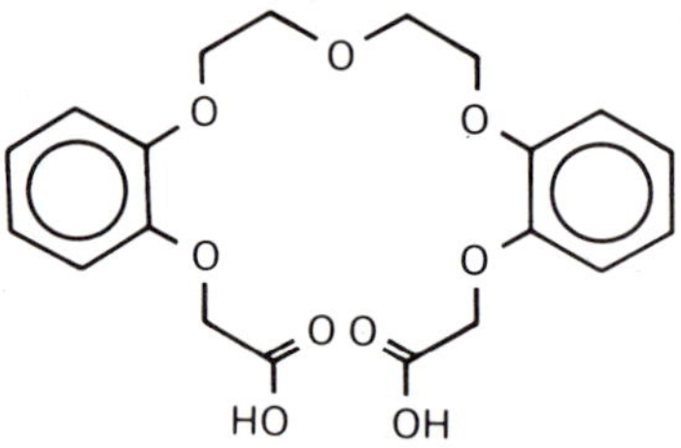

Figure 7 An open-chain ionizable crown ether. Taken from Ref. 109.

solution reaction to produce 1:1 complexes with Na(Pic), whereas with Na(Onp) only a metathetical reaction is noted (109) to obtain OnpH and the sodium salt of the ligand.

For macrobicyclic ligands (discussed in Section 4) the cryptate effect is incorporated (111) during complexation with a cation, which is many orders of magnitude higher than the macrocyclic effect operative during complexation with crown ethers (112, 113). Consequently these ligands complex strongly even calcium and magnesium ions in water as well as in 95% aqueous methanol (Table 3). This suggests that solvation of the small M^{2+} ions with water is not an insurmountable factor during interaction with these ligands. Interestingly the fall in the magnitude of solution stability of the optimal cryptates (underlined in Table 3) is more rapid for the series barium to magnesium than for cesium to lithium, although the M^{2+}/cavity radius ratio is not <1 for any M^{2+}-cryptate. Also in a given medium the interaction of the cryptand [2.1.1] [Figure 3(i), m = 0, n = 1] for magnesium is distinctly weaker than that for the similar-sized lithium. Both the observations indicate that the effect of cation-anion involvement is detectable even in water and that the effect of involvement is more pronounced for M^{2+} ions.

Table 3 Stability Constants (log K_S) of M^+ and M^{2+} Cryptates (K_s in l. mole^{-1} at 25°C) (from Ref. 111)

Ligand[a] (cavity radius, Å)	Medium[b]	Li[c] (0.78)	Na (0.98)	K (1.33)	Rb (1.49)	Cs (1.65)	Mg (0.78)	Ca (1.06)	Sr (1.27)	Ba (1.43)
[2.1.1] (0.8)	W	5.5	3.2	<2.0	<2.0	<2.0	2.5 ± .3	2.5	< 2.0	< 2.0
	M-W	7.58	6.08	2.26	<2.0	<2.0	4.0 ± .8	4.34	2.9	< 2.0
[2.2.1] (1.1)	W	2.5	5.4	3.95	2.55	<2.0	<2.0	6.95	7.35	6.3
	M-W	4.18	8.84	7.45	5.80	3.9	<2.0	9.61	10.65	9.7
[2.2.2] (1.4)	W	<2.0	3.9	5.4	4.35	<2.0	<2.0	4.4	8.0	9.5
	M-W	1.8	7.2	9.75	8.45	3.54	<2.0	7.6	11.5	12.0
[3.2.2] (1.8)	W	<2.0	1.65	2.2	2.05	2.0	<2.0	~2.0	3.4	6.5
	M-W	<2.0	4.57	7.0	7.3	7.06	<2.0	4.74	7.06	10.4

[a]For notation see Figure 3.
[b]W = water, M-W = methanol-water (95:5).
[c]The value in parentheses denotes the ionic radius (Å).

It is worth mentioning that transportation of potassium, employing DB18C6, through a layer of $CHCl_3$ (hence interaction of potassium with this crown ether), is highly dependent on the nature of the anion (110).

2.3.3 Effect of the Anion on Complexation in the Crystal Lattice and on the Stoichiometry of Solid Phase

The complex $NaBr(DB18C6) \cdot 2H_2O$ (43) constitutes an interesting example of the effect of the anion on metal-crown interaction in the crystal lattice. In the solid state this complex exists in two types of molecules, and the cation is coordinated with six oxygen atoms of the donor ring in each molecule. In one type of the molecule both the bipyramidal positions of the cation are occupied by water molecules, and the cation lies roughly in the plane of the donor ring of the crown. However, in the other molecule one of the pyramidal positions is occupied by the bromide ion, and the cation is pulled a distance of 0.07 Å from the donor ring. Such an effect of the anion in the lattice is enhanced as the charge density of the cation and the ligating ability of the anion increase. This point is illustrated by the structural results of the complex $Ca(Dnb)_2(B15C5)_2 \cdot 3H_2O$ (114), where Dnb is 3,5-dinitrobenzoate. This complex is not the expected sandwich. It exists as a pseudosandwich formed with a molecule of B15C5 (Ca-O bond distance, 2.57-2.72 Å) on one side and Dnb anions (Ca-O$^-$, 2.47 Å) on the other. The second molecule of B15C5, while failing to displace Dnb anions from the cation, bonds with the available protons of two chains of water molecules (one chain approaching the crown ether from the adjacent unit cell). The cation is pulled from the donor ring toward Dnb anions by as much as 1.38 Å, whereas similar-sized and comparatively low charge density sodium is pulled out only by 0.75 Å from the cavity of the same crown in the complex $Na(B15C5)I \cdot H_2O$ (43). It is also worth emphasizing that the low charge density potassium ion can be ion separated in the lattice by the second molecule of B15C5 in the complex $K(B15C5)_2I$ (38) and sodium ion by a molecule of water in the complex $Na(B15C5)I \cdot H_2O$ (43). However, ion separation of calcium in the $Ca(Dnb)_2$ complex is not possible, although both water and B15C5 are available in the lattice. Obviously the cation-anion involvement becomes pronounced as the charge density of the cation and the ligating ability of the anion become high. The ion-pairing tendency of calcium is particularly enhanced, even with the self-stabilized Pic. Thus in the lattice of the compound $Ca(Pic)_2(B15C5) \cdot 2H_2O$ the hydrated salt exists as one entity and the B15C5 molecule as the other, so that the compound is not a genuinely coordinated complex (115).

With the s-block cations thiocyanate ion pairs with M more strongly (13, 104) than bromide and iodide ions, whereas a chelating organic anion associates more strongly than any of these inorganic radicals. This property of these anions persists even after complexation of the cation with a crown ether. This is illustrated by X-ray analysis, for it has been revealed that Na(B15C5)I·H$_2$O (43), K(B15C5)$_2$I (38), K(DB30C10)I (98), Na(DB18C6)Br·2H$_2$O (43), and Na(DC18C6)Br·2H$_2$O (41) are ion-separated systems, whereas RbNCS(DB18C6) (39), (KNCS)$_2$(DB24C8) (36), CsNCS(TEMF) (42), RbNCS(18C6) (103), CsNCS(18C6) (102), Ca(Dnb)$_2$(B15C5)$_2$·3H$_2$O (114), and (NaOnp)$_2$(DB24C8) (99) are ion-paired ones. The anion that continues to pair with the cation even after its complexation reduces the crown/metal ratio in the solid product and even contributes to the formation of a bimetallic complex. This explains why for NCS (unlike Br and I) a bimetallic complex of sodium is formed with DB24C8, and why for L this cation forms only bimetallic complexes with the same crown ether (example 8 in Section 2.2). It also explains the formation of an abnormal class of 2:2 complexes for M$^+$NCS$^-$ salts instead of the expected 1:2 sandwiches (example 4 in Section 2.2).

2.4 Metal-Crown Complexation: Charge Density of the Cation

The contribution of charge density of the cation toward the association of the latter with the counteranion, and of the anion on the complexation of the countercation, has been discussed in the foregoing section. Now we discuss how the effect of charge density of the cation operates directly on its complexation with a crown ether, for the electrostatic terms arising from complexation and the conformation changes of the crown ether for complexation are both related to the charge density of the cation. Although calculations for the concerned parameters have not been made with the systems in question, it is possible to determine their contribution by comparing the results of a crown ether with two cations of different charge densities or that of a cation with two crown ethers of different cavity sizes.

2.4.1 Stoichiometry of Small-Cavity Crown Complexes

The low charge density cations—potassium (13, 44, 45, 71), rubidium (32, 77), and cesium (32, 71, 77)—invariably produce 1:2 sandwiches (complete encapsulates) for X as well as L with B15C5, even when no external effort is made to stabilize the counteranion (Table 1). The high charge density magnesium also produces (14) 1:2 complexes with the same crown ether when the self-stabilized anion (Pic) is stabilized further by the use of foreign proton donors in the reaction medium. Except for the sodium salt of the bulky tetraphenylborate (106), the medium charge density cations—calcium, sodium, and lithium—do not (13, 14, 44, 45, 67) yield 1:2 complexes, even when an indirect method of synthesis is adopted to obtain them. Obviously it is just not the size but the charge density of the cation which determines the crown/metal ratio in the solid phase product, and it appears that in this operation the charge density of the low charge density cations operates differently compared to that of the high charge density magnesium.

A *superchelate effect*, which is composed of the well known chelate effect and the macrocyclic effect (112, 113), is incorporated during interaction of a cyclic ligand. Consequently a ligand of this type can easily strip a low charge density cation of its anionic species due to its weak involvement with the latter, and the formation of 1:2 complete encapsulates becomes possible. Obviously in this process the role played by the ligand dominates on thermodynamic grounds (14, 15). This process is termed *ligand encapsulation.* The high charge density magnesium is so involved with the anionic species that the superchelate effect proves too weak to help its complete encapsulation under ordinary conditions, as noted from the formation of a 1:1 complex for $Mg(Pic)_2$–(B15C5) systems in nonhydrated ethanol. However, when the conditions favor ion separation, strong polarization of the ligand with the divalent cation becomes possible and the latter manages its own encapsulation, as noted from the formation of 1:2 complexes from the same systems in the presence of water and/or PicH. In this process, therefore, the all important role is played by the cation on electrostatic grounds; this process is called *self-encapsulation.* During ligand encapsulation the polarization of the cation with donor atoms of the crown is more important; during self-encapsulation the polarization of donor atoms with the cation is more important.

During self-encapsulation the electrostatic contribution of metal-crown interaction is supported by the thermodynamic terms arising from the incorporation of the superchelate effect. The energy so obtained appears to exceed the desolvarion energy requirements of the cation plus the conformation energy changes of the crown [the desolvation energy term for the cyclic ligand is negligible (113)]. During ligand encapsulation the desolvation term for the low charge density cation is already poor. The conformation energy term for the crown ether appears to be met by the thermodynamic terms arising from the successful operation of the superchelate effect, which is aided by the thermodynamic terms of metal-crown interaction.

For the medium charge density calcium, sodium, and lithium, both self-encapsulation and ligand encapsulation are weakly operative, because these cations are weaker polarizers than the high charge density Mg^{2+} and are more strongly involved with anionic species than the low charge density potassium, rubidium, and cesium. Consequently complete encapsulation of such cations is ordinarily not possible. Self-encapsulation for these cations can be made possible under special conditions, namely, (*a*) by the use of bulky and noninteracting anions such as Ph_4B^- in the complex $[Na(B15C5)_2]Ph_4B$ (106) or (*b*) by stabilizing the anion effectively through bonding with the protonated sites such as Cl^- with HOH during the formation of $Na(12\text{-crown-4})_2(Cl \cdot 5H_2O)$ (116) and Br^- with the -NH groups of the antibiotic in the complex $Li(antamanide)Br,CH_3CN$ (95).

Dynamics of Encapsulation in Solutions. The concepts of encapsulation appear valid even in solution. The paper chromatographic results (Table 2), for example, reflect special trends for the lowest charge density potassium and the highest charge density magnesium. Thus migrations of salts in toluene-ethanol (9:1) are enhanced when B15C5 is added to the developer, and the migration sequences change from Li > Na > K and Ca > Mg to K > Na > Li and Mg > Ca. Obviously the solution stabilities of the metal-crown complexes increase in the orders opposite to those of metal-ethanol. Understandably the highest stability for K(B15C5), results from ligand encapsulation operating through B15C5, whereas for Mg(B15C5) it is a result of self-encapsulation exhibited by the cation itself.

For macrobicyclic ligands the cryptate effect is exceptionally strong, and self-encapsulation is exhibitied by high as well as medium charge density cations under ordinary conditions. Complexation with these ligands is strong and selective (Table 3), small cations complexing strongly with small-cavity ligands and large ones with large-cavity ligands (111). The solution stability for each cryptate is higher in methanol-water (95:5) than in water. This indicates that by decreasing the polarity of the medium the effect of cation-anion involvement is over compensated through the strengthening of metal-ligand interaction in the case of these ligands. If the ion-cavity fit is considered to be the main operating factor determining the solution stability and selectivity of the cryptates, the selectivity for each optimal metal-cryptate complex should differ from that of the others by about the same magnitude, because the cavity size of a ligand and the ionic size of the M^+ forming the metal-cryptate complex of optimal stability change roughly by the same amount. The peak selectivity in the case of cryptate [2.2.1]-Na, and especially in the case of cryptate [2.1.1]-Li, indicates that the cryptate effect is aided by the self-encapsulating property of the concerned cations. The high stabilities for the rather loose-fitting [3.2.2]-Rb and [3.3.3]-Cs cryptates involving weakly polarizing cations indicate that ligand encapsulation is operating effectively. In M^{2+}-cryptates self-encapsulation is counteracted by M^{2+}-anion involvement, which is pronounced for the small cations. For the larger cations counteraction is weak, and there is a net high self-encapsulating tendency which aids the operative ligand encapsulation.

2.4.2 *Stoichiometry of Large-Cavity Crown Complexes*

Table 1 shows that DB30C10 produces 1:1 wraparound complete encapsulates either with MX and ML salts of the low charge density cations (potassium, rubidium, and cesium) or with the high charge density magnesium. With NaX, NaI, and Li(Pic) salts this crown always produces 2:1 ion-paired bimetallic encapsulates, even when indirect methods of synthesis are employed (14, 45). The bimetallic products, due to coulombic repulsions between the cations in the cavity, should in principle employ a planar cavity of the crown ether; preliminary X-ray examination of the complex (NaNCS)$_2$(DB30C10) has revealed (44) that the sodium ions should be related by symmetry and that the donor ring must be planar. These results for the three groups of cations are quite comparable with those for B15C5, indicating that complete encapsulation of the cation is basically determined by its own interactive characteristics. Obviously even for DB30C10 complete encapsulation for the low and high charge denstiy cations can be explained by considering the operation of ligand encapsulation and self-encapsulation, respectively. From these processes most of the energy needed for a major conformation change of the crown ether is supplied. For the medium charge density sodium and lithium ions 1:1 complete encapsulates are not formed, probably because (a) either type of encapsulation fails to take place effectively, (b) these cations are known to maintain contacts with anionic species even after complexation, and (c) the reaction energy obtained from complexation of DB30C10 with one such involved monovalent cation is not enough to cause the major conformation changes on the crown needed for a wraparound complexation and to cause charge separation of the ion-paired cation. In conclusion, bimetallic encapsulates are produced because involvement of the cation with anionic species cannot be overcome, a conclusion that was reached above by consideration of the nature of the anion. This also explains the results in point 2 of Section 2.2.

2.5 Differentiation of Na/K and Ca/Mg with Crown Ethers

In the Na/K couple there appears to be a basic difference. The weakly involved potassium ion can be completely stripped of its anionic species under all conditions employing small-cavity (sandwich forming) as well as large-cavity (wraparound forming) crown ethers, whereas the sodium ion behaves comparatively anionphilic and strongly solvating. Consequently potassium can produce dehydrated ion-separated complete encapsulates, whereas for sodium this process is usually uncommon (13). While essentially confirming this conclusion, recent workers (106) claim to have isolated the first dehydrated and ion-separated complete encapsulate [Na(B15C5)$_2$]ClO$_4$ of sodium in 1975. This needs clarification, since such exceptions were already known. The sandwich [Na(12-crown-4)$_2$],(Cl·5H$_2$O) (116) was reported in 1974, whereas dehydrated 1:1 complexes of the type Na(B15C5)L (44) were reported in 1973. The former possibility arises from effective stabilization of Cl with bonding molecules of water: the latter situation arises from the chelating L serving to desolvate the cation. In the Mg/Ca couple generalizations cannot be made as confidently as for the Na/K couple, because these cations have been studied much less by employing multidentate macromolecular ligands. Still, from the work that has been reported *it appears* that these cations can also be differentiated with the help of effects arising from their charge density differences. Calcium ion appears to mimic sodium ion in being comparatively anionphilic (see Section 2.3.3), whereas magnesium, like potassium, tends to be completely encapsulated in neutral environments. Thus a potential ligand like B15C5 fails to cause ion separation of the cation and can itself be observed uncomplexed in the crystal lattice only in Ca^{2+} complexes such as $Ca(Dnb)_2(B15C5)_2 \cdot 3H_2O$ (114) and $Ca(Pic)_2(B15C5) \cdot 2H_2O$ (115). However, magnesium in the complex $Mg(Pic)_2(B15C5)_2 \cdot 2H_2O$ (15) appears to be effectively depolarized in a genuine (15) sandwich. With this information it is possible to place sodium and calcium in one group and potassium and magnesium in the other as noted for these four cations in natural systems (6-9).

3 SENSORS

On the basis of model calculations Simon and co-workers (3) synthesized different neutral lipophilic open-chain multidentates (Figure 2) which function selectively for different cations. Although solid complexes such as $(NaNCS)_2$ (DDAD), KNCS(DDAD), $Ca(NO_3)_2(DDAD)$, $Sr(ClO_4)_2(DDAD)$, and $Ba(ClO_4)_2$ (DDAD), where DDAD is a dioxaoctane dicarboxylic acid diamide, are known (117), these compounds have been developed essentially to function as sensors for ion-selective electrodes. Some of those developed successfully for the selective estimation of lithium (118), sodium (119), calcium (120), strontium (121), and barium (122) are shown in Figure 2 and discussed in detail in Chapter 1.

The selectivity order of a sensor for a cation is greatly determined by its own structural aspects and the solvent medium. Apparently the principles of complexation discovered for metal-crown systems are not applicable to the understanding of cation selectivities of the metal-sensor systems. On the contrary the results of the two systems appear dissimilar. Crown ethers are usually selective for the low charge density cations (K^+, Rb^+, and Cs^+), whereas most sensors select the medium charge density cations (Na^+, Ba^{2+}, Sr^{2+}, and Ca^{2+}); the high charge density Mg^{2+} is not bound by any of the two under ordinary reactions employing organic and aqueous organic media.

4 CRYPTANDS

Cryptands, macropolycyclic ligands developed by Lehn and co-workers, have spheroidal three-dimensional cavities. These ligands are bicyclic or tricyclic, while the cavity of the latter can be cylindrical or spheroidal, giving rise to the four types of related ligands shown in Figure 3. They produce inclusion complexes, termed *cryptates,* with the cations; of various complexes such as

[2.1.1-Li$^+$] I$^-$	[2.2.1-K$^+$] SCN$^-$	[2.2.2-Na$^+$] I$^-$
[2.2.2-K$^+$] I$^-$	[2.2.2-Rb$^+$] SCN$^-$	[2.2.2-Cs$^+$] SCN$^-$
[2.2.2-Ba^{2+}] 2SCN$^-$aq	[3.2.2-Ba^{2+}] 2SCN$^-$aq	[C-2Na] 2I$^-$

(where C is a cylindrical tricyclic cryptand), X-ray structural information is available (89-94).

Interactivity of cryptands with the concerned cations in solution has been investigated extensively (111, 123-125) employing potentiometry, H NMR techniques, and, recently (126, 127), calorimetry. Especially at Michigan State University interaction of the bicyclic ligands has been studied employing NMR techniques for ^{7}Li (128, 129), ^{23}Na (79, 130), ^{39}K (131), and ^{133}Cs (132). During the cryptate formation a powerful cryptate effect is incorporated (111). Calorimetric studies (126, 127) show this to have an enthalpic origin, as was also argued for the macrocyclic effect in the case of macrocycles (133). Since the cryptate effect is much larger than the macrocyclic effect, the complexes of cryptands with spherical cations are the most stable of all the multidentate ligands discussed in this chapter.

4.1 Bicyclic Cryptands

The representative bicyclic ligand [2.2.2] (Figure 3, structure *i*, $m = n = 1$) can solubilize BaSO$_4$ in water (1), and the stability constant of [2.2.2]-K$^+$ in methanol is about 10^4 times (134) the stability value of (18C6)K$^+$. As seen in Table 3, the simple criterion of cavity selectivity determines, roughly, the stability of the cryptates. A detailed examination of the table values, however, indicates that the effect of anion on the metal-cryptand interaction is detectable (Section 2.3.3) and that ligand encapsulations and self-encapsulations are operative (Section 2.4.1). For bicyclic cryptands M^{2+}/M$^+$ selectivity can also be controlled by changing the number of binding sites and the thickness of the ligand; the ligating power and selectivity of the ligands rapidly diminish as donor oxygens are replaced by comparatively less electronegative nitrogen and sulfur atoms (1, 134). Dye and his co-workers (Chapter 2) found that cryptands (and crown ethers) can oxidize alkali metal atoms in THF, alkylamines, and benzene to produce (cryptand-M)$^+$ and the solvated electron (135-138). This obviously occurs because the collective nucleophilicity of the donor system of the multidentate comfortably exceeds that of the electron. Because the rather large (low charge density) cryptate cation is unable to provide electrostatic stabilization to the solvated electron, the latter, usually in solvents of low donicity (139), is received by an uncomplexed alkali cation (or atom) to produce the alkali metal anion or is "crystallized" as an anion to the cryptate cation. Spectacular situations such as these are illustrated (134, 139-142) by the formation of crystalline species like $[2.2.2\text{-}Na^+]Na^-$, $[2.2.2\text{-}K^+]K^-$, and $[2.2.2\text{-}Na^+]e^-$. Also like the crown ethers (77), cryptands cause charge separation of a salt to produce complete encapsulates of a cation. The counteranion becomes destabilized (activated) (141), since it is not adequately stabilized by the large cryptate cation. Such activated anions are attractive for use in organic synthesis.

4.2 Tricyclic Cryptands

These ligands are formed by two macrocycles linked through two bridges, and contain three cavities—two lateral on the top and the bottom, and the third in the center [see Figure 3(*ii*), and (*iv*)]. The cavity sizes can be varied by modifying the size of the macrocycles and the length of the bridges. They produce 1:1 as well as 2:1 complexes (134), the latter especially with the larger tricyclic (cylindrical) ligands. Since the solution stabilities and selectivities of the 1:1 complexes are comparable to those of the macrocycles and to their own 2:1 complexes, the lateral cavities appear to function as macrocyclic units independent of each other. Since the distance between the lateral cavities is sufficiently large, intramolecular cation exchange, as revealed by temperature-dependent NMR studies (142), is operative. Such studies are aimed to understand, ultimately, the "jump from site to site" mechanism during ion transport. The representative spheroidal molecule shown in Figure 3(*iii*) possesses a spherical cavity with 10 binding sites in an octahedrotetrahedral arrangement where the octahedron is completed by the six oxygen atoms and the tetrahedron by four nitrogen atoms (143). Only those 1:1 cryptates that are fairly stable in water are produced; the complex with Cs^+ is the most stable complex of the cation known to date for any macromultidentate. Compared to the bicyclic ligand [3.2.2], which is comparable in cavity size and number of binding sites, this ligand produces more stable complexes that afford a slow cation-exchange rate. A unique feature of the molecule is its ability to produce tetrahedrally bonded inclusion complexes with anions such as Cl^-. The binding in these cases proceeds through the -NH protons, with H_2O in its diprotonated form, and with NH_4^+ in its tetraprotonated form (144). Due to the presence of a closed rigid cavity, a strong cryptate effect is operative which is greater than the one known for macrobicyclic ligands.

5 ANTIBIOTICS

Antibiotics, which are of interest to the chemistry of the spherical cations, have been reviewed elsewhere (2, 3, 145, 146). Antibiotics are broadly of two types:

(a) open-chain monobasic macromolecules of the nigericin group (83-85, 147-154) (MM group), and (b) monocyclic electrically neutral macromolecules of the valinomycin group (155-163) (MCM group). Important members of the MM group, namely, nigericin and monensin, are shown in Figure 4. Antibiotics of this group carry an ionizable carboxyl group at one end, and one (grisorixin, X-206, and X-537 A) or two (nigericin and dianemycin) hydroxyl groups at the other end. In the ionized form the macromolecule behaves as a cyclic compound due to bridging of -COO$^-$ with one or both of the -OH groups. Oxygen atoms belonging to the bonded -OH and/or -COO$^-$, as well as others belonging to 5- and 6-membered rings or to their substituents, act as donor sites for M. Valinomycin (155, 156) and related compounds (Figure 5)—enniatins (157) and beauvericin (158)—are depsipeptides (159). Actins—nonactin, monactin, dinactin, trinactin, and tetranactin—are macrotetrolids (160-162). Antamanide is a 30-membered cyclic decapeptide (163).

5.1 Metal-Antibiotic Complexes

The affinity of the above-mentioned antibiotics for the s-block cations was first evidenced by the observation that the former rendered the salts of the latter soluble in nonpolar solvents (164, 165). Later it was discovered that the antibiotics also stimulate transport of these cations across artificial (21, 24, 25) as well as natural (18, 166, 167) membranes. Despite some arguments to the contrary (168, 169) it has been established that the activity of the antibiotics toward transport of the concerned cations across a membrane (18, 19, 24, 34, 170-176), and toward their lipid solubilization in a nonpolar phase (20), is a consequence of a definite metal-antibiotic interaction in solution. To understand the interactive behavior and the interaction mechanism of antibiotics with these cations, antibiotics and metal-antibiotic systems have been studied extensively in recent years.

The specificities (18-20, 49-62) of various antibiotics have been studied in the solution state, and the subject has been reviewed (177). Conformation aspects (30, 178-185) of these macromolecules in the absence and presence of the cations have been determined. Information regarding the conformation has been obtained with the help of techniques such as NMR (50, 53, 178-191), IR (179, 186-188), and ORD and CD (50, 53, 179, 180, 186).

The stoichiometry of M^+-antibiotic interaction in solution is always 1:1; only the M^+-enniatin B complexes have been found to be 1:2 (sandwiches) (192) in the presence of excess antibiotic. MM-type antibiotics show poor selectivity for M^{2+} ions as compared to M^+ ions (19). In methanol the larger nigericin prefers potassium over sodium (60, 61), monensin prefers sodium over potassium (60, 61), and dianemycin does not quite distinguish among Na, K, and Rb (2). All the MCM-type antibiotics show a high specificity for potassium over sodium

Table 4 Stability Constants (log K) for Some Metal-Antibiotic Complexes (kg mole^{-1}) at 25°C (from Ref. 3)

Antibiotic	Medium	Li	Na	K	Rb	Cs
Nonactin	Methanol	—	5.2×10^2	3.1×10^4	—	—
Valinomycin	Methanol	4	3.7	6.3×10^4	1.4×10^5	2.1×10^4
Enniatin B	Methanol	1.5×10	2.1×10^2	6.6×10^2	4.3×10^2	1.7×10^2
Antamanide	Ethanol	—	2.0×10^3	2.0×10^2	—	—
Nigericin	Methanol	—	1.9×10^4	1.2×10^5	—	—
Monensin	Methanol	—	5.5×10^5	7.5×10^4	—	—

(49-54) (Table 4). The solution stabilities of M^+-actin systems follow the order $K > Rb > Cs > Na > Li$; changes in the sequence with change in environmental conditions, in the case of actins (24) as well as other MCM-type antibiotics (21), can be explained in terms of conformation changes that take place due to changes in environmental conditions. One of the noted exceptions, however, is that antamanide is more selective for sodium than for potassium (55) in spite of the fact that it, like the actins, is also a 30-membered cyclic ligand. Regardless of the nature of the anionic species the solution stabilities of Na^+-antamanide are 1-2 orders of magnitude higher than those of comparable K^+-antamanide (19) complexes; in ethanol antamanide is about 11 times more specific for sodium than for potassium (55), and in a methylene chloride phase containing antamanide, Na(Pic) is extracted from the aqueous phase about 20 times more than K(Pic) (193).

In the solid state the coordinative characteristics and conformation aspects of antibiotics in systems like (Na-nonactin)NCS (194), (K-nonactin)NCS (80), (K-enniatin B)I (81), (K-valinomycin)AuCl$_4$ (82), (Ag^+-monensin$^-$) (83) (and analogous salts of Na, K, and Tl), (M^+-nigericin$^-$) (84) (M = Na, K, or Tl, which produce isomorphous salts), (M^+-grisorixin$^-$) (86, 87) (M = Ag or Tl), [Ba^{2+}(X-537A)$_2^-$]·2H$_2$O (88), and [Ag_2^+(X-537A)$_2^-$] (88) have been investigated with the help of X-ray diffraction technique.

Antibiotics of the MM group form metal salts with M, Ag^+, and Tl^+ which are insoluble in water and soluble in polar organic solvents. An antibiotic in the anionic form acts as a multidentate ligand to fulfill the coordinative requirements of the cation as well as to serve as a charge neutralizer for the latter. The anionic ligand always shows cyclic behavior due to intramolecular bonding, and encapsulates the cation completely. All known complexes of these antibiotics have been shown to be complete encapsulates by single-crystal X-ray analysis;

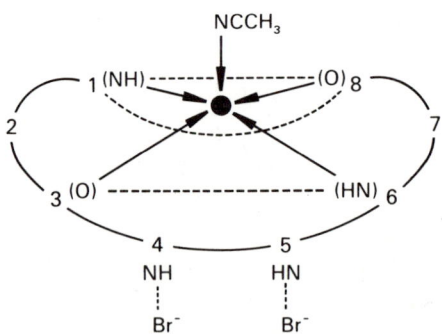

Figure 8 Schematic representation of X-ray molecular structure of (Li-antamanide)Br, CH$_3$CN (95). The closed circle represents Li, and 1, 2, 3,... denote amino acids shown in Figure 5 (*iii*).

in the complex $[Ba^{2+}(X-537A)_2^-] \cdot 2H_2O$ (88) both -COO$^-$ and the bonded -OH act as donor sites, whereas in all the others investigated so far, either of the two sites coordinates with the cation. MCM-type of antibiotics complex with MX; however, only inorganic salts have been investigated so far. The M-MCM complexes are usually 1:1, and those investigated in the solid state have been found to be complete encapsulates. However, (Li-antamanide)Br,CH$_3$CN (95) (Figure 8), which is crystallized from methyl cyanide and carries a molecule of the solvent, is an incomplete encapsulate.

6 CROWNS AND ANTIBIOTICS—SIMILARITY IN PRINCIPLES OF COMPLEXATION

There is a degree of similarity in the behavior of crowns and antibiotics toward the concerned cations (3-6). For both types of macromolecules cation specificities in solution can be measured either by adopting a direct titration technique (46-48) in a homogeneous phase, or by extracting the cation from an aqueous to an organic phase in the presence of the macrocycle (16, 20, 21), or by following transport reactions (27) of the cation across a membrane using the latter as the mediator of transport. Regardless of the mode of determination the specificity sequence for M$^+$ has been found to be K > Rb > Cs > Na > Li for actins (57, 58, 195, 196), enniatins (145), and for those crown ethers in which the cavity size exceeds that of B15C5 (46-48) (Table 5). DC18C6 (cis-syn-cis isomer) (K/Na = 80) (46) and DB30C10 (K/Na = 400) (46) show higher specificity for potassium in methanol, as was also noted for nonactin (K/Na = 16) (49) and valinomycin (K/Na = 17,000) (81) in the same medium.

Both types of macromolecule induce transport of the cations across artificial as well as natural membranes (19, 23, 25, 27, 33-35), although the effect exercized by the crown ethers is comparatively weak (27). For both crown ethers and antibiotics, an approximate proportionality exists (19, 27) between the transport of M$^+$ with a given macrocycle and (*a*) the extraction constant of the corresponding M$^+$-macrocycle species in the two-phase extraction systems (16, 17, 19) as well as (*b*) the stability constant of the species in a homogeneous (methanol) medium (46).

The rate constants for the formation of complexes are high both for the crowns (197) and the antibiotics (198, 199), which indicates that the activation barrier for either molecule is lowered through stepwise dehydration. Various types of equilibrium constants (the constant of extraction into the organic phase, the constant of ion pairing in the organic phase, and the product of the two constants) are dependent on the nature of the organic phase. This indicates that either molecule adopts a conformation to suit the requirements of M$^+$ and that the mechanism of "induced fit" is involved for both during complexation.

Table 5 Stability Constants (log K) of 1:1 Metal-Crown Complexes (1 mole^{-1}) at 25°C (from Ref. 46, 48)

Crown Ether	Medium	Metal					
		Li	Na	K	Rb	Cs	
DB18C6	Methanol	—	4.4	5.0	—	3.6 (log K = 2 for 1:2)	
18C6	Methanol	—	4.3	6.1	—	4.6	
DC18C6 (cis-syn-cis isomer)^a	Water	0.6	1.7	2.2	1.5	1.2	
	Methanol	—	4.1	6.0	—	4.6	
DC18C6 (cis-anti-cis isomer)^b	Water	—	1.4	1.8	0.9	0.9	
	Methanol	—	3.7	5.4	—	3.5	
DB24C8	Methanol	—	—	3.5	—	3.8	
DB30C10	Methanol	—	2.0	4.6	—	—	

^{a} Stability sequence for M^{2+} ions; Mg, Ca, Sr (log K = 3.2), Ba (log K = 3.6).
^{b} Stability sequence for M^{2+} ions; Mg, Ca, Sr (log K = 2.6), Ba (log K = 3.3).

6 Crowns and Antibiotics—Similarity in Principles of Complexation 145

The structural aspects of K(DB30C10)$^+$ (98) and (K-nonactin)$^+$ (80) complexes are comparable in the solid state. Both ligands desolvate potassium ion by wrapping around it to produce cubic systems possessing twofold axes of symmetry. Donor oxygen atoms for both ligands appear as if on the seam of a tennis ball, with the cation located at the center.

6.1 Synthetic and Natural Multidentates—Further Resemblance in Principles of Complexation

The solution stabilities of the alkali metal complexes derived from crowns (32, 46, 176), cyclic antibiotics (24, 200), bicyclic macromolecules (111), and other macromolecular ligands can in general (201) be explained by considering the ion-cavity-radius concept. This is justified because the solution stability of the complex is related to the energy of the reaction, and the latter to the effectiveness of the metal-donor atom contacts. However, the concepts of encapsulation are useful in making qualitative inferences about the stabilities of all the systems. Thus out of the stability sequence K > Rb > Cs > Na > Li, obtained with actins and the crowns (Tables 4 and 5), the lowest stability for the last two cations can be attributed to their strongest involvement with anionic species and to the consequent minimal operation of ligand encapsulation and self-encapsulation. The high stability for the other three cations can result from successful ligand encapsulation, but an enhanced stability in the order Cs, Rb, K appears to be caused by a progressively strong contribution of self-encapsulation even by these cations. This conclusion is confirmed by the results of experiments with the bicyclic ligands (Table 3). The ion/cavity ratios in the case of these ligands (calculated from the values given in Table 3) indicate that the ligand [2.2.2] should be equally suited for potassium and rubidium, and that [3.2.2] should be most suited for cesium. However, the stability values show that the former ligand is more specific for potassium, and the latter for rubidium and even for potassium. Such a selectivity preference for the loose-fitting smaller cation indicates that the self-encapsulating tendency is detectable even for the low charge density cations such as rubidium and cesium.

Successful operation of ligand encapsulation by any cyclic macromolecule on a cation usually leads to the formation of a complete encapsulate, and the stability of the latter should in principle be high due to the absence of cation-anion interactions; the stability of K(B15C5)$_2$Pic is higher than that of K(B15C5)Pic in THF (100). Possibly the state of complete encapsulation also contributes to a high cation specificity of the crown ethers and the actins for potassium, rubidium, and cesium; complete encapsulation of potassium with nonactin (80) and of potassium, rubidium, and cesium with the crowns (38, 42, 46, 77, 100) is known, and the same should be possible for rubidium and cesium with actins. The concepts of ligand encapsulation and complete encapsulation may be

146   Multidentate Macromolecules

collectively put forth in the case of potassium to explain the observed high specificities of actins (49) and valinomycin (51) for this cation over sodium.

From our experience with metal-crown systems we know that, if charge separation of the complexing salt is possible due to effective stabilization of the anion, complexation of the cation is favored by increase in its charge density through its enhanced self-encapsulating tendency. This explains why, due to bonding of Br⁻ with -NH of the antibiotic (Figure 8), a small cation like lithium can be complexed in the presence of a highly nucleophilic bromide by a large-cavity multidentate like antamanide to produce the complex (Li-antamanide)-Br, CH_3CN (95).

6.1.1 Discrimination of Na/K and Ca/Mg in Natural Systems

Encapsulation with electrically neutral macrocycles is possible for low (e.g., K) and high (e.g., Mg) charge density cations with molecules such as B15C5. Encapsulation of the middle charge density cation (e.g., Li, Na, Ca) is usually possible when stabilization of the counteranion takes place, as in the complexing of lithium by antamanide (95). By keeping in view the possibility of two types of the transport mediators—represented by B15C5 and antamanide—in natural membranes, it is possible to visualize two types of membrane (Figure 9). Within each membrane the population of one type of mediator decreases in one direction whereas that of the other decreases in the opposite direction. As observed in most natural systems, a membrane selective for potassium and magnesium from the exterior is expected to be selective for sodium and calcium from the interior [Figure 9(i)] and vice-versa [Figure 9(ii)].

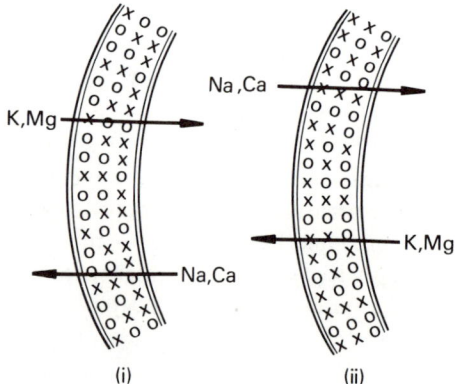

Figure 9 Postulated presentation of natural membranes. Open circles indicate K-Mg selective mediators, represented by B15C5, and crosses indicate Na-Ca selective mediators, represented by antamanide.

6.1.2 General

The preference for smaller cations over larger ones by synthetic (1) and natural (3) macromolecules has been attributed in part to the enhanced site-site repulsions within the ligand molecule. This can be true for the bi- and tricyclic ligands. However, for the flexible macrocycles for which complexes of the small cations have been isolated under special conditions, the discrimination may be attributed primarily to the strong involvement of such cations. The site-site repulsions may only aid the retreat of the macrocycle when the same fails to encapsulate an involved cation. If provisions are made for the stabilization of the counteranion during complexation of the cation, for example, by using strong proton donors in the reaction mixture or by using macrocycles carrying positively polarized sites (such as $-OH^{+\delta}$, $-NH^{+\delta}$ or $-COOH^{+\delta}$), it is believed that high specificity for the small divalent (Ca and Mg) and monovalent (Na and Li) cations may be noted, contrary to the earlier predictions (3, 12). There are numerous examples (2) of complexation of high lattice energy (highly associated) salts with ligands that provide simultaneous stabilization to the counteranion through their positively polarized protons.

To explain the observed discrimination between M^+ and M^{2+} ions and between cations of the same charge, elaborate theoretical treatments exist (3) that invoke the role played by the size of the cation, its coordination number, the number of donor sites carried by the ligand and the distribution of charge on the latter (202), and the geometry of the complex (3). Several of these considerations helped in understanding the variation of the solution stability of metal-macro-multidentate systems with changes in such factors. However, these treatments do not explain certain observations. One example is the total lack of Mg-macrocycle complexes in the solid state versus the formation of M^+-macrocycle complexes, which are common; the reported (5) complex of magnesium is not (67) well characterized. Presently the diversity in the behavior of different cations is being explained by invoking qualitative concepts of encapsulation evolved on the basis of experimental results, and this is being confirmed by carrying out synthesis of the nonexisting metal-crown complexes. Such concepts provide a basic explanation for the formation of complexes, and are complementary to the above-mentioned theoretical considerations for the complete understanding of such systems.

7 SUMMARY

In this chapter coordinative characteristics of the multidentate macromolecules with alkali and alkaline earth cations (M) have been discussed with special reference to crown ethers and natural antibiotics. The crowns discussed include benzo-15-crown-5 (B15C5), dibenzo-18-crown-6 (DB18C6), 18-crown-6 (18C6), dicyclohexano-18-crown-6 (DC18C6), dibenzo-24-crown-8 (DB24C8), and dibenzo-30-crown-10 (DB30C10). The antibiotics under discussion include those of the nigericin group, valinomycin group, and antamanide.

Until now the metal-macrocycle systems have been discussed by considering the size of the metal and the cavity size of the ligand—the ion-cavity-radius concept—which is shown to be incorrect. Instead it has been demonstrated that such systems are better explained by considering the charge density of the metal and the conformation/flexibility aspect of the macrocycle.

For a salt MX the effect of X during complexation of the metal has been discussed with respect to (*a*) metal-macrocycle interaction in solution, (*b*) isolation of metal-macrocycle complexes in the solid state, and (*c*) the stoichiometry and structural aspects of the complex in the lattice; the stoichiometry of the solid complexes (especially of the bimetallic products) and the conformation of the crown ether in the solid complex are shown to be primarily controlled by the nature of the anion and the flexibility characteristics of the ligand. It is illustrated that the thermodynamic solution stability of metal-macrocycle complexes as well as the crystallization possibility of a complex from solution are far more dominantly controlled by the anion than by the solution medium.

Stoichiometries of metal-crown complexes are also covered from the viewpoint of the charge density of metal and, in the case of large-cavity crown ethers, such as (DB30C10), the conformation/flexibility characteristics of the ligand. The dynamics of complexation of small-cavity crown ethers, such as B15C5, are also discussed. The literature information and the experimental results obtained have been analyzed to show that the charge density of the high charge density cations (e.g., magnesium) operates differently during complexation than does that of the low charge density cations (e.g., potassium). During complexation with low charge density cations the crown ether plays a dominant role on thermodynamic grounds, and the process is termed *ligand encapsulation*. In the case of high charge density cations, however, the cation itself plays the dominant role on electrostatic grounds to manage its own encapsulation—*self-encapsulation*. The medium charge density cations (lithium, sodium, and calcium), for which neither phenomenon is effectively operative, are discriminated from potassium and magnesium with the help of crown ethers in the sense that for the former only 1:1 ligation of the ligand is common, whereas for the latter charge separation of the cation (1:2 encapsulation) is normally possible.

By considering the charge density of M, the nucleophilicity of X, and the conformation aspect of the ligand, it becomes possible to predict whether a particular complexation is possible and, if so, the preferred stoichiometry of the resulting complex.

Commonly known natural antibiotics and their complexes with M are briefly treated. Cation selectivity with antibiotics is outlined. Explanations for the observed selectivities have been provided with the help of complexation principles discovered from work with crown ethers. Existing similarities in the principles of complexation with the two types of ligands are outlined, and further similarities have been discussed in the light of the available physicochemical data for the solution and solid states. Reasons for the complexation and selectivity of antamanide with the medium charge density cations are discussed. By postulating the distribution of (a) K-Mg selective and (b) Na-Ca selective transport mediators in natural membranes, the phenomenon of Na/K and Ca/Mg discrimination in natural systems is illustrated.

REFERENCES

1. J. M. Lehn, *Struct. Bonding (Berlin)*, **16**, 1 (1973).
2. M. R. Truter, *Struct. Bonding (Berlin)*, **16**, 71 (1973).
3. W. Simon, W. E. Morf, and P. Ch. Meier, *Struct. Bonding (Berlin)*, **16**, 113 (1973).
4. R. M. Izatt, D. J. Eatough, and J. J. Christensen, *Struct. Bonding (Berlin)*, **16**, 161 (1973).
5. J. J. Christensen, D. J. Eatough, and R. M. Izatt, *Chem. Rev.*, **74**, 351 (1974).
6. A. Kotyk and K. Janacek, *Cell Membrane Transport*, Plenum, London, 1970.
7. E. E. Bittar (Ed.), *Membrane and Ion Transport*, Vol. I, Wiley-Interscience, New York, 1970.
8. R. J. P. Williams, *Q. Rev.*, **24**, 331 (1970).
9. R. J. P. Williams, in *Bio-inorganic Chemistry*, American Chemical Society Publication No. 100, 1971.
10. D. Ammann, E. Pretsch, and W. Simon, *Tetrahedron Lett.*, 2473 (1972).
11. D. Ammann, E. Pretsch, and W. Simon, *Anal. Lett.*, **5**, 843 (1972).
12. W. E. Morf, D. Ammann, E. Pretsch, and W. Simon, *Pure Appl. Chem.*, **36**, 421 (1973).
13. N. S. Poonia, *J. Am. Chem. Soc.*, **96**, 1012 (1974).
14. N. S. Poonia, International Conference on Coordination Chemistry, Hamburg, 1976, p. 343; *J. Am. Chem. Soc.*, communicated.
15. N. S. Poonia, unpublished work.
16. G. Eisenman, S. Ciani, and G. Szabo, *J. Membr. Biol.*, **1**, 294 (1969).
17. H. K. Frensdorff., *J. Am. Chem. Soc.*, **93**, 4684 (1971).
18. B. C. Pressman, *Fed. Proc.*, **27**, 1283 (1968).
19. B. C. Pressman and D. H. Haynes, in *The Molecular Basis of Membrane Function*, D. C. Tosteson, (Ed.), Prentice-Hall, Englewood Cliff, New Jersey, 1970, pp. 221-246.
20. D. H. Haynes and B. C. Pressman, *J. Membr. Biol.*, **18**, 1 (1974).
21. G. Szabo, G. Eisenman, and S. Ciani, *J. Membr. Biol.*, **1**, 346 (1969).
22. R. S. Cockrell, E. J. Harris, and B. C. Pressman, *Biochemistry*, **5**, 2325, (1966).
23. E. J. Harris, G. Catlin, and B. C. Pressman, *Biochemistry*, **6**, 1360 (1967).
24. P. Mueller and D. O. Rudin, *Biochem. Biophys. Res. Commun.*, **26**, 398 (1967).

25. B. C. Pressman, E. J. Harris, E. J. Jagger, and J. H. Johnson, *Proc. Natl. Acad. Sci.*, **58**, 1949 (1967).
26. D. C. Tosteson, P. C. Cook, T. E. Andreoli, and M. Tieffenberg, *J. Gen. Physiol.*, **50**, 2513 (1967).
27. D. H. Haynes, T. Wiens, and B. C. Pressman, *J. Membr. Biol.*, **18**, 23 (1974).
28. E. Eyal and G. A. Rechnitz, *Anal. Chem.*, **43**, 1090 (1971).
29. G. A. Rechnitz and E. Eyal, *Anal. Chem.*, **44**, 370 (1972).
30. D. H. Haynes, Ph.D. Thesis, University of Pennsylvania, Philadelphia, 1970.
31. H. R. Wuhrmann, W. E. Morf, and W. Simon, *Helv. Chim. Acta*, **56**, 1011 (1973).
32. C. J. Pedersen and H. K. Frensdorff, *Angew. Chem., Int. Ed. Engl.*, **11**, 16 (1972).
33. H. Lardy, *Fed. Proc.*, **27**, 1278 (1968).
34. D. C. Tosteson, *Fed. Proc.*, **27**, 1269 (1968).
35. G. Eisenman, *Fed. Proc.*, **27**, 1249 (1968).
36. D. E. Fenton, M. Mercer, N. S. Poonia, and M. R. Truter, *Chem. Commun.*, 66 (1972).
37. M. A. Bush and M. R. Truter, *J. Chem. Soc., Perkin Trans..*, 341 (1972).
38. P. R. Mallinson and M. R. Truter, *J. Chem. Soc., Perkin Trans.*, 1818 (1972).
39. D. Bright and M. R. Truter, *Nature*, **225**, 176 (1970).
40. M. A. Bush and M. R. Truter, *J. Chem. Soc., B*, 1440 (1971).
41. D. E. Fenton, M. Mercer, and M. R. Truter, *Biochem. Biophys. Res. Commun.*, **48**, 10 (1972).
42. P. R. Mallinson, *J. Chem. Soc., Perkin Trans.*, 261 (1975).
43. M. A. Bush and M. R. Truter, *Chem. Commun.*, 1439 (1970).
44. N. S. Poonia and M. R. Truter, *J. Chem. Soc., Dalton Trans.*, 2062 (1972).
45. N. S. Poonia, *J. Inorg. Nucl. Chem.*, **37**, 1855 (1975).
46. H. K. Frensdorff, *J. Am. Chem. Soc.*, **93**, 600 (1971).
47. R. M. Izatt, J. H. Rytting, D. P. Nelson, B. L. Haymore, and J. J. Christensen, *Science*, **164**, 443 (1969); R. M. Izatt, J. J. Christensen, and J. O. Hill, *Science*, **174**, 459 (1971).
48. R. M. Izatt, D. P. Nelson, J. H. Rytting, B. L. Haymore, and J. J. Christensen, *J. Am. Chem. Soc.*, **93**, 1619 (1971).
49. W. E. Morf and W. Simon, *Helv. Chim. Acta*, **54**, 2683 (1971).
50. M. M. Shemyakin, Y. A. Ovchinnikov, V. T. Ivanov, V. K. Antonov, A. M. Shkrob, I. I. Mikhaleva, A. V. Evstratov, and G. G. Malenkov, *Biochem. Biophys. Res. Commun.*, **29**, 834 (1967).
51. M. M. Shemyakin et al., *J. Membr. Biol.*, **1**, 402 (1969).
52. H. K. Wipf, L. A. R. Pioda, Z. Stefanac, and W. Simon, *Helv. Chim. Acta*, **51**, 377 (1968).
53. E. Grell, Th. Funch, and F. Eggers, Symposium on Molecular Mechanism of Antibiotic Action on Protein Biosynthesis and Membrane, University of Granada, Spain, June 1-4, 1971.
54. H. J. Moschler, H. G. Weber, and R. Schwyzer, *Helv. Chim. Acta*, **54**, 1437 (1971).
55. T. Wieland et al. *FEBS Lett.*, **9**, 89 (1970).
56. W. Simon and W. E. Morf, in *Membranes*, G. Eisenman (Ed.), Marcel-Dekker, 1973.
57. Ch. U. Zust, Dissertation ETH, Zurich, 1972.

References

58. W. E. Morf, Ch. U. Zust, and W. Simon, in Ref. 53.
59. V. T. Ivanov et al., *Biochem. Biophys. Res. Commun.,* **42**, 654 (1971).
60. W. K. Lutz, H. K. Wipf, and W. Simon, *Helv. Chim. Acta,* **53**, 1741 (1970).
61. W. K. Lutz, P. U. Fruh, and W. Simon, *Helv. Chim. Acta,* **54**, 2767 (1971).
62. Y. A. Ovchinnikov, V. T. Ivanov, and I. I. Mikhaleva, *Tetrahedron Lett.,* 159 (1971).
63. J. M. Lehn and J. P. Sauvage, *Chem. Commun.,* 440 (1971).
64. J. M. Lehn and F. Montavon, *Tetrahedron Lett.,* 4557 (1972).
65. J. Cheney, J. M. Lehn, J. P. Sauvage, and M. E. Stubbs, *Chem. Commun.,* 1100 (1972).
66. B. Dietrich, J. M. Lehn, and J. P. Sauvage, *Chem. Commun.,* 15 (1973).
67. C. J. Pedersen, *J. Am. Chem. Soc.,* **89**, 7017 (1967).
68. C. J. Pedersen, *Fed. Proc.,* **27**, 1305 (1968).
69. C. J. Pedersen, *J. Am. Chem. Soc.,* **89**, 2495 (1967).
70. C. J. Pedersen, *J. Am. Chem. Soc.,* **92**, 391 (1970).
71. C. J. Pedersen, *J. Am. Chem. Soc.,* **92**, 386 (1970).
72. J. Dale and P. O. Kriastiansen, *Chem. Commun.,* 670 (1971).
73. R. N. Greene, *Tetrahedron Lett.,* 1793 (1972).
74. J. Petranek and O. Ryba, *Collect. Czech. Chem. Cummun.,* **39**, 2033 (1974).
75. M. R. Truter and C. J. Pedersen, *Endeavour,* **30**, 142 (1971).
76. M. R. Truter, *Chem. Br.,* 203 (1971).
77. N. S. Poonia, *J. Sci. Ind. Res.,* **36**, 268 (1977).
78. J. D. Dunitz and P. Seiler, *Acta Crystallogr.,* **30B**, 2750 (1974).
79. E. Shchori, J. J. Grodzinski, and M. Shporer, *J. Am. Chem. Soc.,* **95**, 3842 (1973).
80. H. Dobler, J. D. Dunitz, and B. T. Kilbourn, *Helv. Chim. Acta,* **52**, 2573 (1969).
81. M. Dobler, J. D. Dunitz, and J. Krajewski, *J. Mol. Biol.,* **42**, 603 (1969).
82. M. Pinkerton, L. K. Steinrauf, and P. Dawkins, *Biochem. Biophys. Res. Commun.,* **35**, 512 (1969).
83. A. Agtarap, J. W. Chamberlain, M. Pinkerton, and L. K. Steinrauf, *J. Am. Chem. Soc.,* **89**, 5737 (1967); M. Pinkerton and L. K. Steinrauf, *J. Mol. Biol.,* **49**, 533 (1970).
84. L. K. Steinrauf, M. Pinkerton, and J. W. Chamberlain, *Biochem. Biophys. Res. Commun.,* **33**, 29 (1968).
85. E. W. Czerwinski and L. K. Steinrauf, *Biochem. Biophys. Res. Commun.,* **45**, 1284 (1971).
86. M. Alleaume and D. Hickel, *Chem. Commun.,* 1442 (1970).
87. M. Alleaume and D. Hickel, *Chem. Commun.,* 175 (1972).
88. J. F. Blount and J. W. Westley, *Chem. Commun.,* 927 (1971).
89. D. Moras and R. Weiss, *Acta Crystallogr., Sect. B,* **29**, 396 (1973).
90. D. Moras and R. Weiss, *Acta Crystallogr., Sect. B,* **29**, 400 (1973).
91. D. Moras, B. Metz, and R. Weiss, *Acta Crystallogr., Sect. B,* **29**, 383 (1973).
92. D. Moras, B. Metz, and R. Weiss, *Acta Crystallogr., Sect. B,* **29**, 388 (1973).
93. F. Mathieu, Thesis, Strasbourg, 1972.
94. M. Metz, D. Moras, and R. Weiss, *J. Am. Chem. Soc.,* **93**, 1806 (1971).
95. I. L. Karle, *J. Am. Chem. Soc.,* **96**, 4000 (1974).
96. R. Wiest and R. Weiss, *Chem. Commun.,* 678 (1973).

97. M. Dobler, J. D. Dunitz, and P. Seiler, *Acta Crystallogr.*, **30B**, 2741 (1974).
98. M. A. Bush and M. R. Truter, *J. Chem. Soc., Perkin II*, 345 (1972).
99. D. L. Hughes, *J. Chem. Soc., Dalton Trans.*, 2374 (1975).
100. K. H. Wong, M. Bourgoin, and J. Smid, *Chem. Commun.*, 715 (1974).
101. S. G. A. McLaughlin, G. Szabo, S. Ciani, and G. Eisenman, *J. Membr. Biol.*, **9**, 3 (1972).
102. M. Dobler and R. P. Phizackerley, *Acta Crystallogr.*, **30B**, 2748 (1974).
103. M. Dobler and R. P. Phizackerley, *Acta Crystallogr.*, **30B**, 2746 (1974).
104. M. C. Menard, B. Wojtkowiak, and M. Chabanel, *Bull. Soc. Chim. Belg.*, **81**, 241 (1972).
105. R. T. Sanderson, *Inorganic Chemistry,* Affiliated East-West Press, New Delhi, 1971.
106. D. G. Parsons, M. R. Truter, and J. N. Wingfield, *Inorg. Chim. Acta,* **14**, 45 (1975).
107. N. S. Poonia, A. Banthia, and V. W. Bhagwat, to be published.
108. E. Heftman, *Chromatography,* Reinhold, New York, 1969.
109. D. L. Hughes, C. L. Mortimer, D. G. Parsons, M. R. Truter, and J. N. Wingfield, *Inorg. Chim. Acta Lett.,* **21**, L23 (1977).
110. J. J. Christensen, G. Weed, S. Starr, and R. M. Izatt, paper presented at the First Symposium on Macrocyclic Compounds, August 15-17, 1977, Brigham Young Univ. Provo, Utah.
111. J. M. Lehn and J. P. Sauvage, *J. Am. Chem. Soc.,* **97**, 6700 (1975).
112. D. K. Cabbiness and D. W. Margerum, *J. Am. Chem. Soc.,* **91**, 6540 (1969).
113. F. P. Hinz, and D. W. Margerum, *J. Am. Chem. Soc.,* **96**, 4993 (1974).
114. P. D. Cradwick and N. S. Poonia, *Acta Crytallogr.*, **B33**, 197 (1977).
115. N. S. Poonia, V. W. Bhagwat, and H. Manohar, to be published.
116. F. P. Von Remoorkere and F. P. Boer, *Inorg. Chem.*, **13**, 2071 (1974).
117. N. N. L. Kirsch and W. Simon, *Helv. Chim. Acta,* **59**, 357 (1976).
118. M. G. U. Fiedler, E. Pretsch, and W. Simon, *Anal. Lett.*, **8**, 857 (1975).
119. M. Guggi, M. Oehme, E. Pretsch, and W. Simon, *Helv. Chim. Acta,* **58**, 2417 (1976).
120. P. Viullenmier, P. Gazzotti, E. Carafoli, and W. Simon, *Biochim. Biophys. Acta,* **467**, 12 (1977).
121. W. E. Morf, P. Wuhrmann, and W. Simon, *Anal. Chem.*, **48**, 1031 (1976).
122. M. Guggi, E. Pretsch, and W. Simon, *Anal. Chim. Acta,* **91**, 107 (1977).
123. B. Dietrich, J. M. Lehn, and J. P. Sauvage, *J. Chem. Soc., Chem. Commun.* 15 (1973).
124. B. Dietrich, J. M. Lehn, and J. P. Sauvage, *Tetrahedron Lett.,* 2885, 2889 (1969).
125. J. Cheney, J. M. Lehn, J. P. Sauvage, and M. E. Stubbs, *J. Chem. Soc., Chem. Commun.,* 1100 (1972).
126. G. Anderegg, *Helv. Chim. Acta,* **58**, 1218 (1975).
127. E. Kauffmann, J. M. Lehn, and J. P. Sauvage, *Helv. Chim. Acta,* 1099 (1976).
128. Y. M. Cahen, J. L. Dye, and A. I. Popov, *J. Phys. Chem.*, **79**, 1292 (1975).
129. Y. M. Cahen, J. L. Dye, and A. I. Popov, *J. Phys. Chem.*, **79**, 1289 (1975).
130. J. M. Ceraso and J. L. Dye, *J. Am. Chem. Soc.,* **95**, 4432 (1973).
131. M. Shporer and Z. Luz, *J. Am. Chem. Soc.,* **97**, 665 (1975).
132. E. Mei, J. L. Dye, and A. I. Popov, *J. Am. Chem. Soc.,* **98**, 1619 (1976).

References

133. R. D. Hancock and G. J. McDougall, First Symposium on Macrocyclic Compounds, Brigham Young University, August 1977, Provo, Utah.
134. J. M. Lehn, *Pure Appl. Chem.*, **49**, 857 (1977).
135. J. L. Dye, *J. Chem. Educ.*, **54**, 332 (1977).
136. J. L. Dye, J. M. Ceraso, M. T. Lok, B. L. Barnett, and F. J. Tehan, *J. Am. Chem. Soc.*, **96**, 608 (1974).
137. J. L. Dye, C. W. Andrews, and S. E. Mathews, *J. Phys. Chem.*, **79**, 3065 (1975).
138. F. J. Tehan, B. L. Barnett, and J. L. Dye, *J. Am. Chem. Soc.*, **96**, 7203 (1974).
139. J. L. Dye, *Pure Appl. Chem.*, **49**, 3 (1977).
140. J. L. Dye, C. W. Andrew, and S. E. Mathews, *J. Phys. Chem.*, **79**, 3065 (1975).
141. D. Clement, F. Damm, and J. M. Lehn, *Heterocycles*, **5**, 477 (1976).
142. J. M. Lehn, J. Simon, and J. Wagner, *Angew. Chem., Int. Ed. Engl.*, **12**, 578, 579 (1973).
143. E. Graf and J. M. Lehn, *J. Am. Chem. Soc.*, **97**, 5022 (1975).
144. B. Metz, J. M. Rozalky, and R. Weiss, *J. Chem. Soc., Chem. Commun.*, 533 (1976).
145. B. C. Pressman, in *Inorganic Biochemistry*, (G. L. Eichhorn, Ed.), Elsevier, London, 1973.
146. N. S. Poonia and B. P. Yadav, *Chem. Rev. Rajasthan Hindi Granth Acad. Jaipur*, **1974**, (4), 370.
147. T. Kubota, S. Matsutani, M. Shiro, and H. Koyama, *Chem. Commun.*, 1541 (1968).
148. P. Gachon, A. Kergomard, and H. Veschambre, *Chem. Commun.*, 1421 (1970).
149. M. Alleaume and D. Hickel, *Chem. Commun.*, 1422 (1970).
150. M. Pinkerton and L. K. Steinrauf, *J. Mol. Biol.*, **49**, 533 (1970).
151. J. F. Blount and J. W. Westley, *Chem. Commun.*, 927 (1971).
152. S. M. Johnson, J. L. Herrin, S. J. Liu, and I. C. Paul, *Chem. Commun.*, 72 (1970).
153. S. M. Johnson, J. L. Herrin, S. J. Liu, and I. C. Paul, *J. Am. Chem. Soc.*, **92**, 4428 (1970).
154. C. A. Maier and I. C. Paul, *Chem. Commun.*, 181 (1971).
155. H. Brockmann and G. Schmidt-Kastner, *Chem. Ber.*, **88**, 57 (1955).
156. H. Brockmann, M. Springourum, G. Traxler, and I. Hofer, *Naturwissenschaften*, **50**, 689 (1963).
157. P. A. Plattner, U. Nager, and A. Boller, *Helv. Chim. Acta*, **31**, 594 (1948).
158. R. L. Hamill, C. E. Higgens, H. E. Boaz, and M. Gorman, *Tetrahedron Lett.*, 4255 (1969).
159. M. M. Shemyakin, *Angew. Chem.*, **72**, 342 (1960).
160. W. Keller-Schierlein and H. Gerlach, *Fortschr. Chem. Org. Naturst.*, **26**, 161 (1968).
161. K. Ando, H. Oishi, S. Hirano, T. Okutomi, K. Suzuku, H. Akazaki, M. Sawada, and T. Sagawa, *J. Antibiot. (Tokyo)*, **24**, 347 (1971).
162. K. Ando, Y. Murakami, and Y. Nawata, *J. Antibiot. (Tokyo)*, **24**, 418 (1971).
163. T. Weiland, G. Luben, H. Ottenheym, J. Faesel, J. X. de Vries, W. Konz, A. Prox, and J. Schmid, *Angew. Chem.*, **80**, 209 (1968).
164. R. L. Harned, P. H. Harter, C. J. Corum, and K. L. Jones, *Antibiot. Chemother.*, **1**, 592 (1951).

165. J. Berger, A. I. Rachin, W. E. Scott, L. H. Sternbach, and M. W. Goldberg, *J. Am. Chem. Soc.*, **73**, 5295 (1951).
166. D. C. Tosteson, P. C. Cook, T. E. Andreoli, and M. Tieffenberg, *J. Gen Physiol.*, **50**, 2513 (1967).
167. C. Moore and B. C. Pressman, *Biochem. Biophys. Res. Commun.*, **15**, 562 (1964).
168. B. C. Pressman and E. J. Harris, *Abstracts of the Seventh International Congress on Biochemistry*, H-79, 1967.
169. S. N. Graven, H. A. Lardy, and O. S. Estrada, *Biochemistry*, **6**, 365 (1967).
170. Z. Stefanac and W. Simon, *Chimia*, **20**, 436 (1966).
171. Z. Stefanac and W. Simon, *Microchem. J.*, **12**, 125 (1967).
172. D. C. Tosteson, T. E. Andreoli, M. Tieffenberg, and P. Cook, *J. Gen. Physiol.*, **51**, 373 (1968).
173. H. K. Wipf, W. Pache, P. Jordan, H. Zahner, W. Keller-Schierlein, and W. Simon, *Biochem. Biophys. Res. Commun.*, **36**, 387 (1969).
174. H. K. Wipf, A. Olivier, and W. Simon, *Helv. Chim. Acta*, **53**, 1605 (1970).
175. Y. A. Ovchinnikov, V. T. Ivanov, and A. M. Shkrob, in Ref. 53.
176. R. Winkler, *Struct. Bonding (Berlin)*, **10**, 1 (1972).
177. Y. A. Ovchinnikov, V. T. Ivanov, and A. M. Shkrob, *Membrane Active Complexones*, Elsevier, Amsterdam, 1974.
178. D. H. Haynes, A. Kowalsky, and B. C. Pressman, *J. Biol. Chem.*, **240**, 502 (1969).
179. V. T. Ivanov, I. A. Laine, N. D. Abdulaev, L. B. Senyavina, E. M. Popov, Y. A. Ovchinnikov, and M. M. Shemyakin, *Biochem. Biophys. Res. Commun.*, **34**, 803 (1969).
180. Y. A. Ovchinnikov, V. T. Ivanov, V. F. Bystrov, N. D. Abdullaev, E. M. Popov, G. M. Lipkind, S. F. Arkhipova, E. S. Efremov, and M. M. Shemyakin, *Biochem. Biophys. Res. Commun.*, **37**, 668 (1969).
181. J. H. Prestegard and S. I. Chan, *Biochemistry*, **8**, 3921 (1969).
182. J. H. Prestegard and S. I. Chan, *J. Am. Chem. Soc.*, **92**, 4440 (1970).
183. D. J. Patel, *Biochemistry*, **12**, 677 (1973).
184. A. E. Tonelli, D. J. Patel, M. Goodman, F. Neider, H. Faulstich, and T. Wieland, *Biochemistry*, **10**, 3211 (1971).
185. A. E. Tonelli, *Biochemistry*, **12**, 689 (1973).
186. V. F. Bystrov, S. L. Portonva, V. T. Ivanov, and Y. Ovchinnikov, *Tetrahedron*, **25**, 493 (1969).
187. Y. A. Ovchinnikov, V. T. Ivanov, V. F. Bystrov, A. I. Miroshnikov, E. N. Shepel, N. D. Abdullaev, E. S. Ffromov, and L. B. Senyavina, *Biochem. Biophys. Res. Commun.*, **39**, 217 (1970).
188. E. Pretsch, M. Vasak, and W. Simon, *Helv. Chim. Acta*, **55**, 1098 (1972).
189. D. F. Mayers and D. W. Urry, *J. Am. Chem. Soc.*, **94**, 77 (1972).
190. M. Ohnishi, and D. W. Urry, *Biochem. Biophys. Res. Commun.*, **36**, 194 (1969).
191. M. Ohnishi, M. C. Fedarko, J. D. Baldeschwieler, and L. F. Johnson, *Biochem. Biophys. Res. Commun.*, **46**, 312 (1972).
192. V. T. Ivanov, E. V. Evstratov, L. V. Sumskaya, E. I. Melnik, T. S. Chumburidz, S. L. Portnova, T. L. Balashova, and Y. U. Ovchinnikov, *FEBS Lett.*, **36**, 72 (1973).
193. T. Wieland, H. Faulstich, and W. Burgermeister, *Biochem. Biophys. Res. Commun.*, **47**, 984 (1972).

References

194. M. Dobler and P. R. Phizackerley, *Helv. Chim. Acta,* 57, 664 (1974).
195. B. C. Pressman, *Proc. Natl. Acad. Sci.,* 53, 1076 (1965).
196. S. N. Graven, H. A. Lardy, and A. Rutter, *Biochemistry,* 5, 1735 (1966).
197. P. B. Chock, *Proc. Natl. Acad. Sci.,* 69, 1939 (1972).
198. H. Diebler, M. Eigen, G. Ilgenfritz, G. Mass, and R. Winkler, *Pure Appl. Chem.,* 20, 93 (1969).
199. R. Winkler, Dissertation, Vienna-Göttingen, 1969; Ref. 139.
200. J. B. Chappell and A. R. Crofts, *Biochem. J.,* 95, 393 (1965).
201. L. Y. Martin, L. J. Dehayes, L. J. Zompa, and D. H. Busch, *J. Am. Chem. Soc.,* 96, 4046 (1974).
202. S. Krasne and G. Eisenman, in *Membranes,* Vol. 2, (G. Eisenman, Ed.), Marcel-Dekker, New York, 1973.

CHAPTER FOUR
CROWN ETHERS AND RELATED MACROCYCLES WITH BIS(METHYLENE)AROMATIC OR -HETEROAROMATIC SUBUNITS: THEIR SYNTHESIS AND COMPLEXATION

D. N. REINHOUDT AND F. DE JONG

Koninklijke/Shell-Laboratorium
(Shell Research B.V.)
Amsterdam, The Netherlands

1 Introduction	158
2 Synthesis and Properties	160
2.1 Synthesis, 160	
2.2 Templated Synthesis, 176	
2.3 NMR Spectra, 179	
2.4 Mass Spectra, 180	
3 Complexation with Salts of Alkali and Alkaline Earth Metals	181
3.1 Complexation in Polar Solvents, 181	
3.2 Complexation in Apolar Solvents, 182	
4 Complexation with Ammonium and Alkylammonium Salts	186
4.1 Complexation with Ammonium Salts, 188	
4.2 Complexation with Water, 189	
4.3 Complexation with Alkylammonium Salts under Anhydrous Conditions, 193	
4.4 Complexation with Alkylammonium Salts in the Presence of Water, 195	
4.5 Kinetics of Complexation with Alkylammonium Salts, 201	
References	213

1 INTRODUCTION

In the past 10 years, after the first publications by Pedersen (1, 2), the chemistry of crown ethers has developed rapidly, and in many areas of chemistry and biology these compounds have started to play an increasingly important role. They have been successfully used for complexation, where they show a high degree of cationic specificity (3), and for the dissolution of salts in apolar solvents, where they give rise to an increased nucleophilicity of the weakly solvated anions (4-6). Furthermore the optical resolution of racemic protonated amines by chiral crown ethers (7, 8) and the mimicry of enzyme systems (9, 10) have widened the prospects for the synthesis of chiral compounds and stereoselective catalysis.

Thus there is a wide variety of possible applications, but the number of synthesis routes remains small. The first synthesis of crown ethers, via reaction of catechol and α,ω-dihalides or α,ω-ditosylates in the presence of a base, was reported in 1967 (1, 2). Since then a large number of crown ethers and related macrocycles have been prepared that differ greatly in ring size and in the number and type of heteroatoms (O, N, S) and additional functional groups. The macrocyclic polyethers have exclusively been obtained by formation of a carbon-oxygen bond in the cyclization step. This kind of synthesis nearly always comprises intramolecular replacement of a halide or tosylate group by an alkoxide nucleophile (Williamson synthesis). Only a few crown ethers have also been synthesized by other reactions such as cyclooligomerization of ethylene oxide

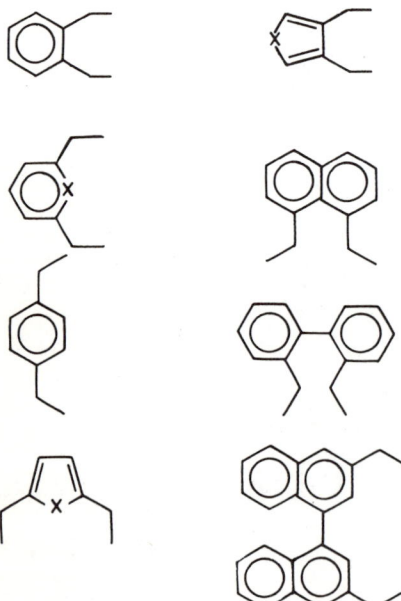

Figure 1 Macrocyclic subunits discussed in this chapter.

1 Introduction

(11-13) and reaction of 1,3-dioxane with alkenes (14). For the synthesis of macrocycles in which other hetero atoms are present in addition to oxygen, several alternative methods of ring closure have been used. Thus acylation of amino groups followed by reduction of the amides, and nucleophilic substitution at sulfur have been employed by, for example, Lehn and co-workers (15) and Vögtle's group (16).

The major drawbacks of ether formation via the Williamson synthesis are the slow rate of reaction of alkoxides with alkyl halides or tosylates (17) and the concurrent base-catalyzed elimination of acid (18).

For various applications these macrocycles must have very specific structures. The structural requirements relate not only to ring size, type and number of donor atoms, and regularity of the structure, but often also to the presence of one or more functional groups. Because of this a great deal of work has been done on the synthesis of macrocycles in which substituents are incorporated, preferentially in close proximity to the macroring.

In view of the limitations of the Williamson synthesis, the lack of alternative reaction routes, and the requirement of additional functionality, a large number of crown ethers and related macrocycles have been designed with bis(methylene)-aromatic or -heteroaromatic moieties as the main building blocks.* In the first place such a design eliminates the two drawbacks of the Williamson ether synthesis because (*a*) benzyl halides are very reactive in solvolysis, and (*b*) the absence of β-protons eliminates the possibility of the undesired 1,2-elimination (19). In the second place the incorporated aryl or heteroaryl moieties can be used as potential points of attachment for functional groups (7, 8).

This chapter deals with crown ethers and related macrocycles in which one or more bis(methylene)aromatic or -heteroaromatic moities are incorporated, their synthesis and properties, and their capacity to form complexes with salts. Only a small part of the work that is discussed here has been described in other review articles that cover topics such as the synthesis and chemistry of crown ethers (20-23), macrocyclic sulfides (24), chiral recognition by crown ethers (7, 8), and complexation of salts (3, 25, 26).

The following section covers the synthesis and properties of these macrocycles, with the emphasis on yield, the role of the template cation in the cyclization reaction, and spectroscopic data. The next two sections deal with the formation

---

Note on nomenclature. Because the IUPAC nomenclature of these macrocyclic compounds is complicated, we have adopted the "crown ether" nomenclature that was suggested by Pedersen (2). Apart from this the subunits present in the macrocycles (see Figure 1) have been named bis(methylene)aromatic or -heteroaromatic moieties. Although it is recognized that this convention again involves a deviation from the IUPAC nomenclature, it clearly describes this particular group of macrocylic compounds. The term *crown ethers and related compounds* thus refers to all macrocyclic compounds that contain at least one oxygen atom as a donor ligand, and can consequently be regarded as being derived from a crown ether by replacement of one or more, but not all, oxygen atoms by other heteroatoms.

of complexes with salts of alkali and alkaline earth metals and with protonated amines; in these sections special attention is paid to the effects of xylenyl, furyl, pyridyl, and thienyl moieties on complexation. In this context it seems appropriate to include a review of the methods used for the determination of association constants in polar and apolar solvents. The importance of the work summarized in these sections goes beyond the scope of the particular macrocycles reviewed here because it has led to the development of methods for the determination of complexation constants in general. The same holds for the last section, which deals with the rates of decomplexation of complexed macrocycles: investigations into the mechanism of cation exchange for crown ethers having bis(methylene)aromatic and -heteroaromatic subunits have provided a better insight into the phenomena taking place during decomplexation and association of crown ether complexes with ammonium salts in general.

2 SYNTHESIS AND PROPERTIES

2.1 Synthesis

Macrocyclic compounds with one or more bis(methylene)aromatic or -heteroaromatic subunits have been synthesized with oxygen as the only type of hetero atom, but also with nitrogen and oxygen, sulfur and oxygen, and nitrogen, sulfur, and oxygen. As these groups of compounds differ in properties, particularly in their complexing abilities, they are discussed under separate headings. This section also deals with the role of cations in the templated (nonhigh-dilution) synthesis of such macrocycles.* Finally, the results of some relevant spectroscopic studies are summarized. Emphasis has been placed on mass spectrometry and PMR spectroscopy, each of which has proved to be a valuable tool, the former in the elucidation of structures and the latter in detailed studies on complexation with salts.

2.1.1 *Macrocyclic Polyethers (O_x)*

The first examples of macrocyclic polyethers with 1,2-bis(methylene)benzene subunits were reported by Vögtle and Zuber in 1972 (27). These investigators reacted the disodium salt of 1,2-bis(hydroxymethyl)benzene with 1,2-bis-(bromomethyl)benzene under carefully controlled conditions of high dilution, and obtained mixtures of two macrocyclic polyethers (**1** and **2**). The individual yields of these compounds depended greatly on the dilution conditions used, varying from 15 and 55% for **1** and from 7 to 40% for **2**. The analogous reaction of 1,2-bis(hydroxymethyl)benzene with 1,3-bis(bromomethyl)benzene produced only the cyclic 2:2 reaction product **3**. Recently (28) these authors reported the formation of **4** from the reaction of 1,2-bis(bromomethyl)benzene and 1,5-bis(2′-hydroxyphenoxy)-3-oxapentane. Although this compound too was pre-

---
*Whenever high-dilution conditions were used, this is explicitly stated in the text.

pared under conditions of high dilution, its yield amounted to no more than 8% It should be emphasized that, although by structure these macrocyclic polyethers (**1-4**) belong to the class of compounds reviewed in this chapter, they do not form complexes with salts.

Macrocyclic polyethers derived from 1,2-bis(halomethyl)benzene that do form complexes with salts were prepared by Reinhoudt and co-workers (19) via reaction with dilithium, disodium, or dipotassium salts of polyethylene glycols (di-octa). Even under conditions of nonhigh dilution high yields of macrocyclic polyethers (**5** and **6**) were obtained. The ratio of the yields of cyclic 1:1 and 2:2 reaction products varied with the type of glycol and base cation used (Table 1). The absolute yields of **5** varied from 1 to 53%, and a relation was found between the size of the cavity of the crown ether formed and the size of the base cation. The maximum yields were obtained when the crown ether cavity was slightly larger than the cation (Table 2). Crown ethers of type **6** could only be isolated

Table 1 Yields of Crown Ethers 5 and 6 Obtained using Different Alkoxide Cations of Metal Alkoxides (Ref. 19)

Crown Ether	Yield (%)		
	Li^+	Na^+	K^+
5 (n = 0)	<1	1	—
5 (n = 1)	31	10	8
5 (n = 2)	23	34	24
5 (n = 3)	13	29	53
5 (n = 4)	10	18	50
5 (n = 5)	—	22	43
5 (n = 6)	—	16	28
6 (n = 1)	—	17	—
6 (n = 2)	—	14	24
6 (n = 3)	—	<1	8

4

5

6

7

8, R = H, X = O

10, CH_3, X = S

9

11

12a A−B = C_6H_5 N

12b A−B = $CH_3OOC\,C{=}C\,COOCH_3$

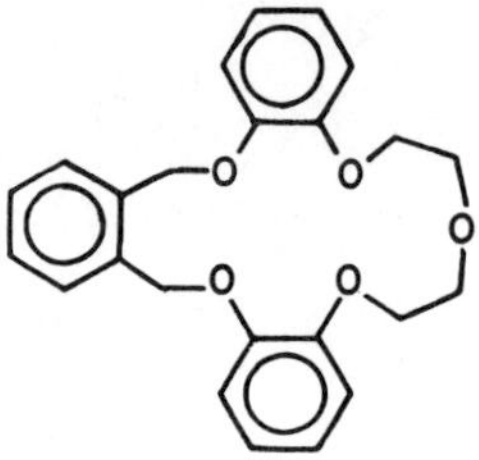

from reactions with di-, tri-, and tetraethylene glycol. A recent publication dealt with the use and synthesis of a similar 2,3-bis(methylene)naphthyl crown ether (7), obtained by reaction of 2,3-bis(hydroxymethyl)naphthalene with pentaethylene glycol ditosylate; no yield was reported (29). When a similar procedure was used for the synthesis of **5** and **6**, the yields were lower than those obtained by reaction of bis(bromomethyl)benzenes and polyethylene glycols (19). Analogous series of crown ethers (**8** and **9**) were obtained by reaction of 3,4-bis(chloromethyl)furan with polyethylene glycols. From the reaction mixtures of 4,5-bis(chloromethyl)benzo-1,3-dioxole and 3,4-bis(chloromethyl)-2,5-dimethylthiophene only the 1:1 reaction products (**10** and **11**) could be isolated in a pure state (19). The furan ring in **8** was used for further modification via Diels-Alder reactions with *N*-phenylmaleimide and with dimethyl acetylenedicarboxylate, giving **12a** and **12b**, respectively.

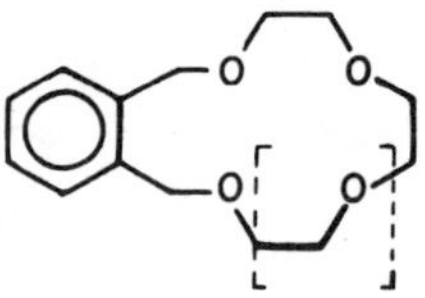

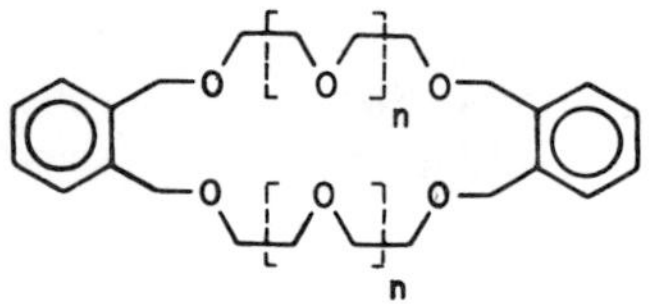

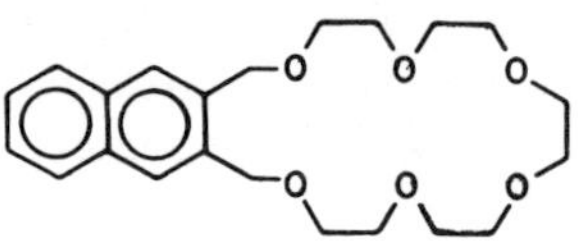

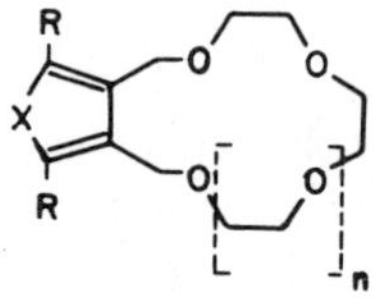

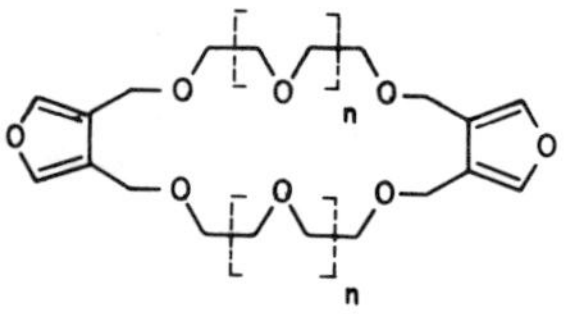

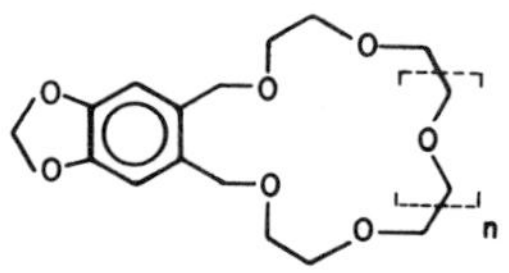

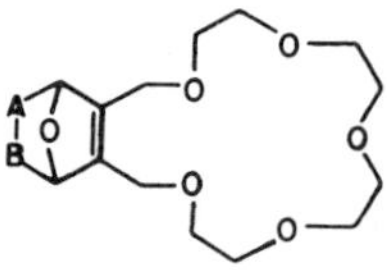

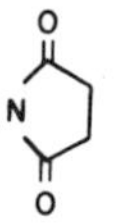

Crown ether *esters* (**13**) containing a 1,2-bis(methylene)benzene moiety have been obtained from 1,2-bis(bromomethyl)benzene and dipotassium dicar-

Table 2a Maximum Yields of Crown Ethers 5, 8, and 10 (Ref. 19)

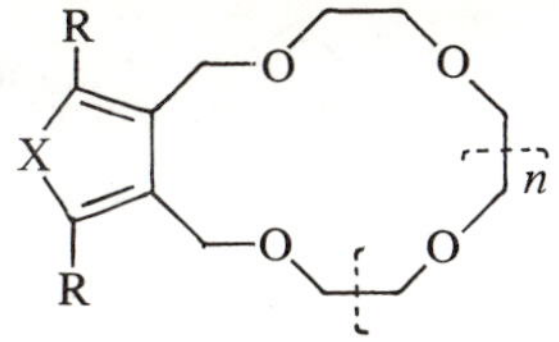

	Yield (%)			
n	5 (X = CH = CH) (R = H)	8 (X = O) (R = H)	10 (X = S) (R = CH$_3$)	Base Cation
1	10^a	8	7	Na$^+$
2	34	31	35	Na$^+$
3	53	30	38	K$^+$
4	50	44	43	K$^+$
5	43	43	42	K$^+$
6	28	—	22	K$^+$

^{a}The yield was 31% with lithium as the template cation.

Table 2b Maximum Yields of Crown Ethers 6 and 9 (Ref. 19)

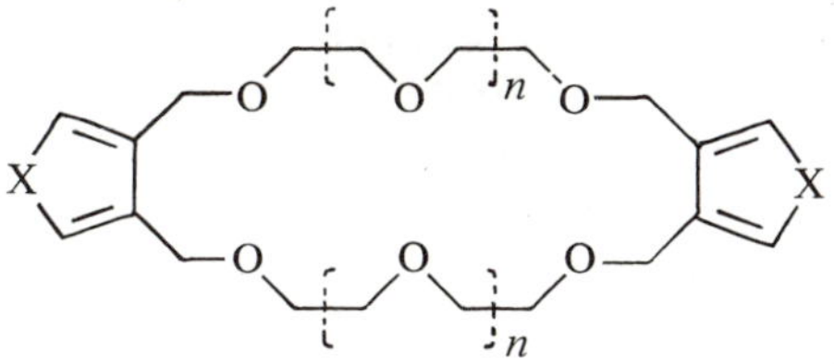

	Yield (%)		Base Cation
n	6 (X = CH = CH)	9 (X = O)	
1	17	18	Na$^+$
2	24	8	K$^+$
3	8	—	K$^+$

boxylates in dimethylformamide (30). The yields were low, with the exception of those of compounds with $n = 0$ and 2 (35 and 16%, respectively). Attempts to form complexes with salts failed.

There are two types of crown ethers with 1,3-bis(methylene)aromatic subunits. In the first type the macrocyclic ring is linked to a benzene (or naphthalene) ring via the 1- and 3-positions (or the 1- and 8-positions); in the second type it is attached to a furan ring via the 2- and 5-positions. In the latter type of compound the furan oxygen atom (or atoms) form(s) part of the macrocyclic array of oxygen atoms.

Reactions of 1,3-bis(bromomethyl)benzene with disalts of polyethylene glycols in apolar solvens yielded two types of crown ether (**14** and **15**, $R^1 = R^2 = H$) as the result of 1:1 or 2:2 reaction (31-33). The yields of **15** were of practical value (30 and 9%) in two cases only ($n = 1$ or 2); the yields of **14** were very much dependent on the ratio of the size of the crown ether ring to that of the base cation present in the cyclization step (31, 32). The alternative method of synthesis of **14** and **15**, that is, reaction of 1,5-bis(hydroxymethyl)benzene with polyethylene glycol ditosylates, gave much lower yields (32). Crown ethers of type **14** and **15** with substituents (OR, COOR, etc.) at C(2) and/or C(4) of the aryl group(s) have been obtained either by reaction of substituted 1,3-bromomethylbenzenes (Table 3) or by modification of a substituent already present (Table 4) (33-36). Intraannular substituents are of particular interest, as their reactivity and/or other properties may differ from those of extraannular substituents. Newcomb and Cram (33) studied the acidity of carboxyl groups at C(2) in crown ethers **14**. The reported pK_a values are higher than that of 2,6-bis(methyoxymethyl)benzoic acid ($pK_a = 3.3$). This decreased acidity of the intraannular carboxyl groups was ascribed to transannular hydrogen bonding of the proton to crown ether oxygen atoms. The effect is most pronounced in the smallest, most rigid rings [pK_a ($n = 2$) = 4.8 ~ pK_a ($n = 3$) = 4.8 > pK_a ($n = 4$) = 3.8 > pK_a ($n = 7$) = 3.4]. Crown ethers **14** ($R^1 = OCH_3$, $R^2 = H$) with intra-annular methoxy groups, obtained by reaction of 2,6-bis(bromomethyl)anisole with disodium polyethylene glycolates, have been readily converted into the

Table 3 Crown Ethers Containing a 1,3-Bis(methylene)benzene Subunit (14, 15) Obtained by Cyclization Reactions

Crown Ether	n	Ring Size	R^1	R^2	Yield (%)	M.p. (°C)	Ref.
14	1	12	H	H	2	oil	32
14	2	15	H	H	16	oil	32
14	2	15	COOCH$_3$	H	34	oil	33
14	2	15	OCH$_3$	CH$_3$	58	oil	34
14	2	15	OCH$_3$	H	45	oil	35
14	3	18	H	H	67 (60)	45-48	32 (33)
14	3	18	COOCH$_3$	H	82	oil	33
14	3	18	H	Br	46	oil	36
14	3	18	H	NPhth	37	78-79	36
14	3	18	H	COOCH$_3$	11	62-64	36
14	3	18	Br	H	7	oil	33
14	3	18	Cl	H	53	oil	33
14	3	18	CN	H	10	oil	33
14	3	18	OCH$_3$	CH$_3$	49	70-72	34
14	3	18	OCH$_3$	H	58	oil	35
14	4	21	H	H	49	oil	32
14	4	21	COOCH$_3$	H	68	oil	33
14	4	21	OCH$_3$	CH$_3$	59	71-73	34
14	5	24	H	H	18	oil	32
14	6	27	H	H	21	oil	32
14	7	30	H	H	21	oil	32
14	7	30	COOCH$_3$	H	34	oil	33
15	0	18	OCH$_3$	CH$_3$	14	148-151	34
15	1	24	H	H	30	oil	32
15	1	24	OCH$_3$	CH$_3$	26	88-90	34
15	2	30	H	H	9	84-85	32

Table 4 Crown Ethers Containing a 1,3-Bis(methylene)benzene Subunit (14) Obtained by Substituent Interconversion

n	Ring Size	R^1	R^2	Yield (%)	M.p. (°C)	Ref.
2	15	COOH	H	~98	106-112	33
2	15	OH	H	>90	66-66.5	35
2	15	OH	NO_2	–	105-106	35
3	18	COOH	H	~98	100-101	33
3	18	CH_2OH	H	80	oil	33
3	18	CH_2OCH_3	H	50	70-71	33
3	18	H	CN	68	oil	36
3	18	OH	H	>90	48-49	35
3	18	OH	NO_2	–	91.0-91.5	35
4	21	COOH	H	~98	86-95	33
7	30	COOH	H	~98	oil	33

corresponding hydroxyl crown ethers (14, R^1 = OH, R^2 = H) by demethylation with lithium iodide in pyridine (35). This facile conversion has been attributed to complex formation, because an attempt to demethylate 2,6-dimethylanisole under the same conditions was unsuccessful. Nitration of the phenols yielded the corresponding 4'-nitro derivatives (14, R^1 = OH, R^2 = NO_2). All four phenols were reported to exhibit strong intramolecular hydrogen bonding, but this was not reflected in their pK_a values (10.6-10.8).

Similar crown ethers with intraannular substituents (16) were obtained by reaction of 1,5-bis(2'-hydroxyphenoxy)-3-oxapentane with various 2-substituted 1,3-bis(bromomethyl)benzenes (28). The yields (see Table 5) varied from 3 to 36%; the latter, notably high, yield was reported for the 2-nitro-substituted crown ether.

Other crown ethers with 1,3-bis(methylene)aromatic groups are those derived from 1,8-substituted naphthalenes (17 and 18) (28, 29). The yield of 18 was 10%; that of 17 was not reported.

Compounds that resemble these crown ethers, with respect to both the structure of the 1,3-bis(methylene)aromatic subunit and their complexing

2 Synthesis and Properties

Table 5 Crown Ethers (16) Containing 1,3-Bis(methylene)benzene and Catechol Subunits (Ref. 28)

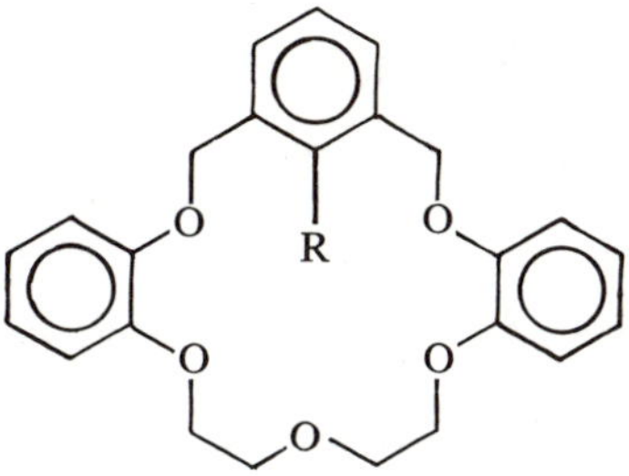

R	Yield (%)	M.p. (°C)
H	5	158-160
F	3	178-179
OCH_3	9	215-216
NO_2	36	179-180
$SOCH_3$	8	273-274

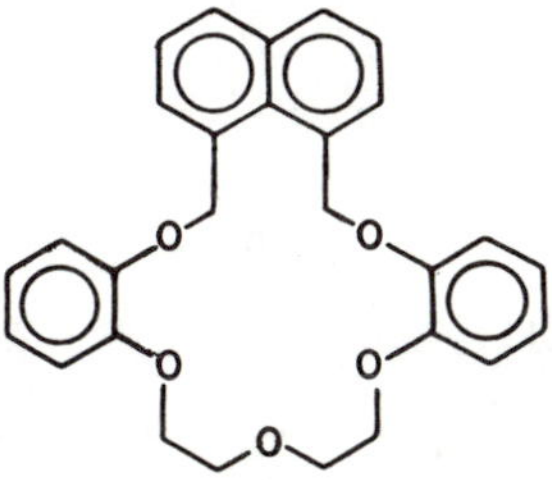

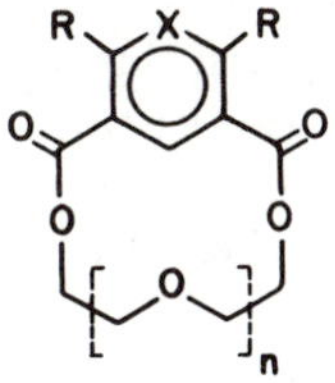

19, X = CH, N
R = H, CH_3

18

abilities, are the crown ether esters (19, X = CR^1, R = H) that were recently reported (37).

Incorporation of a 2,5-bis(methylene)furan subunit into macrocyclic polyethers has been accomplished in two different ways, namely, by reaction of 2,5-bis(chloromethyl)furan with polyethylene glycolates, and by reaction of 2,5-bis(hydroxymethyl)furan with polyethylene glycol ditosylates (31, 32, 38, 39). The second method gives the better yields of **20** (Table 6), probably because 2,5-

Table 6 Crown Ethers with 2,5-Bis(methylene)furan Subunits

Crown Ether	n	Yield (%)	M.p. (b.p./mm Hg) (°C)	Ref.
20	0	10	(150/0.01)	32
20	1	36 (27)	~0 (150/0.01)	39 (32)
20	3	15	(230/0.01)	32
21	1	11	109-111	38, 39
21	3	7	(250/0.01)	32
22	–	35	69-70	38, 39
23	–	40 (10)	124-126	39 (38)

bis(chloromethyl)furan eliminates hydrogen chloride under the prevailing basic conditions. By-products formed by dimerization of such an elimination product have been isolated (38, 39). Crown ethers with more than one 2,5-bis-(methylene)furan subunit have been obtained either by cyclization of two molecules of 2,5-bis(hydroxymethyl)furan and glycols to give **21** (31, 32) or by a stepwise synthesis using other furan derivatives (38, 39). Reaction of two molecules of 2-(2'-chloroethoxymethyl)-5-hydroxymethylfuran with sodium hydride in dimethylformamide gave 11% of **21** (n = 1). Crown ether **22** was obtained from diol **24** (X = OH) by reaction with diethylene glycol ditosylate, and crown ether **23** by reaction of the dichloride (**24**, X = Cl) with 2,5-bis(hydroxymethyl)-furan. Chemical transformation of these crown ethers has been reported: hydrogenation of the furan ring in **20** (n = 1) and also the Diels-Alder reaction of **20** with dimethyl acetylenedicarboxylate affords **25** (38, 39). The temperature-dependent PMR spectrum of **25** indicated that the ring inversion of this bicyclic system is slow on the PMR time scale at 25°C (E_a = 21 kcal mole^{-1}).

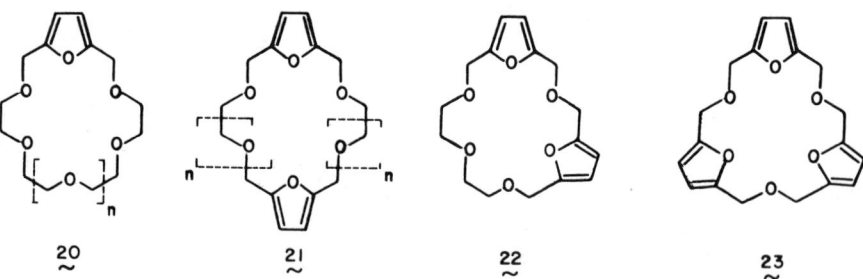

20   21   22   23

The incorporation of 1,4-bis(methylene)benzene subunits into crown ethers has also been reported (32). Compounds **26** were obtained by reaction of 1,4-bis(bromomethyl)benzene and disodium or dipotassium polyethylene glycolates.

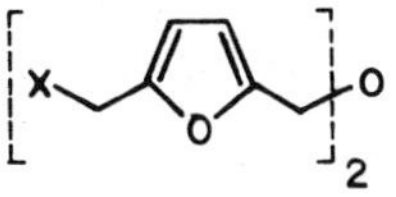

2 Synthesis and Properties

24 X = OH
~ X = CL

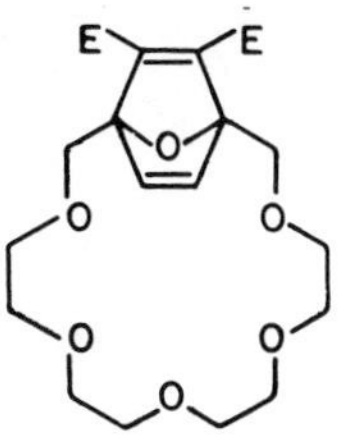

25
~
E = COO CH$_3$

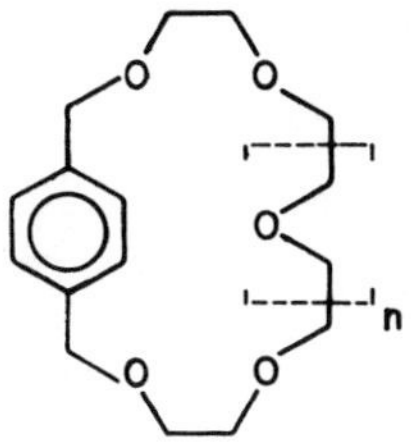

26
~

The yields varied from 4 to 35%, the maximum yield being that of the 19-membered macrocyclic polyether (Table 7). The same type of crown ether (**27**) was obtained from a reaction of 1,4-bis(bromomethyl)benzene with 1,5-bis(2′-hydroxyphenoxy)-3-oxapentane (**28**). Despite the fact that this reaction was carried out in high dilution, the yield was only 2%.

Table 7 Crown Ethers with 1,n-Bis(methylene)aromatic Subunits

Crown Ether	n	R	Yield (%)	M.p. (°C)	Ref.
26	0	—	4	oil	32
26	1	—	35	55-56	32
26	2	—	16	oil	32
26	3	—	6	oil	32
26	4	—	7	oil	32
26	5	—	13	oil	32
27	—	—	2	121-122	28
28	—	—	3	112-115	28
29	2	CH$_3$	40	glass	34
29	2	H	40	214-215	34
29	3	CH$_3$	40	glass	34
29	3	H	30	188-189	34
29	4	CH$_3$	20	glass	34
29	4	H	15	162-163	34

Reaction of the same diol with 2,2′-bis(bromomethyl)-1,1′-biphenyl afforded crown ether **28** in 3% yield (28). Cram and co-workers (7, 8), who prepared many crown ethers derived from 2,2′-dihydroxy-1,1′-binaphthyls, have used the

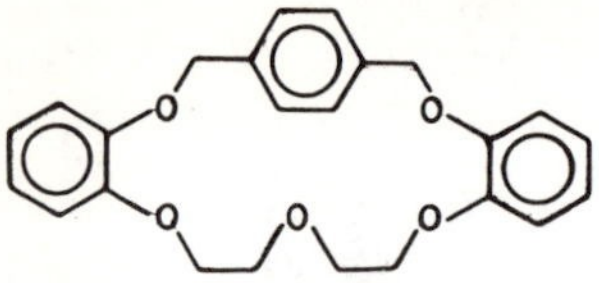

                                27 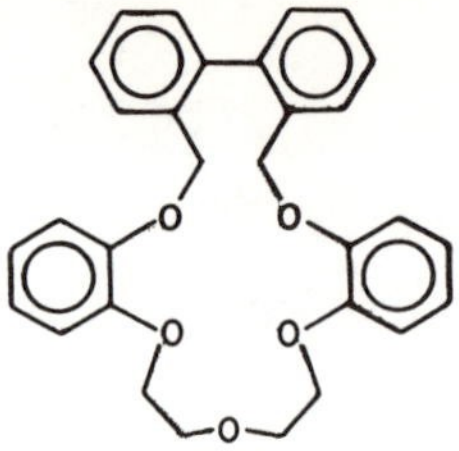   28 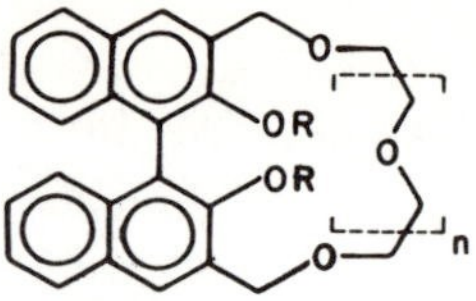   29

same type of building block for another class of crown ethers with bis-(methylene)aromatic subunits (34). They reacted disodium 3,3'-bis(hydroxymethyl)-2,2'-dimethoxy-1,1'-binaphthyl with polyethylene glycol ditosylates and obtained crown ethers **29** ($R = CH_3$) in good yields (Table 7). Similarly they prepared the corresponding dihydroxy crown ethers **29** (R = H) by reaction of 3,3'-bis(hydroxymethyl)-2,2'-bis(methoxymethoxy)-1,1'-binaphthyl with polyethylene glycol ditosylates in the presence of sodium hydride, followed by hydrolysis of the acetal groups with acid.

2.1.2 Macrocyclic Azapolyethers (O_xN_y)

The nitrogen atoms in the macroring of (aza)polyethers are either sp^3- or sp^2-hybridized, and the bis(methylene)aromatic subunit can be either a bis(methylene)benzene ring or a 2,6-bis(methylene)pyridine moiety.

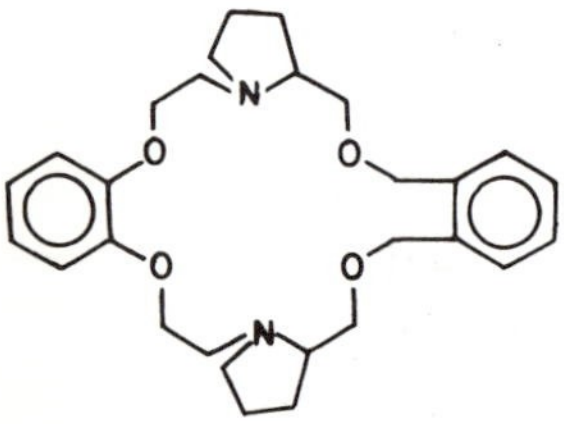

                                30 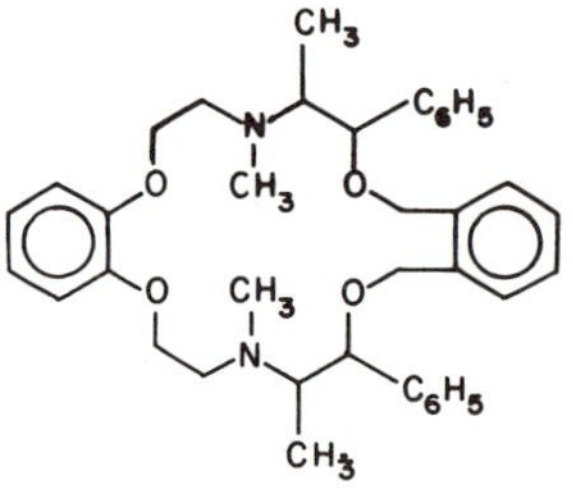   31 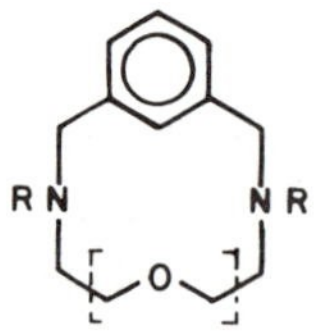   32

Wudl and Gaeta reported the synthesis of two chiral macrocyclic diazapolyethers (**30** and **31**), using L-proline and D-Ψ-ephedrine as the chiral starting materials (40). The overall yield of **31**, after three steps, was 40%. Sutherland's group (41-43) synthesized a number of crown ethers in which a 1,3-bis(methylene)benzene subunit is present and two of the oxygen atoms of the polyether ring are replaced by urethane or methylamino groups (**32** and **33**). The

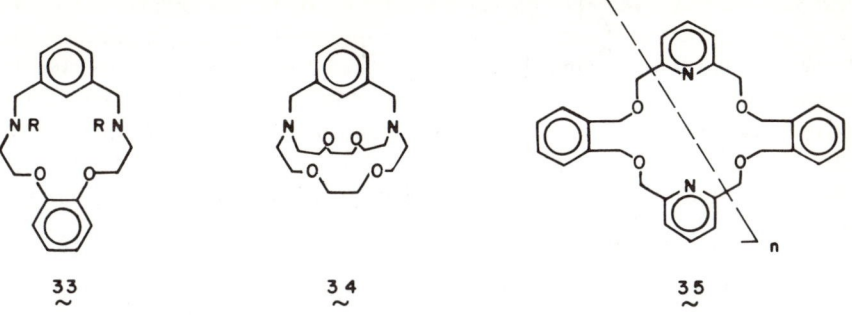

33

34

35

synthesis of **32** (R = COOC$_2$H$_5$) and **33** (R = COOC$_2$H$_5$) involves reaction of α,ω-dihalides or α,ω-ditosylates with dianions of α,ω-bis(urethanes). Although high-dilution conditions were not used, the yields were approximately 40%. The conversion of the urethane groups in **32** and **33** (R = COOCH$_3$) into N-methyl groups was achieved by reduction with lithium aluminium hydride. These macrocycles have been used in detailed PMR studies to investigate the mechanism of complex formation with ammonium salts (Section 4.5). A bicyclic azapolyether (**34**) has been obtained by reduction of the corresponding macrocyclic bisamide (**44**). The ring-closure step in the synthesis of this bisamide was effected in high dilution according to Lehn's procedure of the cryptate synthesis (15).

Incorporation of a 2,6-bis(methylene)pyridine subunit was first reported by Newkome and Robinson (45), who reacted disodium 2,6-bis(hydroxymethyl)-pyridine with 1,2-bis(bromomethyl)benzene to obtain mixtures of products (**35**, n = 1-4). The 22- and 33-membered rings (n = 1 or 2) were isolated in yields of 40 and 9%, respectively. They claimed to have isolated the two other ring compounds (n = 3 or 4), but no definite proof of their structure could be given. The 1:1 reaction product (n = 0) could not be obtained, even under high-dilution conditions. Other macrocycles with a 2,6-bis(methylene)pyridyl group have been prepared by reactions of 2,6-bis(halomethyl)- or 2,6-bis(hydroxymethyl)pyridine with polyethylene glycols or the corresponding ditosylates (46, 47), and with 1,5-bis(2'-hydroxyphenoxy)-3-oxapentane (28). The yields of the

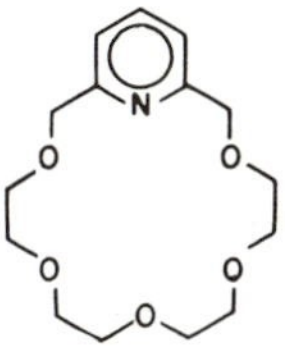

36

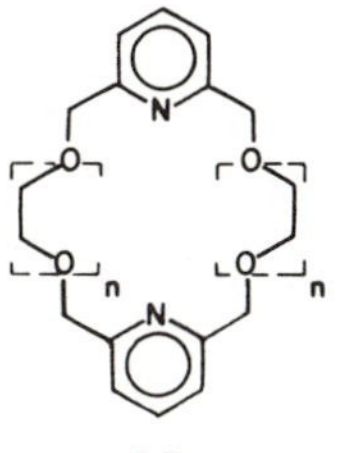

37

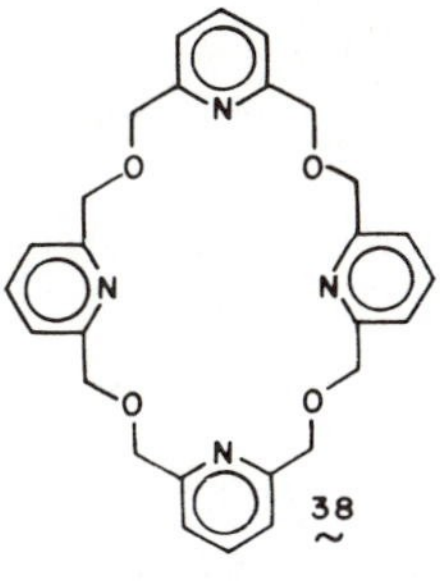

38

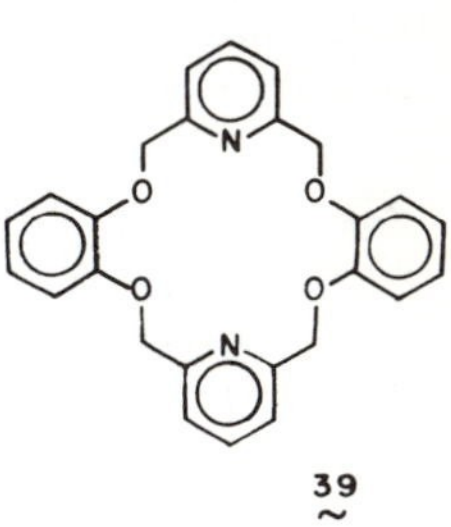

39

macrocyclic azapolyethers from these reactions (36-42) are summarized in Table 8. The pK_a values of some mono- and diprotonated macrocycles have been measured, and it was found that monoprotonated 37 ($n = 0$) is more stable than monoprotonated 36 or 38 by about 4 kcal mole^{-1}. The relatively high stability of monoprotonated 37 ($n = 0$) has been ascribed to the fact that the proton in this structure is linked to both nitrogen and oxygen atoms. This is in line with the observation that diprotonated 37 ($n = 0$) is less stable than diprotonated 38.

Table 8 Crown Ethers with 2,6-Bis(methylene)pyridine Subunits

Crown Ether	Yield (%)	M.p. (°C)	Ref.
36	29	40-41	46
37 ($n = 0$)	6	172-175	46
37 ($n = 1$)	18	147-148	46
38	20	173-176	46
39	9	184-186	46
40	32	125-128	46
41	—	—	46
42 (X = N)	30	132 (dec.)	28
42 (X = NO)	25	159 (dec.)	28
43 (SS)-	26	288-292[a]	46
44 (SS)-	43	—	48
45 (SS)-	29	—	48
46 -	15	224	49
47 (DD)-	7.5	147-149	50
48 ($n = 1$, R = Tos)	34[b]	175-177	28
48 ($n = 2$, R = Tos)	32[b]	184-185	28
48 ($n = 3$, R = Tos)	27[b]	163-165	28
49	63[c]	95-96	44

[a] Bis(tetrahydrofuran)clathrate.
[b] High-dilution conditions; Tos = p-toluenesulfonyl.
[c] Reduction of the corresponding bisamide.

Reaction of 2,6-bis(bromomethyl)pyridine or the corresponding N-oxide with 1,5-(2'-hydroxyphenoxy)-3-oxapentane in high dilution yielded macrocycles 42. The complexes of 42 with a number of salts are listed in Table 11 (28) (see Section 3.1).

The two types of crown ether having 2,6-bis(methylene)pyridine subunits and a chiral cavity have been prepared. In one type of compound (43-45) the chiral-

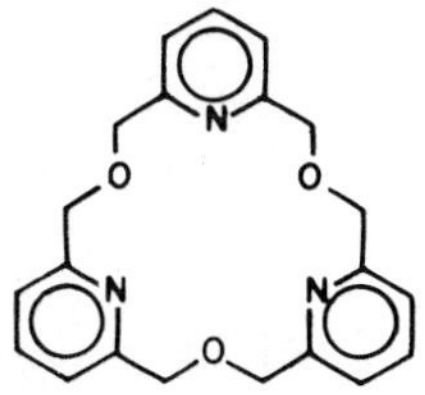

40

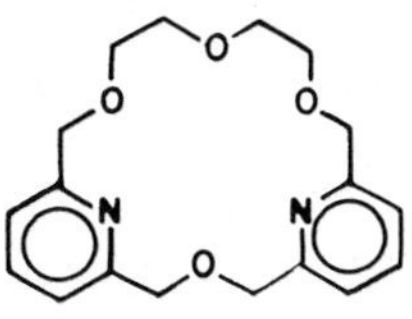

41

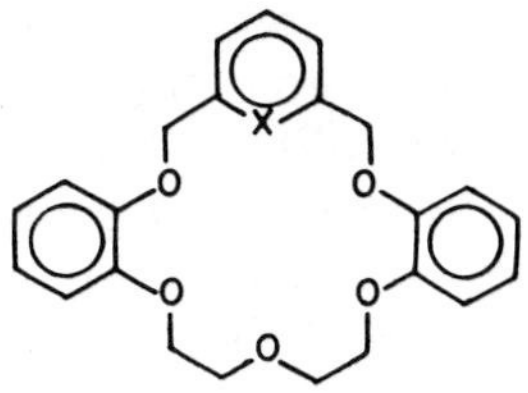

42 (X=N, NO)

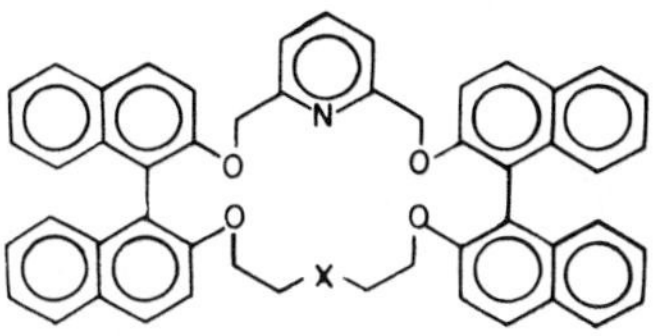

43; X = 2,6 - C$_5$H$_3$N
44; X = 2,6 - C$_6$H$_4$
45; X = CH$_2$

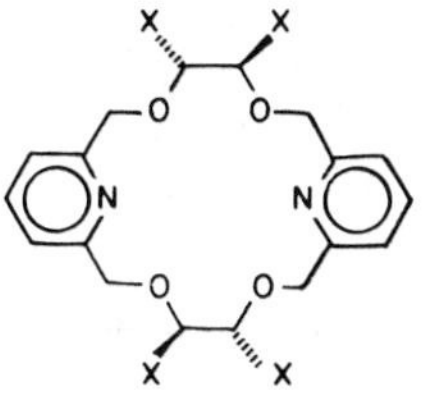

46; X = CO N(CH$_3$)$_2$

47; X =
   -O CH$_3$
     ><
   -O CH$_3$
   H

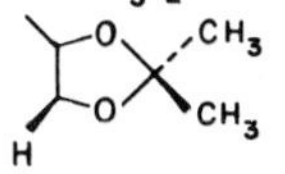

ity arises from the steric barrier in 2,2'-disubstituted-1,1'-binaphthyls (48). In the other type (46 and 47) the chirality results from subunits derived from an L(+)-tartaric acid bis(N,N-dimethylamide) moiety (46) or 1,2:5,6-di-O-isopropylidene-D-mannitol (47) (49, 50). The stereoselective complexation of enantiomeric protonated amines such as S-phenylglycine methyl ester and R-valine methyl ester hexafluorophosphate has been reported (48) (see Section 4.4).

Several azapolyethers have been synthesized by formation of a carbon-nitrogen bond in the cyclization step (28). Compounds 48 (R = p-toluenesulfonyl) were obtained by reaction of 2,6-bis(bromomethyl)pyridine with the disodium salts of 1,ω-bis(p-toluenesulphonamides) in high dilution (Table 8). A macrobicyclic polyether with a 2,6-bis(methylene)pyridine subunit (49) was prepared in the

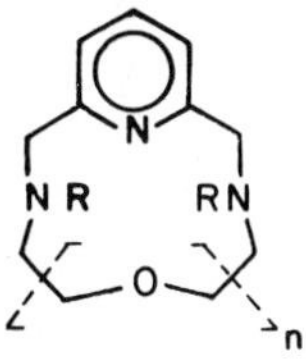

48

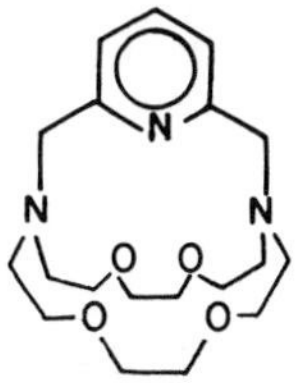

49

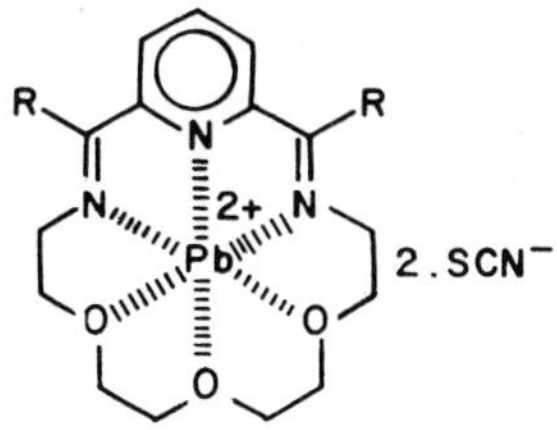

50 2 . SCN$^-$

same way as **34** (44). It reacts with sodium tetrafluoroborate to form a complex that is sufficiently stable to be detected in the mass spectrometer. Complexation of **49** with salts is further discussed in Section 3.1.

Other macrocycles with 2,6-pyridyl subunits that are related to the crown ethers described in this chapter are the macrocyclic ether esters (10, 37), the macrocyclic ether-amides (28), macrocyclic polyethers linked with 2,6-pyridyl groups (51, 52) via oxygen rather than carbon, and compounds **50**. The last-mentioned compounds were obtained (53) as the lead thiocyanate complexes on reaction of 1,11-diamino-3,6,9-trioxaundecane with 2,6-diacetylpyridine or pyridine-2,6-dicarbaldehyde in the presence of lead(II) cations. Structurally related noncyclic compounds with 2,6-bis-(methylene)pyridyl subunits that form complexes with salts have also been reported (54).

2.1.3 Macrocyclic Thiapolyethers ($O_x S_y$)

Macrocyclic polyethers in which one or more of the oxygen atoms are replaced by sulfur have been synthesized by Vögtle and co-workers (28, 55, 56) via reaction of dihalides or ditosylates with dithiols in high dilution. Pronounced template effects were not observed. The bis(methylene)aromatic subunit may be 2,5-bis(methylene)thiophene, as in **51**, or 1,3-bis(methylene)benzene, as in **52**.

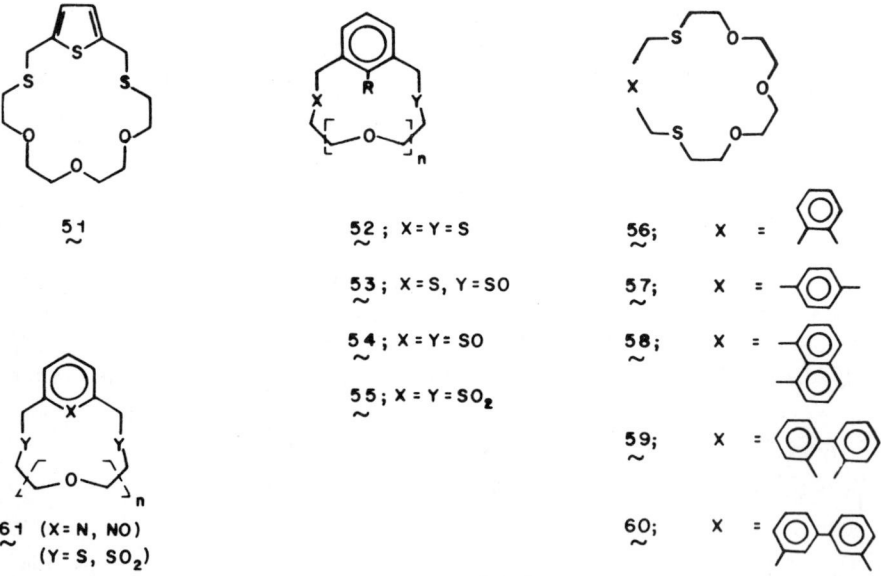

Examples of other subunits are 1,2- and 1,4-bis(methylene)benzene, 1,8-bis-(methylene)benzene, and 2,2'- and 3,3'-bis(methylene)-1,1'-biphenyl (**56-60**).

Compounds 52 in particular were prepared with a wide variety of intraannular substituents and ring sizes (Table 9). Oxidation of one or both of the sulfur atoms in compounds 52 (n = 1-3) yielded the corresponding mono- and disulfoxides (53 and 54) and the sulfones (55), respectively. The sulfoxide groups in these macrocycles constitute one or more centers of chirality, but attempts to resolve these crown ethers into their enantiomers have not been reported (28).

Table 9 Macrocyclic Thiapolyethers with 1,n-Bis(methylene)aromatic Subunits

Crown Ether	n	R	Yield (%)	M.p. (°C)	Ref.
51	–	–	–	--	16
52	1	SOCH$_3$	27	194-196	55
52	1	F	44	89-91	56
52	1	NO$_2$	47	126-127	56
52	1	NH$_2$	74^a	88-89	56
52	2	Cl	45	58-59	56
52	2	NO$_2$	31	82-83	56
52	2	NH$_2$	40^a	73-74	56
52	3	NO$_2$	–	–	56
52	3	NH$_2$	42^a	143-145	56
52	3	H	24	56-57	28
52	3	C$_6$H$_5$	27	110-111	28
52	3	CN	46	72-74	28
52	3	OCH$_3$	29	oil	28
52	3	SOCH$_3$	48	oil	28
52	3	COOCH$_3$	34	oil	28
53	3	H	35^b	103-105	28
54	3	H	29^b	148-151	28
55	3	H	46^b	251-253	28
55	3	OCH$_3$	38^b	218-220	28
55	3	SO$_2$CH$_3$	35^b	229-231	28
56	–	–	53	47-49	28
57	–	–	15	67-68	28
58	–	–	19	76-78	28
59	–	–	31	67-69	28
60	–	–	36	50-52	28

^{a}Reduction of the corresponding nitro compound (52, R = NO$_2$).
^{b}Oxidation of the corresponding polysulfide (52).

2.1.4 Macrocyclic Thiazapolyethers $(O_xN_yS_z)$

Of the macrocyclic compounds in which all three kinds of donor atoms (O, N, and S) are incorporated, only those with 2,6-bis(methylene)pyridine subunits have been reported. They were obtained by reaction of 2,6-bis(bromomethyl)-pyridine and α,ω-dithiols in high dilution, in one case followed by oxidation of the sulfide groups (16, 28) (see Table 10). A structurally related macrocyclic thioester ether that forms complexes with alkali metal salts has also been reported (37).

Table 10 Macrocyclic Thiazapolyethers with 1,n-Bis(methylene)aromatic Subunits

61

X	Y	n	Yield (%)	M.p. (°C)	Ref.
N	S	1	46	133-135	28
N	S	2	36	90-91	28
N	S	3	49	58-59	28
NO	S	3	47	oil	28
NO	SO$_2$	3	61[a]	198-201	28

[a] Oxidation of the corresponding macrocyclic polysulfide.

2.2 Templated Synthesis

The macrocyclic compounds described in the preceding section have been obtained in yields ranging from less than 1% to nearly 70%, depending on ring size, type and number of heteroatoms, and reaction conditions. Two types of reaction procedure have been used. In the first type the cyclization is effected with the linear reactant(s) in concentrations of 0.1-1.0 M, whereas in the second type they are reacted under carefully controlled conditions of high dilution during the entire reaction. The latter type of reaction has been employed mainly by Vögtle and co-workers for the synthesis of macrocycles in which both "soft"

2 Synthesis and Properties

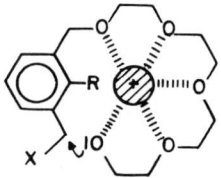

Figure 2 Representation of the cyclization of a linear bifunctional precursor using a metal alkoxide.

heteroatoms (S, N) and "hard" oxygen atoms are incorporated, and also for the synthesis of the crown ether-type compounds ("all oxygen") (27, 28). However, most of the crown ethers containing oxygen as the only heteroatom, as well as crown ethers in which one or more of the oxygen atoms are replaced by a (pyridine) nitrogen atom, were prepared by reactions at much higher concentrations (19, 31-35, 38-43, 45-49). The fact that these compounds were formed in remarkably high yields has been attributed by the various workers to the so-called template effect of the base cation that is present during the cyclization step, which comprises a nucleophilic displacement of halide or tosylate by alkoxide groups (22, 23, 45). High yields were reported not only for crown ethers derived from bis(methylene)aromatics but also for other crown ethers (22, 23). The explanation given (57, 58) is that the cyclization of the linear bifunctional precursor is made possible by a cyclic conformation in which the alkoxide cation is coordinated to the oxygen atom of the precursor, thus bringing the two ends of the molecule in close proximity (see Figure 2). This means that generally the effect has been related to a favorable entropy contribution.

Long before crown ethers were known, it had been found that cations could influence the yield of macrocycles such as porphyrins and corrins (59, 60), but these cations were of the "soft" type, like nickel, cobalt, and copper ions. In some other cases high yields of macrocyclic compounds were obtained under nonhigh-dilution conditions in the absence of cations that might act as a template ion (61, 62). The high yields in such reactions have been ascribed to the low internal entropy of the precursors (63). Only a few detailed studies have been made of the nature of the template effect in crown ether synthesis. Only indirect evidence of the occurrence of such an effect has been obtained, for example, in the form of the observed relationship between yields and size of base cations employed in the synthesis of 18-crown-6 (K^+), 15-crown-5 (Na^+), and 12-crown-4 (Li^+) (57, 58, 64). Similar salt effects have been noted in the cyclocotetramerization of furan and acetone, in which lithium salts in particular proved to have a beneficial effect on the yield of macrocyclic polyethers (65-67).

As part of their work on the synthesis of crown ethers with a 1,2-bis-(methylene)benzene subunit, Reinhoudt and co-workers (19) (see Section 2.1)

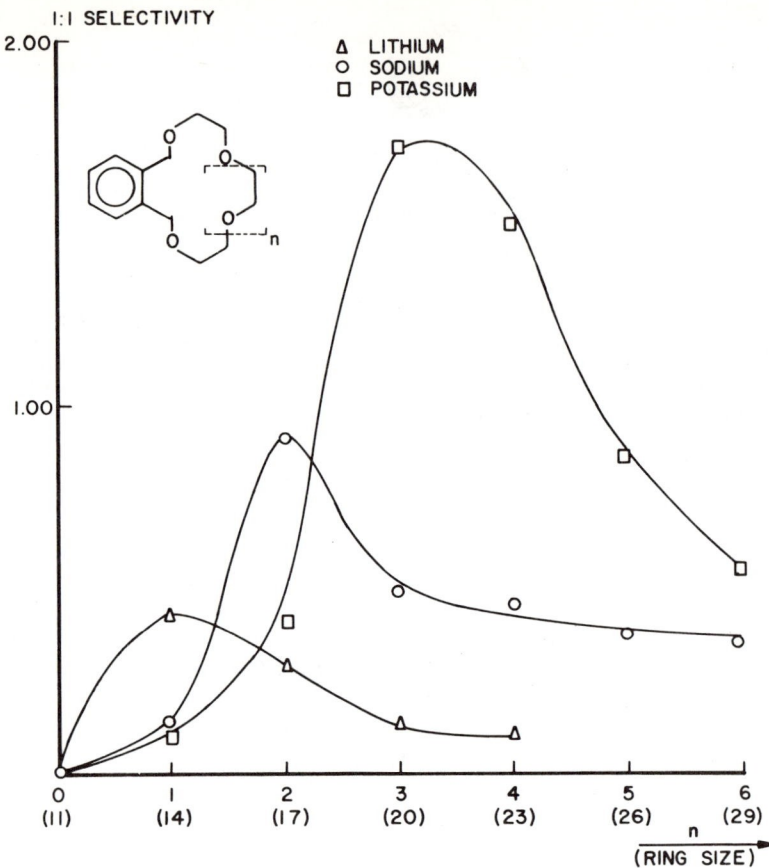

Figure 3 1:1 selectivity of the reactions of 1,2-bis(bromomethyl)benzene and polyethylene glycolates (Li$^+$, Na$^+$, K$^+$).

made a special study of the effect of various base cations on the yield of cyclic products. Their results (see Figure 3) indicate that the selectivity to ring formation is highest when the crown ether can just barely accommodate the base cation. The optimal fit was determined independently via complexation studies of the crown ethers with MPtCl$_3 \cdot$C$_2$H$_4 \cdot$H$_2$O (M = Na$^+$, K$^+$, Rb$^+$, or Cs$^+$) in apolar solvents (68) (see Section 3.2). The results clearly showed that macrocyclic polyethers are obtained in low yields when the crown ether formed is too small to accommodate the base cation and that the yields are surprisingly high when the ring cavity is much larger than the cation. These high yields have been related to a "twisted" conformation of the polyether chain of the complexed precursor. Complexation studies indeed revealed this type of conformation in complexes with cations where the ratio between the radii of the crown ether cavity and the cation is considerably larger than 1 (68). In their interpretation of the nature of the template effect, Reinhoudt's group considered the differences in both entropy and enthalpy between cyclization of the linear precursor in which the alkoxide forms a contact ion pair and cyclization of a precursor that has a cyclic conformation with the alkoxide as a partly charge-separated ion pair. It has been well established that the nucleophilicity of anions, in this case an alkoxide anion, is enhanced on complexation of the positive counterion (4, 5, 69).

2.3 NMR Spectra

NMR spectroscopy is the most important technique by far for the elucidation of structures, as well as for studies of conformation in solution, qualitative complexation, and the mechanism of exchange processes of crown ethers. It is particularly suitable for studying the crown ethers reviewed in this chapter because of the "benzylic" methylene protons in the bis(methylene)aromatic and -heteroaromatic subunits. The reason is twofold. First, in contrast to the absorptions of the OCH_2CH_2O groups in most other crown ethers—which consist of multiplets or broad singlets, and are rather insensitive to variations in structure, as a result of which appear in a relatively narrow "band" of the PMR spectrum—the absorptions of the benzylic CH_2O protons are shifted by 0.6-1.2 ppm to lower field. Second, these protons lie in close proximity to the aromatic groups, with the magnetic anisotropy arising from the ring current. Therefore on complexation and in rigid conformations the absorptions of the "benzylic" protons are the most sensitive to small steric differences (70).

^{13}C NMR data on these crown ethers have only incidentally been reported. Spectral data on crown ethers **5-10, 14, 15** ($R^1 = R^2 = H$), **20**, and **26** revealed that the differences between the absorptions of "benzylic" and polyether carbon atoms are small but consistent (19, 32). The values are only slightly dependent on ring size: **5, 6** (71.5 ± 0.5 ppm); **8, 9** (63.7 ± 0.3 ppm); **10** (65.0 ± 0.2 ppm); **14, 15** (72.6 ± 0.6 ppm); **20** (64.9 ± 0.3 ppm); and **26** (72.8 ± 0.7 ppm).

PMR spectroscopy has been used for various purposes such as the analysis of crude mixtures of crown ethers (19). Thus in the reaction mixture of 1,2-bis-(bromomethyl)benzene and triethylene glycol, the 14-membered ring **5** ($n = 1$), the 28-membered ring **6** ($n = 1$), and higher-membered cyclic polyethers could be identified directly.

The absorptions of benzylic protons were also used to study the conformation of azapolyethers (41). The temperature-dependent PMR spectra of **32** ($R = CH_3$, $n = 1$) revealed that at $-40°C$ the conformational inversion of the macroring is slow on the PMR time scale ($\Delta G^{\neq} = 11.0$ kcal mole^{-1}). As mentioned in Section 2.1, PMR spectroscopy indicated that the ring-inversion process in the bicyclic crown ether (**25**) has an activation energy of more than 21 kcal mole^{-1} (38, 39). The effect of intraannular substituents in **52** on ring inversion was also studied by dynamic PMR spectroscopy, and it was found that the nitro substituent in particular decreased the rate of ring inversion (56).

In addition to the study of free macrocycles PMR spectroscopy has been used for studying the complexes that they form with salts. Both qualitative studies and the determination of complexation constants (39, 47, 68, 71, 72) have been reported. PMR spectroscopy also revealed the mechanism of complexation and decomplexation in apolar solvents (41-43, 50, 73, 74). For these studies mainly the signals of the benzylic protons were used, in addition to those of the intra-annular aryl protons, for example, those in **14** and the furan protons in **8**. This work is reviewed in detail in the sections on complexation with alkali and alkaline earth metal salts (Section 3.2) and with ammonium salts (Section 4.5).

^{13}C NMR spectroscopy is less suitable for studying complexation, as the incidentally reported results point to only small differences in absorption between the benzylic carbon atoms in the complex and in the free ligand (68). This observation seems to hold for most complexes of crown ethers (75).

2.4 Mass Spectra

Mass spectrometry has been widely used for the elucidation of structures of crown ethers with one or more bis(methylene)aromatic or -heteroaromatic subunits. However, in most cases only the molecular compositions are mentioned, and detailed fragmentation patterns have rarely been reported, if at all. This is a general feature in crown ether chemistry, and only a single publication deals specifically with the electron-impact-induced fragmentation of a number of benzo-3n-crown-n compounds (76).

Gray and co-workers (77) have reported details of the mass spectra of crown ethers **5-10**, **14**, **15**, **20**, and **26**. They observed that these compounds exhibit unusually intense molecular ions, which allows identification of crown ethers with a molecular weight of up to about 1000 (**62**, n = 3) and 56 ring atoms. The fragmentation patterns proved to be relatively independent of the type of (hetero)aromatic nucleus, and a characteristic series of polyether fragments was observed.

62 (n = 2, 3)

It has been reported that complexes of some crown ethers, for example, **49**, with salts exhibit fragment ions that contain the cation (Na^+, K^+) (44).

3 COMPLEXATION WITH SALTS OF ALKALI AND ALKALINE EARTH METALS

The great interest in the chemistry of macrocyclic polyethers and their aza and thia analogues is largely a result of their ability to form complexes with salts both in polar and in apolar solvents. Several reviews have been devoted to this aspect, but most of them are restricted to complexation in polar solvents (3, 25, 26).

The subject of this section is the complexation of macrocycles containing bis(methylene)aromatic and -heteroaromatic subunits with salts of alkali and alkaline earth metals in polar and in apolar solvents. In the next section (4) complexation with ammonium salts is discussed.

3.1 Complexation in Polar Solvents

Weber and Vögtle (28) reported on the complexation of a number of crown ethers having 1,3-bis(methylene)aromatic and -heteroaromatic subunits with several salts in methanol. As a qualitative test they used the dissolution of various thiocyanate salts in methanol (28). For instance, the addition of potassium thiocyanate to a suspension of the crown ethers in methanol resulted in clear solutions. A number of such complexes were obtained as crystalline solids (see Table 11). Similarly several macrocycles were reported to form com-

Table 11 Complexes (1:1) Isolated from Methanol-Ethyl Acetate (Ref. 28)

Crown ether	Salt
16	NaSCN
28	NaSCN
42 (X = N)	NaSCN, KSCN, NH_4SCN, $Hg(SCN)_2$, $Co(SCN)_2$, RbI, BaI_2, $Pr(NO_3)_3 \cdot 6H_2O$, $AgNO_3$
42 (X = NO)	KSCN
61 (X = N, Y = S, n = 3)	NaSCN, KSCN, $Ba(SCN)_2$, $Co(SCN)_2$, $CuCl_2 \cdot 2H_2O$[a]
61 (X = NO, Y = S, n = 3)	NaSCN,[b] KSCN, NH_4SCN, $Ba(SCN)_2$

[a] Ligand/salt ratio: 4:5.
[b] Ligand/salt ratio: 2:3.

plexes with Co^{2+}, Ag^+, Hg^{2+}, and Cu^{2+} salts. The selectivity of complex formation was related to the "hardness" and "softness" of the donor atoms, the steric influence of intraannular groups in these macrocycles, and the effect of additional donor abilities of such groups. Replacement of, for instance, the 2,6-pyridyl group in **61** (X = N, Y = S) by a 1,3-phenylyl group resulted in complete suppression of complex formation with alkali metal salts. Oxidation of the nitrogen atom, on the other hand, increased the complex stability of complexes with these salts. Another effect observed on oxidation was the increased selectivity toward sodium as compared to potassium, as a result of the smaller crown ether cavity.

The bicyclic compound (**49**) showed a pronounced cation selectivity in complexation with thiocyanate salts in water (44). The stability constants (Table 12) clearly indicate that **49** has a strong preference for the monovalent sodium and the divalent calcium, strontium, and barium cations.

Table 12 Stability Constants of Complexes of **49** with $M(SCN)_n$ in Water (Ref. 44)

$M(SCN)_n$	Log K
Li	3.28
Na	5.28
K	3.44
Rb	2.60
Cs	<2.00
Mg	<2.00
Ca	7.82
Sr	8.60
Ba	7.90

3.2 Complexation in Apolar Solvents

In several publications the dissolution of potassium or sodium permanganate in chloroform or benzene by a macrocyclic polyether is mentioned as an indication of complex formation (16, 28, 37). One of the first more or less quantitative methods that was developed for measuring the complexing abilities is the so-called *picrate extraction* method. Originally developed by Pedersen (78) and Frensdorff (79), this method utilizes the equilibration of an aqueous picrate salt solution with a crown ether dissolved in a water-immiscible organic solvent. The distribution of the picrate at equilibrium, determined by UV spectrometry, is taken as a measure of the complexing ability of the crown ether. One of the drawbacks of this method is that the distribution depends not only on the degree of complexation but also on the solubilities and the partition coefficients of the various components over the two phases. On the other hand the method is simple and often sufficiently accurate to provide some indication of the relative complexing abilities of a series of crown ethers having small variations in structure.

Cram and co-workers (34) used this method to study the complexation of crown ethers in which one or two methoxy or phenoxy groups are present as intraannular substituents (**14** and **15**, R^1 = OCH_3, R^2 = CH_3; **29**, R = H or CH_3) with alkali metal salts and with ammonium ions (Table 13). One of their conclusions is

Table 13 Association Constants (K_a) of Crown Ether Picrate Salt Complexes ($CHCl_3$, 25°C) (Ref. 34)

Crown Ether	n	R,R^1	R^2	\multicolumn{5}{c}{Log K_a of Picrate Complexes}				
				Li^+	Na^+	K^+	Rb^+	Cs^+
14	2	OCH_3	CH_3	3.58	4.15	4.32	–	3.70
14	3	OCH_3	CH_3	4.00	4.67	6.26	–	5.06
14	4	OCH_3	CH_3	3.46	4.66	5.26	–	5.82
14	3	H	H	3.08	3.38	4.90	–	4.58
14	3	$COOCH_3$	H	5.96	5.95	6.09	–	5.68
15	0	OCH_3	CH_3	2.93	3.30	3.59	3.23	3.28
15	1	OCH_3	CH_3	2.75	3.70	3.94	3.63	3.78
29	2	CH_3	–	4.21	6.09	6.57	–	5.75
29	3	CH_3	–	4.27	4.63	5.98	5.99	5.98
29	4	CH_3	–	4.11	4.52	5.36	5.29	5.56
29	2	H	–	3.34	3.90	4.61	–	4.40
29	3	H	–	3.85	4.19	5.37	5.35	5.30
29	4	H	–	3.69	4.49	5.02	4.94	5.35

that methoxy groups are better additional donor groups than hydroxy groups. This has been attributed to the strong intramolecular hydrogen bonding of the hydroxyl proton to the crown ether oxygen atoms. These hydrogen bonds have to be broken before complexation with the cation can take place in an optimal manner. Another publication (33) deals with the complexing properties of crown ethers bearing intraannular carboxyl groups (**14**, R^1 = COOH, R^2 = H).

The amount of salt transferred, at equilibrium, from an aqueous solution of the Li^+, Na^+, K^+, or Ca^{2+} salts of the crown ether carboxylate to a dichloromethane phase was measured (Table 14). Each of the individual cations requires a specific ring size for maximum lipophilization: Li^+, an 18-membered ring; Na^+, a 21-membered ring; K^+, a 30-membered ring, and Ca^{2+}, an 18-membered ring.

Table 14 Lipophilization of Cations by Crown Ethers (14, R^1 = COOH, R^2 = H) (Ref. 33)

Crown Ether 14, R^1 = COOH, R^2 = H	Extracted Salt (%) in CH_2Cl_2			
	Li^+	Na^+	K^+	Ca^{2+}
$n = 2$	1.4	1.5	1.4	1.1
$n = 3$	7.2	7.9	6.7	4.8
$n = 4$	6.1	8.7	6.8	1.8
$n = 7$	3.4	5.2	8.0	2.9
(2,6-dimethoxybenzoic acid)	3.3	3.2	2.8	3.4

Another method for the study of complexation of alkali metal salts by crown ethers in apolar solvents was reported by Reinhoudt's group (68, 80). This method is more direct than the one discussed above, as it requires no partition over two immiscible liquid phases. Crown ethers in chloroform, or other apolar solvents, were equilibrated with solid alkali metal trichloroethyleneplatinum(II) salts ($MPtCl_3 \cdot C_2H_4 \cdot nH_2O$), of which the potassium salt is the well-known Zeise's salt. Equilibrium was reached virtually instantaneously. The degree of complexation, that is, the ratio between complexed and free crown ether, was determined by PMR spectroscopy. The ratio between the intensities of ethylene (salt) protons and crown ether protons gave directly the value of $K_a \cdot$ [salt]. Furthermore the assumption was made that the concentration of free salt in solution equals the solubility of the salt in the absence of a crown ether as long as an excess of solid salt ($MPtCl_3 \cdot C_2H_4 \cdot nH_2O$) is present. These solubilities were determined independently by atomic absorption spectrometry ($\sim 5 \times 10^{-6}$ M, 27°C in chloroform). Combining these values with those obtained for the ratios between free and complexed crown ether gave the values of the association constants (K_a). Such constants have been determined for a variety of crown ethers and cations (Table 15). The accuracy of this method depends on the accuracy of the determination of the ratio of free to complexed crown ether, which ratio should be within the 0.05-20 range.

From PMR studies of solutions of these complexes it was concluded that the ligand exchange is fast on the PMR time scale at temperatures higher than

Table 15 Association Constants (K_a) of Crown Ether-MPtCl$_3 \cdot$C$_2$H$_4 \cdot n$H$_2$O Complexes (0.1 M, 27°C, in CDCl$_3$) (Ref. 68)

Crown Ether	Log K		
	K^+	Rb^+	Cs^+
5 (n = 1)	4.97	3.84	<3.30
5 (n = 2)	5.85	5.30	4.89
5 (n = 3)	>6.70	5.88	6.18
8 (n = 1)	5.59	3.74	3.65
8 (n = 2)	>6.70	5.67	3.52
8 (n = 3)	>6.70	>6.40	6.41
10 (n = 1)	6.00	4.04	4.23
10 (n = 2)	>6.70	5.36	5.63
10 (n = 3)	>6.70	6.11	5.98
14 (R^1 = R^2 = H, n = 2)	4.28	3.84	<3.30
14 (R^1 = R^2 = H, n = 3)	6.23	5.32	5.38
14 (R^1 = R^2 = H, n = 4)	>6.70	5.34	5.48

−40°C. A second conclusion of this work was that only one type of complex with a 1:1 stoichiometry is formed. These studies also provided valuable information about the conformation of these complexes in solution. For complexes of **5**, **8**, **10**, and **14** the following three types of conformation could be distinguished:

1. A conformation in which the polyether ring and the aromatic ring are almost perpendicular to each other, observed when the crown ether cavity is small compared with the cation.
2. A conformation of nearly flat complexes, observed when the crown ether can just accommodate the cation.
3. A conformation with a twisted polyether ring, observed when the crown ether cavity is *substantially* larger than the cation in the complex. Analysis of the PMR spectra of such complexes of crown ethers **5** revealed that this twisting (see Figure 4) occurs above a certain value of the ratio of crown ether cavity size to cation size. This was concluded from the nonequivalence of the polyether ring protons in the PMR spectra (see Figure 5). The twisted conformation has been related to the relatively high yields of "large" crown ethers (e.g., **5**, $n \geqslant 4$), as discussed in Section 2.3. A similar twisted conformation of

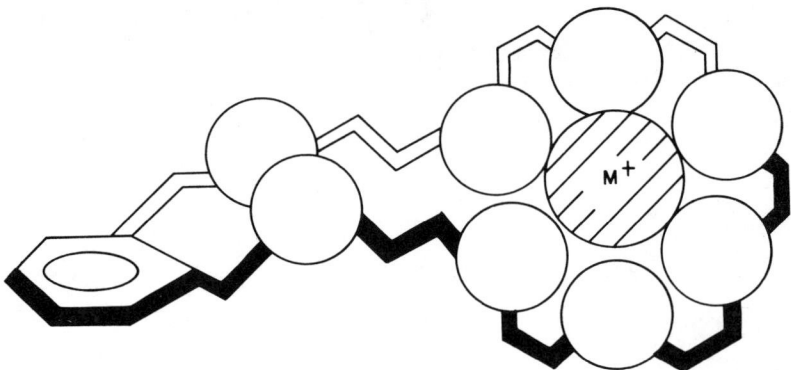

Figure 4 Twisted conformation of the polyether chain in complexes of crown ether 5 ($n = 5$) and $MPtCl_3 \cdot C_2H_4$ in solution.

the polyether chain in the solid potassium thiocyanate complex of dibenzo-30-crown-10 has been demonstrated by X-ray analysis (81). A recent PMR study of dibenzo-30-crown-10 has pointed to the same type of conformation in solution (70).

4. COMPLEXATION WITH AMMONIUM AND ALKYLAMMONIUM SALTS

The formation of complexes between crown ethers and ammonium, alkylammonium, and guanidinium salts in methanol has already been reported by Pedersen (2). From a symmetry point of view the tetrahedral NH_4^+ cation (Section 4.1) bears a closer resemblance to the spherical metal cations (Section 3) than the alkylammonium salts. The complexation of alkylammonium salts offers the possibility of bringing two organic molecules together in a complex of predictable structure. This possibility has been exploited by Cram and co-workers, and has resulted in a vast body of host-guest chemistry (8).

The presence of the alkyl group in the cation allows the use of ^1H and ^{13}C NMR to study complexation and determine the association constants (both relative and absolute) for crown ether complexes of *tert*-butylammonium salts (Sections 4.3-4.4). The study of fast chemical exchange reactions by ^1H NMR spectroscopy has a long and well-documented history. Applied to alkylammonium complexes of crown ethers, this technique has yielded valuable information concerning the kinetics of complexation (Section 4.5).

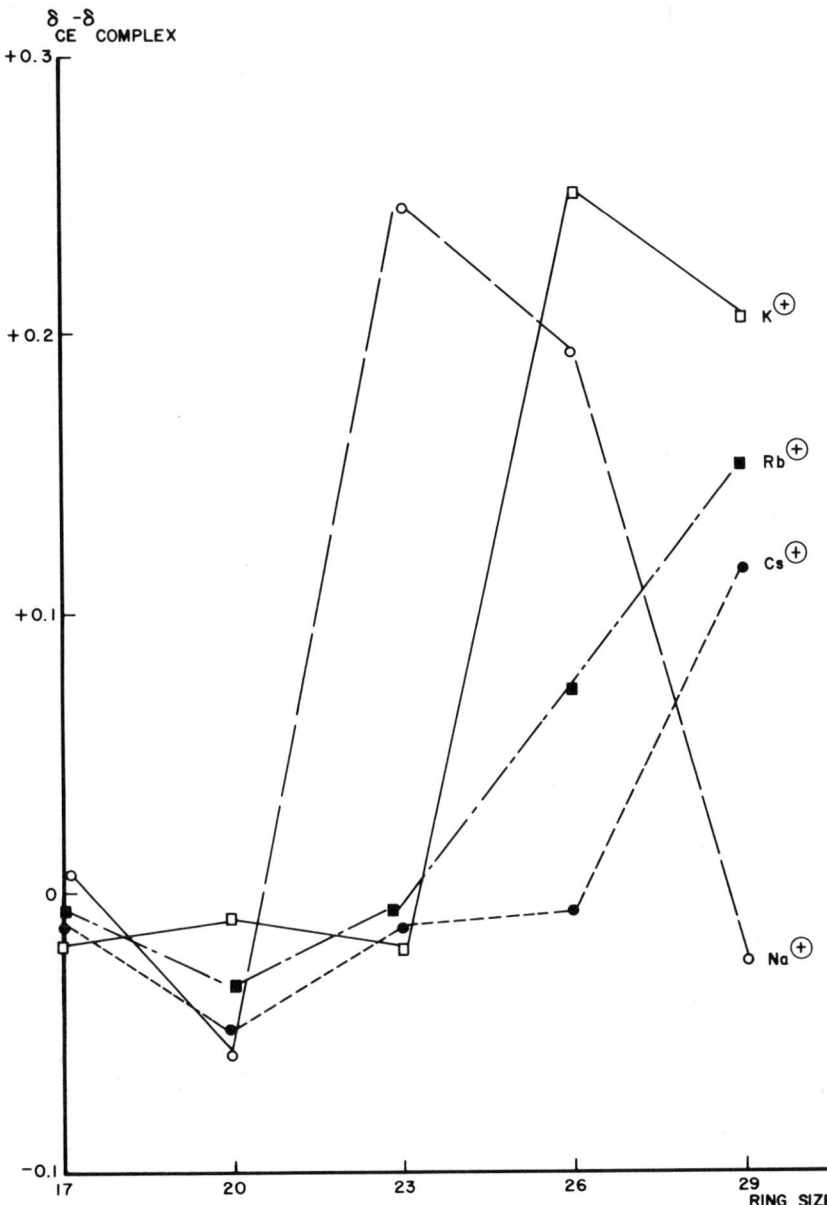

Figure 5 Chemical shift differences of the polyether protons in crown ethers (5)/MPtCl$_3$·-C$_2$H$_4$.

4.1 Complexation with Ammonium Salts

Since the ionic diameters of NH_4^+ (2.86 Å), Rb^+ (2.94 Å), and Tl^+ (2.80 Å) are quite similar (3), their binding constants with crown ethers might be expected to lie within a narrow range from a purely electrostatic point of view. However, in the case of Rb^+ (noble gas configuration) and Tl^+ (noble gas + ns^2-configuration) this expectation proved unjustified: complexes between crown ethers and Tl^+ in water were found to be much more stable (i.e., by about 1 kcal mole^{-1}) than expected on the basis of the ionic radius of Tl^+ (3, 82), and this discrepancy has been attributed to the high polarizability of Tl^+ compared to that of Rb^+ (82).

With NH_4^+ complexes of crown ethers in water the situation is less straightforward. Whereas NH_4^+ complexes of 18-crown-6 (**63**, $n = 3$) and dicyclohexano-18-crown-6 (**65**) are less stable (by about 0.3 kcal mole^{-1}) than the corresponding Rb^+ complexes, the NH_4^+ complex of 15-crown-5 (**63**, $n = 2$). is much more stable (by 1.4 kcal mole^{-1}) than its Rb^+ counterpart (82, 83). The extra stabilization of the NH_4^+ complex was shown to be due to the entropy and not to the enthalpy of binding.

63 (n= 1, 2, 3)   **64** (n= 1, 2, 3)

In apolar solution the binding constants of NH_4^+ and crown ethers **14**, **15** and **29** have been determined by partitioning experiments using ammonium picrate (34, 84). Comparison of the results, presented in Table 16, with the data in Table 13 reveals that the NH_4^+ complex of 1,3-xylyl-18-crown-5 (**14**, $n = 3$, $R^1 =$ H) is about 0.3 kcal mole^{-1} more stable than the Rb^+ complex. The order is reversed for **15** and **29**: here the Rb^+ complexes are 0.3 kcal mole^{-1} more stable. The exceptional behavior of 1,3-xylyl-18-crown-5 compounds (**14**) was attributed to the $N^+ \cdots \pi$-aryl interaction being more favorable than the $M^+ \cdots \pi$-aryl interaction (84).

The beneficial effect of a three-dimensional arrangement of binding sites is evident from the results obtained with compounds **14** and **29**. The intraannular H-bonding acceptor sites (OCH_3, CO_2CH_3) lie above the plane of the crown ether ring and stabilize the NH_4^+ complex by an additional 1 kcal mole^{-1}. Phenolic intraannular substituents decrease the stability of the complex because of intramolecular hydrogen bonding to transannularly located oxygen atoms of the ring (34, 85). These hydrogen bonds must be broken to make room for the

4 Complexation with Ammonium and Alkylammonium Salts

Table 16 Association Constants (log K_a) for Complexes with Ammonium Picrate in $CDCl_3$ at 24°C

Crown Ether	n	R^1	R^2	Log K_a	Ref.
14	2	OCH_3	CH_3	3.63	34
	3	OCH_3	CH_3	5.56	34
	4	OCH_3	CH_3	5.13	34
	3	CO_2CH_3	H	5.70	34
	3	H	H	5.06, 4.87	34, 84
	3	H	Br	4.50	84
	3	H	$C(CH_3)_3$	5.08	84
	3	H	OCH_3	5.09	84
	3	H	CN	4.03	84
15	0	OCH_3	CH_3	3.28	34
	1	OCH_3	CH_3	3.93	34
29	2	CH_3	—	5.90	34
	3	CH_3	—	5.68	34
	4	CH_3	—	5.06	34
	2	H	—	3.77	34
	3	H	—	4.85	34
	4	H	—	4.71	34

cation. The breaking of the hydrogen bond with simultaneous formation of the cation was observed for complexes of **14**, R = OH, with NH_3 (85). Compound **14**, n = 3, R = OH, was completely converted into the ammonium phenoxide, whereas only low conversions were observed for crown ethers such as **14** (n = 2, R = OH), whose cavity is too small to accomodate the NH_4^+ cation.

The distinctive binding features of NH_4^+ compared to those of metal cations have also been noted in the preparation of solid crown ether complexes (86). A theoretical basis has been provided by ab initio calculations. First, it was shown that the contribution of hydrogen bonding to the overall binding enthalpy was as high as 70% (39). Second, a tetrahedral arrangement of hydrogen bonding acceptor sites was found to be the most stable (87). These facts imply that strong complexing agents for ammonium salts must have strong acceptor sites properly arranged for hydrogen bonding.

4.2 Complexation with Water

It is well known that water and dialkyl ethers form hydrogen-bonded complexes in apolar solutions (88). The enthalpy of complex formation between water and dioxane in carbon tetrachloride is about 3.5 kcal mole^{-1} (89). Therefore crown ethers are expected to form stable complexes with water as well. With a few exceptions this property of crown ethers has so far received little attention. In the synthesis of 12-crown-4 (**63**, $n = 1$) and 15-crown-5 (**63**, $n = 2$) the crude compounds were isolated as hydrates (64). NMR evidence has been presented for hydrogen bonding between water and benzo-18-crown-6 (**64**, $n = 2$) in aqueous solution (70). Open-chain counterparts of crown ethers sometimes form hydrates (78), and the hydrogen bonding acceptor properties of polyoxyethylene ethers have been used to solubilize water in apolar solutions (90).

The complexation of water by 1,3-xylyl crown ethers (**14**, $R^1 = R^2 = H, n = 2\text{-}7$) has been studied using the following method (71): C_0 molar solutions of the crown ether (CE) in CDCl$_3$ were equilibrated with water, and the water/crown ether ratio (R_w) in the organic phase was determined by NMR spectroscopy. If it is assumed that only 1:1 complexes are formed, the association constant K_w can be obtained from the following equation:

$$R_w C_0 = S_w + C_0 [K_w S_w/(1 + K_w S_w)] \qquad (1)$$

in which S_w stands for the solubility of water in neat CDCl$_3$. Plots of $R_w C_0$ against C_0 gave linear relationships for all crown ethers, and S_w and K_w were obtained from the slope.

The results, presented in Table 17, clearly show that crown ethers are effective agents for solubilizing water in chloroform, and that their complexing capacity increases with their ring size. Comparison of crown ether **14**, $n = 4$, with its open-chain counterpart **66** reveals that the cyclic structure is only slightly more effective. Hence it must be concluded that the water-complexing ability is main-

Table 17 Complexation of Water by Crown Ethers in CDCl$_3$ at 22 ± 1°C (71)

Crown Ether	K_w (ℓ mole^{-1})	$R_w - S_w/C_0$
14, $n = 2$	8	0.27
$n = 3$	14	0.38
$n = 4$	14	0.39
$n = 5$	20	0.47
$n = 6$	23	0.51
$n = 7$	47	0.68
63, $n = 3^a$	30	0.60
66	7	0.24

^{a}Corrected for the water solubility.

4 Complexation with Ammonium and Alkylammonium Salts

65 **66** **67**

ly determined by the number of ethyleneoxy units, the size of the cavity being less important.

Further information about the structure of the complexes was obtained from the chemical shifts of water protons in the free and complexed state. Since

$$\delta = \delta_{CE \cdot H_2O} - (\delta_{CE \cdot H_2O} - \delta_{H_2O}) \frac{S_w}{R_w C_0} \quad (2)$$

the chemical shift of complexed water, $\delta_{CE \cdot H_2O}$, can be found from a plot of the averaged water signal δ against $1/R_w C_0$ (Figure 6). Within experimental error the points for all crown ethers **14** are on the same line. The downfield shift of the signal of complexed water ($\delta_{CE \cdot H_2O}$ = 2.97 ± 0.05 ppm) relative to that of "free" water (δ_{H_2O} = 1.56 ppm) is in agreement with the formation of a hydrogen-bonded complex (91). The constancy of the chemical shift indicates that the water molecule is not located in the crown ether cavity, as this would result in a sensitive shift-structure relationship, as was observed for both the alkyl and the NH_3^+ signals in *tert*-buylammonium salts. From the chemical shifts of the crown ether signals it was concluded that the changes in crown ether conformation arising on complexation with water are very much smaller than the changes observed on complexation with metal ions or *tert*-butylammonium ions (68, 72).

The formation of complexes between crown ethers and *p*-toluene sulfonic and picric acid in 1,2-dichloroethane was recently reported (92). Association constants increased in the same order as the acidity of the acids in water: picH < TosH. For 18-crown-6 the free energy of association is much higher with TosH (1.5 kcal mole^{-1}) than with picric acid, which in turn is slightly higher (0.7 kcal mole^{-1}) than with water.

Hydrogen bonding of crown ethers to neutral molecules is not limited to compounds with O-H donor groups. Complex formation between dicyclohexano-18-crown-6 and thiols in acetonitrile has been reported by Nakabayashi and co-workers (93). C-H groups, although much weaker donors than OH, NH, and SH (94), can also interact with crown ethers through hydrogen bonding. This is evident from the isolation of solid nitrile and dinitrile complexes of 18-crown-6 and other crown ethers (95-98). X-Ray studies of these complexes provided

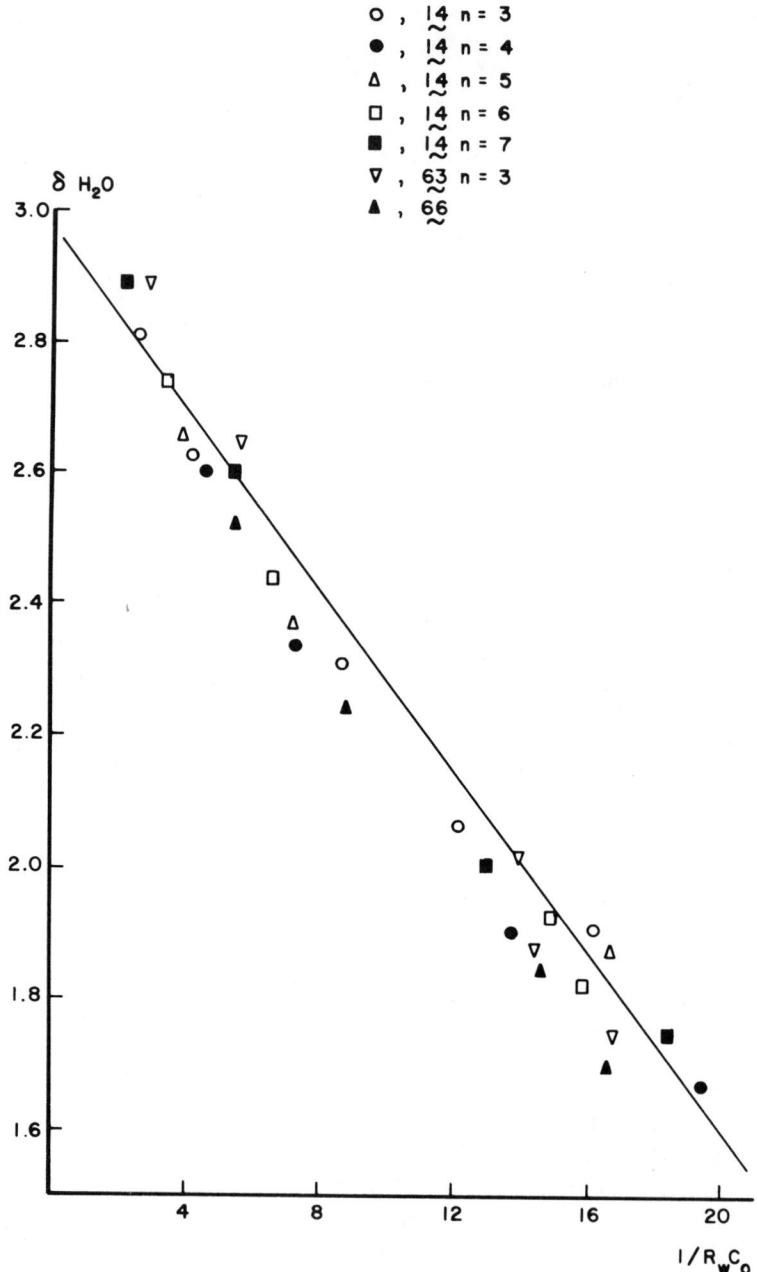

Figure 6 Relation between the chemical shift of water (δH_2O) and reciprocal water concentration ($1/R_wC_o$) for 1,3-xylyl crown ethers.

definite proof of C-H···O bonding (99). Hydrogen bonding of this type was also demonstrated in complexes between 18-crown-6 and dimethyl acetylenedicarboxylate (100).

4.3 Complexation with Alkylammonium Salts Under Anhydrous Conditions

The relative association constants for a number of crown ethers with *tert*-butylammonium hexafluorophosphate under strictly anhydrous conditions have been determined by a PMR titration technique (72), in which small portions of a $CDCl_3$ solution of a crown ether X are added to a solution containing the t-$BuNH_3 \cdot PF_6$ (AM) complex of crown ether Y. Since

$$AM \cdot Y + X \xrightleftharpoons{K_{rel}} AM \cdot X + Y \qquad (3)$$

the relative association $K_{rel} = K_Y/K_X$ can be calculated from the observed shift of the *tert*-butyl group (δ) and

$$\frac{\delta - \delta_Y}{\delta_X - \delta_Y} = \frac{-B + \sqrt{B^2 + 4AC}}{2A} \qquad (4)$$

in which $R_a = [[AMX] + [AMY]]/[Y]_t$, $A = R_a (K_{rel} - 1)$, $C = [X]_t/[Y]_t$, and $B = K_{rel} + C - A$; $[X]_t$ and $[Y]_t$ refer to the total concentrations of (free and complexed) X and Y, and δ_X and δ_Y are the shifts of the *tert*-butyl groups of the 1:1 complexes of X and Y, respectively.

In the 1,3-xylyl series **14** ($R^1 = R^2 = H$) the 18-membered macrocycle shows a sharp maximum in complexing ability (Figure 7). This is in line with expectations based on CPK models. The chemical shifts of the *tert*-butyl protons of compounds **14** increase regularly with increasing ring size: 0.825 ($n = 3$), 0.997 ($n = 4$), 1.154 ($n = 5$), 1.258 ($n = 6$), and 1.304 ppm ($n = 7$). Compared to the *tert*-butyl signal in the t-$BuNH_3^+ \cdot PF_6^-$ complex of 18-crown-6 ($\delta_{Bu} = 1.356$ ppm) they are all upfield. These results point to a conformation of the complex in which the *tert*-butyl group is in the shielding cone of the aryl ring (Figure 8). With increasing size of the cavity the average distance between cation and aryl group, and consequently the shielding effect, decreases. Evidence for this conformation is the fact that the close proximity of the aryl proton in the 2-position and the anion results in an upfield shift (0.16 ppm for $n = 3$).

The precise location of the anion in complexes of alkylammonium salts and crown ethers is not known yet. X-Ray crystal studies of a tetramethylenediammonium hexafluorophosphate complex showed that the PF_6^- anion and the ammonium cation are located on opposite sides of the crown ether (101). In a benzylammonium thiocyanate complex the cation and anion were reported to

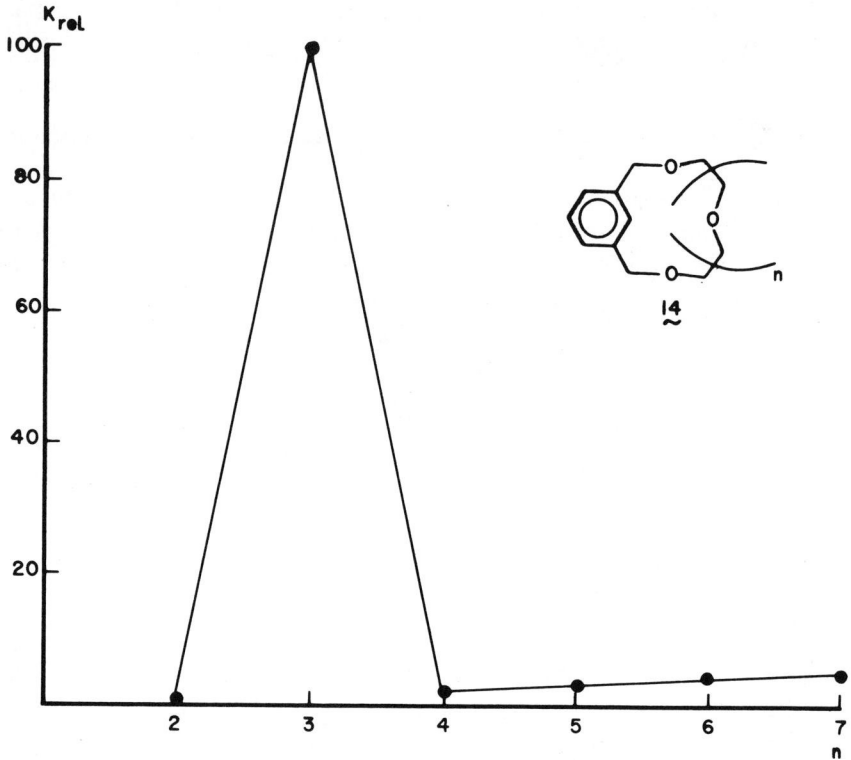

Figure 7 Relative binding constants for t-BuNH$_3$·PF$_6$ complexes of 1,3-xylyl crown ethers **14** in anhydrous CDCl$_3$ at 20°C.

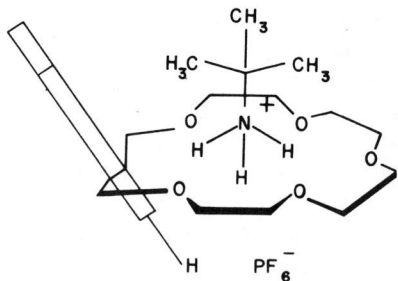

Figure 8 Most likely conformation of the t-BuNH$_3$·PF$_6$ complex of 1,3-xylyl-18-crown-5.

be on the same side of the crown ether (102). Further study is necessary to decide whether these facts reflect the differences in hydrogen-bonding ability between PF_6^- and SCN^-.

4.4 Complexation with Alkylammonium Salts in the Presence of Water

The effect of water on the stability of complexes between crown ethers and salts in solution has received little attention. Molecular models have been used to predict complex stability on the basis of the "best fit" between unsolvated crown ethers and unsolvated cations, and this approach has proved to be very fruitful (3, 8).

Yet there is overwhelming evidence that water molecules are incorporated into many crown ether complexes in the solid state. The water molecule can contribute to the lattice energy in different ways:

1. By hydrogen bonding to the anion, the crown ether, or the cation.
2. By saturating the coordination sphere of the cation.

Examples of these types of interaction have been provided by X-ray studies. Hydrogen bonding to the anion was observed in complexes of 12-crown-4 with NaOH (103) and NaCl(104), and in complexes of [2.2.2]-cryptate with RbSCN (105). Hydrogen bonding to crown ether oxygens only was found in the complex of benzo-15-crown-5 with calcium 3,5-dinitrobenzoate (106). Hydrogen bonding to both anion and crown ether was reported for a quadricyclic cryptate and NH_4I (107). Binding to both the cation and the anion was observed in several cases, for instance, in NaBr complexes of dibenzo-18-crown-6 (108), NaI complexes of benzo-15-crown-5 (109), and $Ba(SCN)_2$ complexes of cryptates (110). Interaction with only the cation was proposed for $NaB(C_6H_5)_4$ complexes of dibenzo-30-crown-10 (111). In all these examples the crown ether still makes the main contribution to the coordination of the cation. In some complexes, however, the crown ether does not contribute at all to the coordination of the cation. In complexes of $UO_2(NO_3)_2$ with 18-crown-6 (112) and of $MgCl_2$ with 12-crown-4 (113) the water molecules constitute the total coordination sphere of the cation and interact with the crown ether through hydrogen bonding.

The same interactions occurring between water and the different species in the complex in the solid state may also be operative in solution. Partition experiments with aqueous solutions of alkali metal salts of hexanitrodiphenylamine and nitrobenzene solutions of dibenzo-18-crown-6 have shown that, on solubilization in the organic phase, the Li^+, Na^+, and K^+ salts retain 2.0, 1.1, and 0.1 molecules of water, respectively (114). This shows that a crown ether does not completely displace the aqueous solvation shell when the cation has a high energy of hydration (15) and/or is too small in the "naked" state to fit into the cavity.

The beneficial effect of water on the stability of t-BuNH$_3$·PF$_6$ complexes of 1,3-xylyl crown ethers **14** is apparent from a comparison of the relative complexing abilities determined under anhydrous conditions with those obtained from partition experiments (71). At very low crown ether concentrations (0.01 M) the relative complexing power in the presence of water is the same as that in the absence of water (Figure 9). At high crown ether concentrations (0.5 M) the apparent binding constants of the larger macrocycles increase, with 1,3-xylyl-24-crown-7 (**14**, $n = 5$) showing the largest increase (by a factor of 20). This apparent increase in binding ability parallels the amount of water coextracted with the cation (Figure 9). The formation of hydrated dimers may well explain the synergistic effect of water and t-BuNH$_3$·PF$_6$ observed at high ligand concentrations.

The chemical shift of water in these hydrates ($\delta \approx 3.7$ ppm) is 2.2 ppm downfield from that of free water in CDCl$_3$, and 0.7 ppm downfield from that of the water–crown ether complexes. Since the PF$_6^-$ anion shows little tendency to interact through hydrogen bonding (115), the water is probably bound to the ammonium cation. The hydrated and anhydrous complexes of the 1,3-xylyl crown ethers have a different conformation in solution, as is evident from the chemical shift of the *tert*-butyl groups. Removing the water of hydration was reported to result in downfield shifts of 0.005, 0.023, 0.186, 0.104, and 0.101 ppm for the 18-, 21-, 24-, 27-, and 30-membered crown ether complexes, respectively (72).

Since the presence of water has very little effect on the stability of complexes in which the size of the crown ether cavity is close to that of the completely dehydrated cation, it may be expected that partition experiments will give reliable binding constants for these compounds (Table 18).

Cram's group (39, 47) reported that association constants (K_a) in CDCl$_3$ can be obtained from a distribution experiment in which a CDCl$_3$ solution of the crown ether (C_0 molar, V_{CDCl_3} ml) is equilibrated with an aqueous solution of the *tert*-butylammonium salt (B_0 molar, $V_{\text{D}_2\text{O}}$ ml). The molar ratio of salt to crown ether ($R = [t\text{-BuNH}_3\text{X}]/C_0$) in the CDCl$_3$ phase can be derived from the ^{1}H NMR spectral data, and K_a follows from

$$K_a = \frac{R}{K_d (1 - R) [B_0 - RC_0 \, V_{\text{CDCl}_3}/V_{\text{H}_2\text{O}}]^2} \tag{5}$$

in which the distribution constant K_d, defined by

$$[^t\text{BuNH}_3^{\oplus} + \text{X}^{\ominus}]_{\text{H}_2\text{O}} \xrightleftharpoons{K_d} [^t\text{BuNH}_3\text{X}]_{\text{CHCl}_3} \tag{6}$$

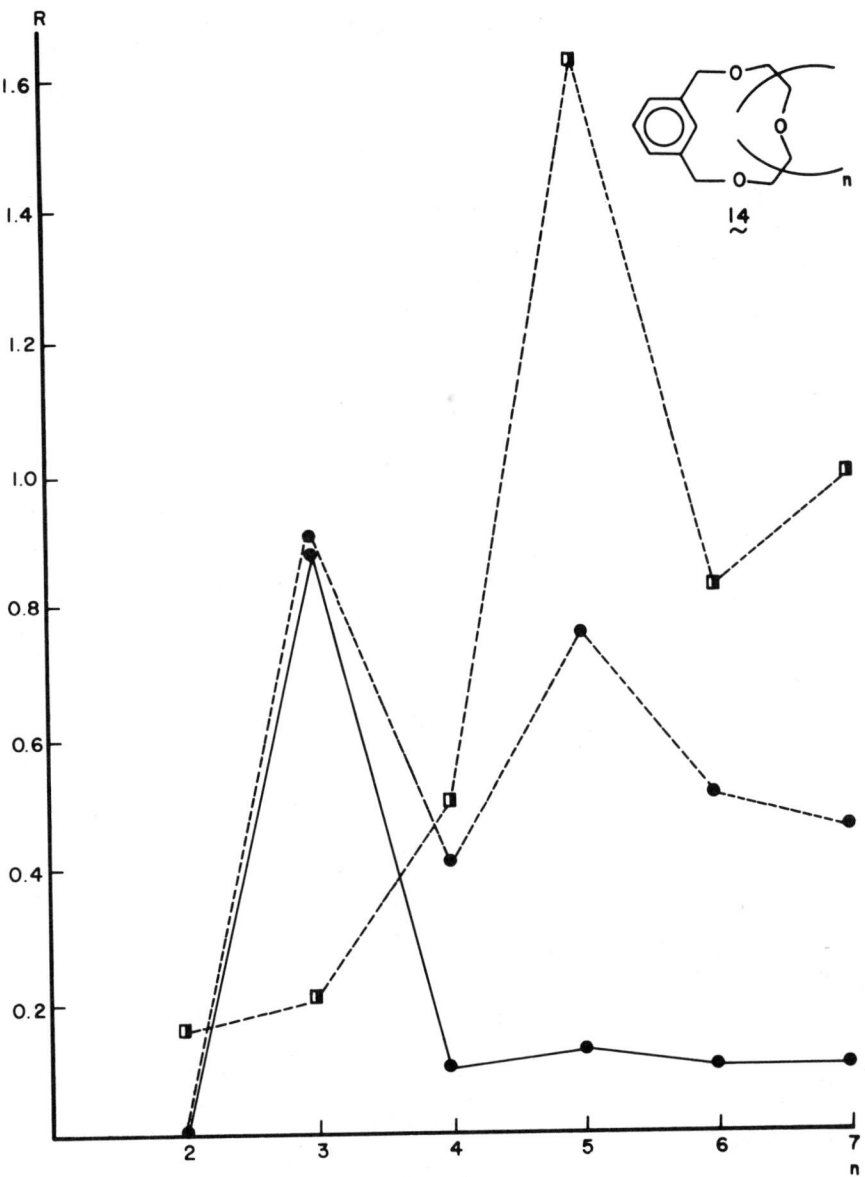

Figure 9 Fraction (R) of 1,3-xylyl crown ethers **14** complexed by t-BuNH$_3$·PF$_6$ (●) and water (□) at low (——, 0.01 M) and high (– – –, 0.5 M) ligand concentration.

Table 18 Association Constants (log K_a) of Crown Ether Complexes of t-BuNH$_3^+ \cdot$ X$^-$ in CDCl$_3$ at $25 \pm 1°$C

Crown Ether				Log K_a				
				$X = PF_6^{-a}$	$X = ClO_4^{-b}$	$X = SCN^{-c}$	$X = pic^{-d}$	$X = Cl^{-e}$
5, $n = 2$				5.2				
5, $n = 3$				5.2				
8, R = H, X = O, $n = 2$				5.0				
14								
	R^1	R^2	n					
	H	H	2	<2				
	H	H	3	5.3	5.3	3.5	3.5	
	H	H	4	3.4				
	H	H	5	3.5				
	H	H	6	3.8				
	H	H	7	4.0				
	H	Br	3		4.4	2.5	2.9	
	H	C(CH$_3$)$_3$	3		5.6	3.7	3.7	
	H	CO$_2$C$_2$H$_5$	3			2.8		
	H	OCH$_3$	3		5.4	3.5		
	H	SCH$_3$	3		4.8			
	H	CN	3		3.8			
	CO$_2$CH$_3$	H	2			≤2.7		
	H	H	3			4.8		
	H	H	4			3.5		

H	H	7		2.9	
CO₂H	H	3		3.1	
CH₂OH	H	3		3.4	
CH₂OCH₃	H	3		3.4	
CN	H	3		3.1	
20, $n = 1$			6.1	5.3	4.0
21, $n = 1$				2.4	
22				4.1	
23				3.0	
36				6.7	5.1
37, $n = 1$				6.3	4.7
37, $n = 0$				2.9	
38				2.8	
47				2.4	
63, $n = 2$			5.0	6.5	5.1
$n = 3$			7.4		
64, $n = 1$			3.9		
$n = 2$			6.7	5.8	4.3
$n = 3$			4.8		
67			6.1	4.9	

[a] Taken from Refs. 73 and 74.
[b] Taken from Ref. 84.
[c] Taken from Refs. 39, 50, 84, 116, and 117.
[d] Taken from Ref. 84.
[e] Taken from Ref. 116.

is determined from separate experiments (39).

The complexing ability toward t-BuNH$_3^+$ (Table 18) as a function of cavity size shows a maximum for 18- to 20-membered crown ethers. Removal of one ethyleneoxy unit from the optimal structure results in a loss in binding energy of at least 3 kcal mole^{-1}. Adding one ethyleneoxy unit to the optimal structure causes a smaller decrease in stability (about 2 kcal mole^{-1}).

Replacement of a CH$_2$CH$_2$OCH$_2$CH$_2$ unit in 18-crown-6 by a bis(methylene)- aromatic or -heteroaromatic unit results in the following order of relative binding abilities toward t-BuNH$_3^+$:

[pyridyl] > [furan-2,5-diyl-like] > [furan] > [1,2-phenylene] ≈ [1,3-phenylene] ≈

≈ [3,4-furyl] ≫ [1,4-phenylene (xylyl)]

The first three units contain a binding site for the cation, in contrast to the xylyl and 3,4-furyl units.

Cram and co-workers (39) have dissected the free energies of association between t-BuNH$_3^+$ and crown ethers into six contact-site parameters, which proved to be additive and were profitably used to predict the free energies of association of a number of crown ethers (39, 116).

It has been reported that for 1,3-xylyl-18-crown-5 electron-donating substituents in the 5-position increase the complex stability, whereas electron-withdrawing substituents have a destabilizing effect. This result was rationalized by assuming an interaction between the positive charge of the cation and the π-binding site at the 2-position (84).

Transannular substituents decrease the stability of t-BuNH$_3^+$ complexes of 1,3-xylyl-18-crown-5, except for CO$_2$CH$_3$, whose extra binding site compensates for the unfavorable steric interactions (117). Steric interactions with a chiral barrier have been successfully used to induce chiral recognition toward alkylammonium salts (8). These chiral barriers cause a large decrease in the free energy of binding. Replacement of a CH$_2$CH$_2$ unit in 18-crown-6 by the 2,2'-binaphthyl unit results in a loss in binding energy of 4.3 kcal mole^{-1} (47). An even larger decrease in binding energy (5.1 kcal mole^{-1}) is observed on introduction of four 2,2-dimethyl-1,3-dioxolanyl groups (compound 47) into the dipyridyl crown ether 37, n = 1, or into 18-crown-6 (50, 118).

4 Complexation with Ammonium and Alkylammonium Salts

The bulkiness of the alkyl groups in the ammonium salt also has a large effect on the complex stability. Decreasing the steric demands of the cation by replacing *tert*-butylammonium by benzylammonium results in an increase in free energy of association of at least 6.8 kcal mole^{-1} for compound **47** (50). Similar large differences were observed between complexes of t-BuNH$_3\cdot$SCN and C$_6$H$_5$CH$_2$NH$_3\cdot$SCN with crown ethers having the same skeleton as 18-crown-6 (118, 119).

Another important factor affecting the association constants is the type of anion (Table 18). Cram and co-workers obtained good linear free energy relationships (slope 0.99) when comparing t-BuNH$_3\cdot$SCN as a standard with the corresponding perchlorate and picrate (84). This suggests that the anion scales are linearly related:

$$\Delta G(X^-) = \Delta G(Y^-) + \text{constant}$$

A rough estimate of the difference in free energy between the scales can be obtained from Table 18:

```
         pic⁻
Cl⁻      SCN⁻   PF₆⁻   ClO₄⁻
|         |      |       |          → -ΔG₀
0        2.0    3.2     4.5          (kcal mole⁻¹)
```

4.5 Kinetics of Complexation with Alkylammonium Salts

Natural antibiotics, such as the valinomycins, and synthetic macrocyclic ligands share the capacity to induce the transport of cations through lipophilic membranes (120). In order to identify the rate-determining step in the overall ion-transport process, several workers have studied the kinetics of complexation between alkali ions and natural cation carriers (121-123). Crown ethers have been frequently used as model systems, and the kinetics of complexation, in aqueous solution, between alkali cations and 18-crown-6 (122, 123), dibenzo-18-crown-6 (126-128), and dibenzo-30-crown-10 (129) have been determined.

In general the kinetics of complexation between salts and crown ethers becomes important in all processes that require several loading and unloading cycles, such as transfer of salts through bilayer (120) and solvent (130) membranes, phase-transfer catalysis (131), and catalysis of organic reactions (23). In many of these applications nonpolar aprotic solvents are used. Because of their good solubility complexes of crown ethers and alkylammonium salts are very suitable for studying the kinetics of complexation in such solvents.

The kinetics of complexation between a crown ether and an alkylammonium salt can be studied by means of NMR line-shape methods using the ^{1}H signals of either the crown ether or the salt. Since only a very small fraction of the 1:1 complex is dissociated in apolar solvents, an equimolar amount of either the crown ether or the salt has to be added to the pure 1:1 complex to ensure an equal concentration of exchanging sites. Adding a surplus of alkylammonium salt to the complex has the disadvantage that under these conditions the 1:1 complex may bind a second salt molecule to form a 1:2 complex. Evidence of the formation of such complexes was obtained when excess solid t-BuNH$_3$·SCN was equilibrated with chloroform solutions containing 1,3-xylyl-18-crown-5 (36). The chemical shift of the *tert*-butyl group (δ = 1.230 ppm) in the 1:2 complex formed was different from that of the free salt (δ = 1.486 ppm) as well as from that of the 1:1 complex (δ = 0.893 ppm) obtained in equilibration experiments with aqueous salt solutions.

To avoid the complication that may arise on addition of an excess of free salt, de Jong's group created equal concentrations of exchanging sites by adding one equivalent of a t-BuNH$_3$·PF$_6$ complex of a different crown ether (73, 74). Under these conditions the *tert*-butyl groups are present in three different species (eqs. 4.7 and 4.8): complexed with crown ether X, complexed with crown ether Y, and as free salt (AM), the latter in undetectably (by NMR) low concentration:

$$X \cdot AM \underset{k_{-1}}{\overset{k_1}{\rightleftharpoons}} AM + X \qquad (7)$$

$$Y \cdot AM \underset{k_{-2}}{\overset{k_2}{\rightleftharpoons}} AM + Y \qquad (8)$$

When the complex of crown ether Y is chosen in such a way that (*a*) there is a large difference between the chemical shifts of the *tert*-butyl groups in the complexes with X and Y, and (*b*) the complex of Y is kinetically far less stable than that of X, the exchange of t-BuNH$_3^+$ between crown ethers X and Y can be treated as a two-site problem, in which the decomplexation of complex X·AM constitutes the rate-limiting step.

In addition to exchange occurring via complexation and decomplexation (Eqs. 7 and 8), the following bimolecular exchange reactions may contribute to the overall cation-exchange process:

$$XAM + AM^* \xrightarrow{k_3} XAM^* + AM \qquad (9)$$

$$XAM + Y \xrightarrow{k_4} YAM + X \qquad (10)$$

… 4 Complexation with Ammonium and Alkylammonium Salts

$$XAM + YAM^* \xrightarrow{k_s} XAM^* + YAM \tag{11}$$

It was found that the dependence of the reciprocal lifetime of XAM (τ_{XAM}^{-1}) on the concentrations of the different species is represented by

$$\tau_{XAM}^{-1} = k_1 + k_3[AM] \tag{12}$$

Hence bimolecular exchange between XAM and free crown ether Y (Eq. 10) and between complexes XAM and YAM (Eq. 11) was not observable.

The rate of decomplexation (k_1) for the t-BuNH$_3$·PF$_6$ complex of 18-crown-6 was obtained from Eq. 12 using two different methods:

1. For a CDCl$_3$ solution containing equimolar amounts of the t-BuNH$_3$·PF$_6$ complexes of 18-crown-6 (X) and 1,3-xylyl-18-crown-6 (14, $n = 3$), the concentration of free salt is approximated by

$$[AM] = \left\{ \frac{[Y\ AM]}{K_Y} \right\}^{1/2} \tag{13}$$

in which K_Y stands for the association constant of 1,3-xylyl-18-crown-5. A plot of τ_{XAM}^{-1} [calculated from the observed line broadening with the aid of the commonly used formulas (132)] against $[YAM]^{1/2}$ gave a good linear relationship (Figure 10), with $k_1 = 70\ \text{s}^{-1}$ and $\frac{k_3}{\sqrt{K_Y}} = 2.65 \times 10^3$ at 20°C.

2. For a CDCl$_3$ solution containing equimolar amounts of complexes together with free crown ether 14 ($[Y]_e$), the concentration of free salt is given by

$$[AM] = \frac{[YAM]}{K_Y[Y]_e} \tag{14}$$

A plot of τ_{XAM}^{-1} against $[YAM]/[Y]_e$ gave a linear relationship (Figure 10), with $k_1 = 65\ \text{s}^{-1}$ and $k_3/K_Y = 5.63$ at 20°C.

Thus the two methods gave the same rate of decomplexation for the t-BuNH$_3$·PF$_6$ complex of 18-crown-6 ($k_1 = 70 \pm 10\text{s}^{-1}$). Furthermore, from the slopes of the curves obtained in these two methods, it was derived that $k_3 = 1.3 \times 10^6\ \text{\textsterling mole}^{-1}\text{s}^{-1}$ and $K_Y = 2.2 \times 10^5$ mole $\text{\textsterling}^{-1}$ at 20°C. The absolute value of the association constant of 1,3-xylyl-18-crown-5 (14, $n = 3$) has been used to quantify the scale of relative complexing abilities found from competition experiments (Table 18). The known association constants of a series of crown ethers have in turn been used to determine the rate of decomplexation of the

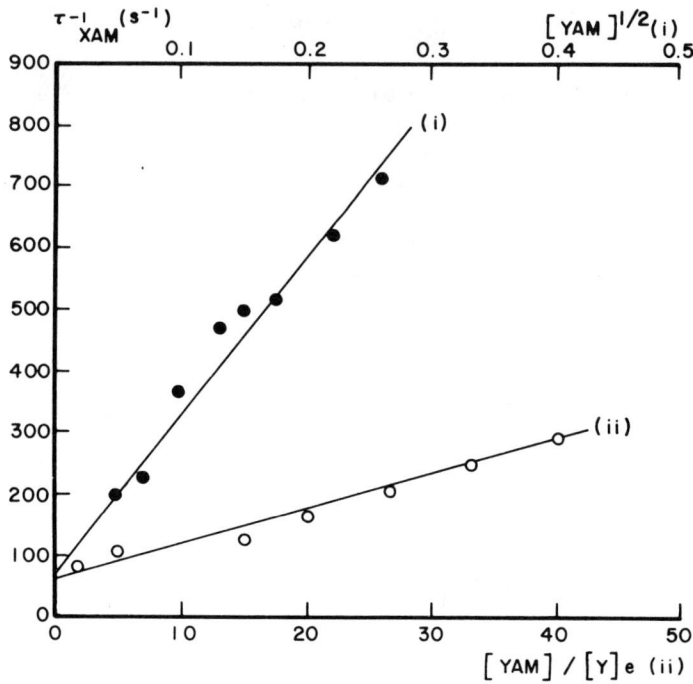

Figure 10 Dependence of the reciprocal lifetime of the t-BuNH$_3$·PF$_6$ complex of 18-crown-6 (XAM) on the concentration of free ammonium salt in CDCl$_3$ at 20°C as obtained from exchange with the t-BuNH$_3$·PF$_6$ complex of 1,3-xylyl-18-crown-5.

t-BuNH$_3$·PF$_6$ complex of 18-crown-6 by a third method (133):

3. For a CDCl$_3$ solution containing equimolar amounts of the t-BuNH$_3^+$·PF$_6^-$ complex of 18-crown-6 and another crown ether Y, the concentration of free salt is given by Eq. 13. When crown ether Y was varied, a plot of τ_{XAM}^{-1} against $K_Y^{-1/2}$ gave a good linear relationship (Figure 11). The values of k_1 = 55 s^{-1} and k_3 = 1.5 × 10^6 ℓ mole^{-1} s^{-1} are in good agreement with those obtained using the first two methods.

An approximate value of the Arrhenius energy of decomplexation for the t-BuNH$_3$·PF$_6$ complex of 18-crown-6 was obtained from a plot of ln τ_{XAM}^{-1} versus $1/T$ for conditions under which the concentration of uncomplexed salt was very low. In one experiment this was accomplished by using an equimolar mixture of 1,3-xylyl-18-crown-5 and its complex, and in another experiment by using the t-BuNH$_3$·PF$_6$ complex of benzo-18-crown-6 only. These two experiments gave about the same value for E_a: 18 ± 1 kcal mole^{-1} (Figure 12), which is

4 Complexation with Ammonium and Alkylammonium Salts

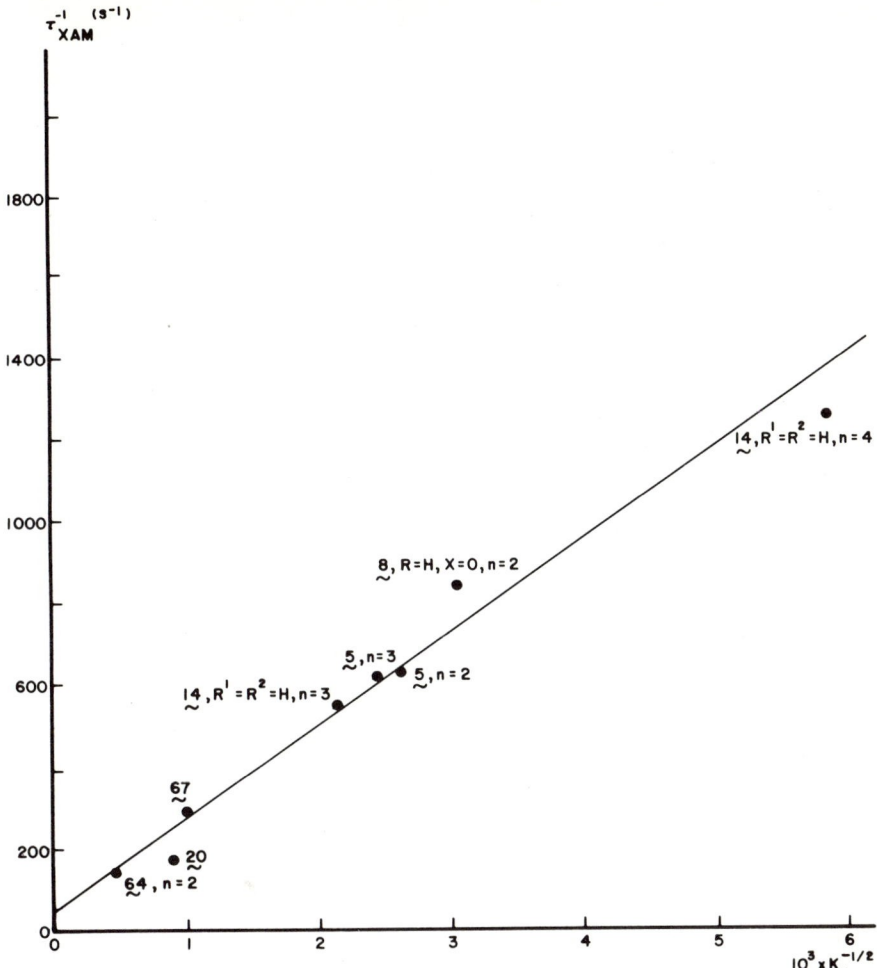

Figure 11 Dependence of the reciprocal lifetime of the t-BuNH$_3$·PF$_6$ complex of 18-crown-6 (XAM) on the binding constant K_y of the second crown ether complex as obtained from equimolar (0.025 M) solutions in CDCl$_3$ at 20°C.

unexpectedly high, considering that most crown ether complexes of alkali cations were reported to give activation energies for decomplexation ranging from 10 and 14 kcal mole^{-1} (126-128).

The method of cation exchange between t-BuNH$_3$·PF$_6$ complexes of different crown ethers has been applied to a number of crown ethers to obtain information about the effect of crown ether structures on the kinetics of complexation. For a series of crown ethers a linear relationship between

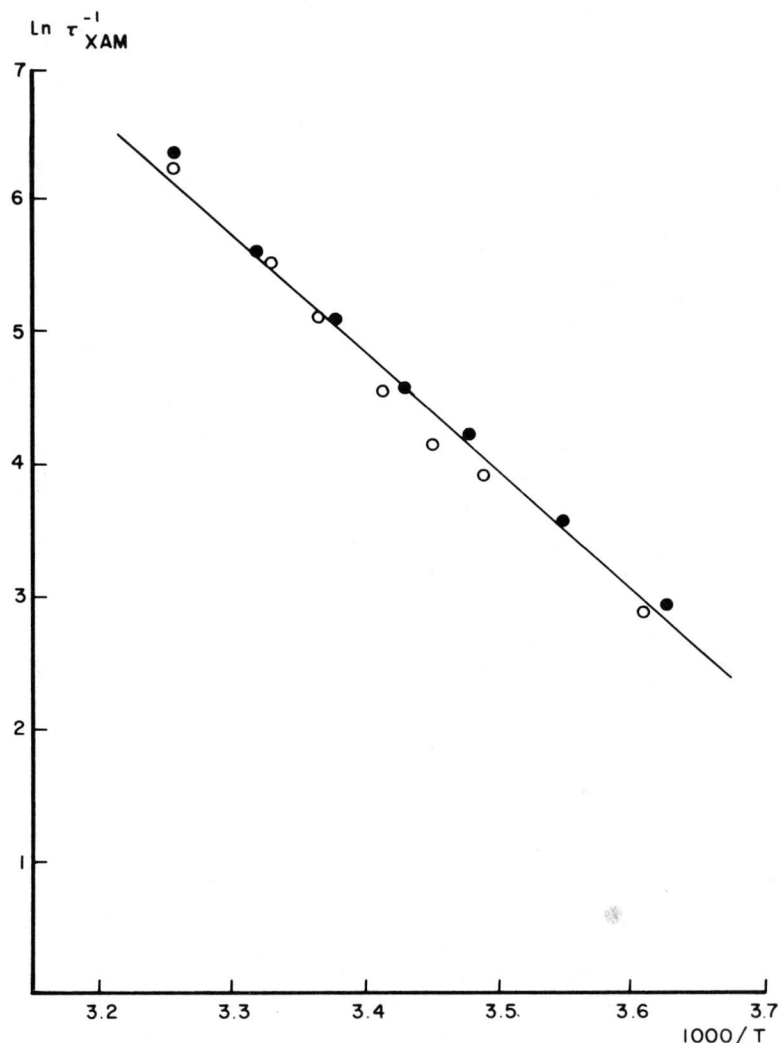

Figure 12 Arrhenius plot of the reciprocal lifetime τ^{-1}_{XAM} of the t-BuNH$_3$·PF$_6$ complex of 18-crown-6 under conditions of low free salt concentration: equimolar (0.02 M) amounts of free and complexed 1,3-xylyl-18-crown-5 (●); equimolar (0.02 M) amount of benzo-18-crown-6 complex (○).

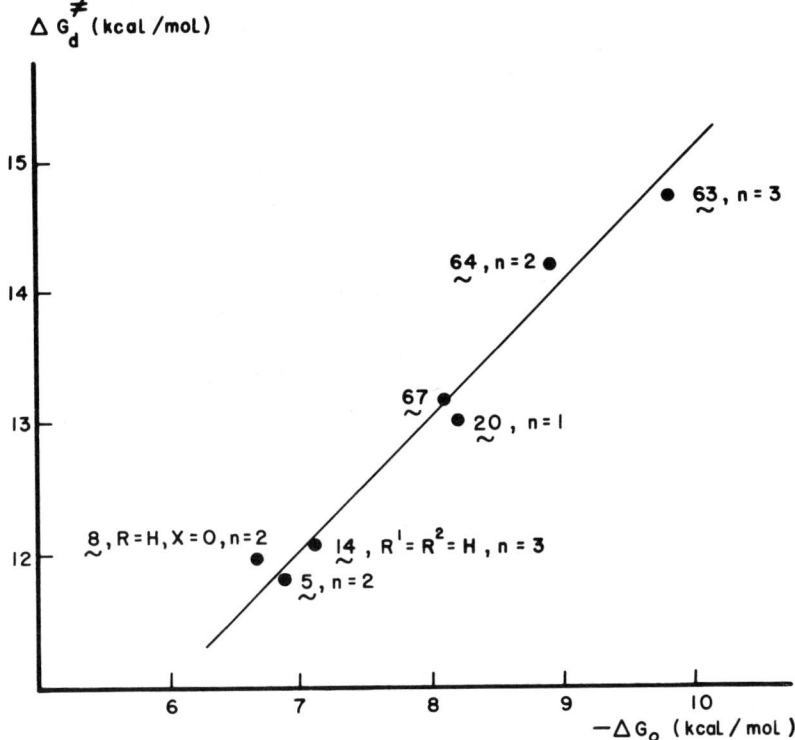

Figure 13 Relation between free energy of activation for decomplexation ($\Delta G_d^{\neq}$) and free energy of association (ΔG_0) for t-BuNH$_3$·PF$_6$ complexes of crown ethers in t-BuNH$_3$·PF$_6$ complexes of crown ethers in CDCl$_3$ at 20°C.

the free energy of activation for decomplexation ($\Delta G_d^{\neq}$) and the free energy of complex formation (ΔG_0) was found (Figure 13). For these crown ethers the free energy of activation for complexation was about the same: $\Delta G_c^{\neq} \approx 5$ kcal mole^{-1}, indicating that the rate of complex formation is independent of the crown ether structure and is probably diffusion controlled ($k_c = 1.1 \times 10^9$ ℓ mole^{-1} s^{-1} at 20°C).

Occasionally the difference in ^{1}H chemical shifts between the free and complexed crown ether is large enough to permit a study of the kinetics of complexation. In dilute (0.01 M) CDCl$_3$ solutions containing equimolar amounts of complex and free crown ether, cation exchange occurs via decomplexation-complexation (Eq. 7). Thus free energies of activation for decomplexation have been obtained for 3,4-furyl-17-crown-5 (**8**, R = H, X = O, n = 2), 1,3-xylyl-18-crown-5 (**14**, R^1 = R^2 = H, n = 3), and dibenzo-18-crown-6 (**67**). The $\Delta G_d^{\neq}$ values at the coalescence temperature of the signals for the aromatic protons

Table 19 Free Energies of Association (ΔG_0) and Free Energies of Activation (kcal mole^{-1}) for Decomplexation ($\Delta G_d^{\neq}$) of t-BuNH$_3 \cdot$PF$_6$ Complexes of Crown Ethers in CDCl$_3$

Crown Ether	$-\Delta G_0$ (at 20°C)	$\Delta G_d^{\neq}$ (temperature)	Method[a]
5, $n = 2$	6.9	11.8 (20)	A
		11.4 (−34)	B
		11.7 (−34)	
5, $n = 3$	7.0		
8, R = H, X = O, $n = 2$	6.7	12.0 (20)	B
		11.1 (−48)	
14, R^1 = R^2 = H, $n = 3$	7.1	12.1 (20)	A
		12.0 (−28)	C
		11.4 (−50)	A
		12.3 (−28)	A
		11.4 (−50)	B
		11.8 (−50)	C
20, $n = 1$	8.2	13.0 (20)	A
		13.2 (−14)	B
		12.9 (−14)	
		11.5 (−50)	
63, $n = 3$	9.8	14.7 (20)	A
64, $n = 2$	8.9	14.2 (20)	A
67	8.1	13.2 (20)	A
		12.8 (−20)	A
		13.1 (−20)	C

[a] A) Exchange between two different complexes; B) exchange between the two faces of the same complex; C) exchange between free and complexed ligand.

were in good agreement (± 0.3 kcal mole^{-1}) with the values obtained from exchange between complexes of different crown ethers (Table 19).

The C_2 axis of symmetry present in many crown ethers is absent in their t-BuNH$_3$·PF$_6$ complexes; consequently protons facing the cation and protons facing the anion have different chemical shifts. The rate of cation exchange between the two faces of the crown ether

$$t\text{-BuNH}_3^+ \quad \overset{H_a}{\underset{PF_6^-}{\bigvee}}\overset{H_b}{} \rightleftharpoons t\text{-BuNH}_3\cdot\text{PF}_6 + \overset{H_a}{\underset{}{\bigvee}}\overset{H_b}{}\overset{H_a}{\underset{PF_6^-}{\bigvee}}\overset{H_b}{} \rightleftharpoons \text{PF}_6^- \quad \text{H}_3\text{N}^+t\text{-Bu} \quad (15)$$

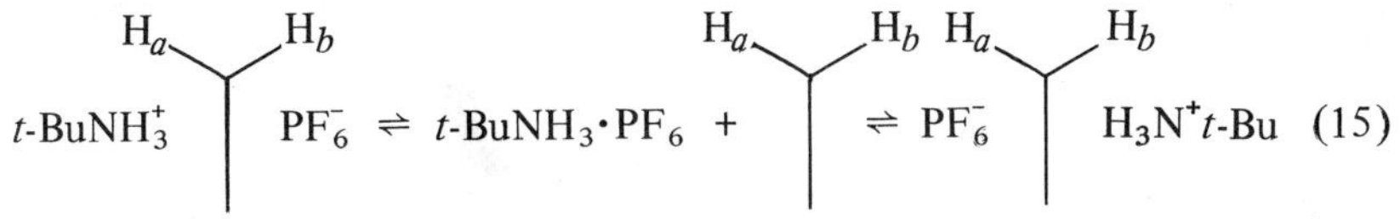

can be determined at the coalescence temperature of the signals of H$_a$ and H$_b$ with the aid of the usual formulas (130). The results obtained by using the benzylic protons of crown ethers containing the xylyl (**5, 14**) and furyl unit (**8, 20**) are given in Table 19. The agreement with the $\Delta G_d^{\neq}$ values obtained from exchange between different complexes is very satisfactory, particularly in view of the inaccuracy introduced by the use of the rather small differences in chemical shift (10-15 Hz) between the benzylic protons.

Sutherland's group (41-43) has studied the kinetics of complexation of diazacrown ethers **32, 33**, and **68-70** with alkylammonium thiocyanates. The signal of

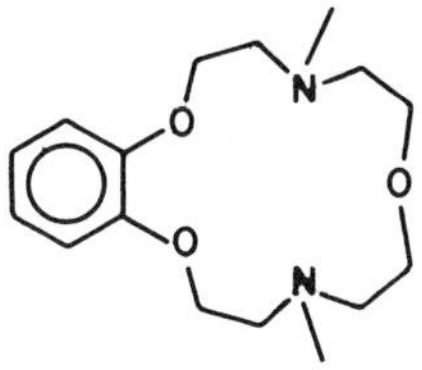

68 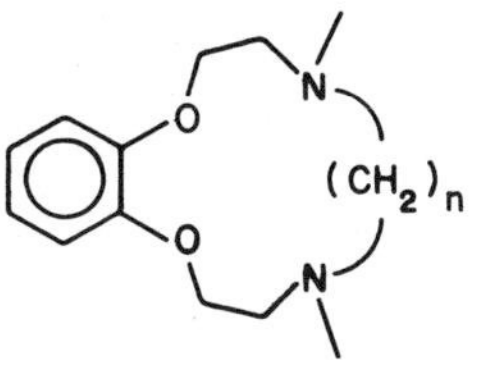 **69** 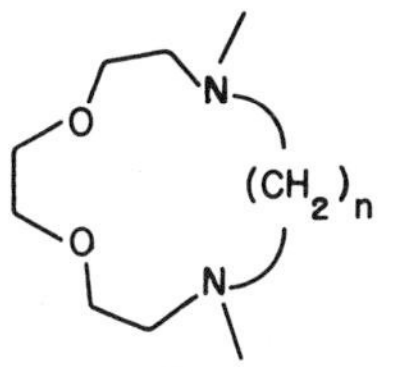 **70**

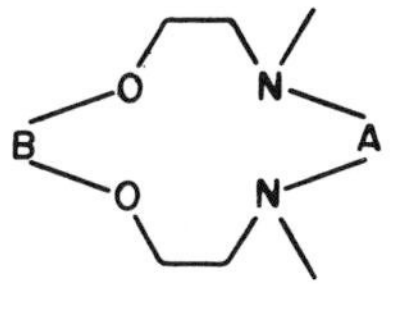

71

the NCH$_2$ protons was used to determine the exchange between the two faces of the crown ether (Eq. 15). The free energy of activation for decomplexation decreased with increasing degree of substitution at the α- or β-position in the alkylammonium salt. This is consistent with the fact that nonbonded inter-

Table 20 Free Energies of Activation ($\Delta G_d^{\neq}$) for the Decomplexation of $R_1NH_3 \cdot SCN$ Complexes of Crown Ethers in $CDCl_3$

Crown Ether	R_1	$\Delta G_d^{\neq}$ (kcal mole^{-1})	T_c (°C)	Ref.
32, R = CH$_3$, n = 2	H	8.7	−90	42
	CH$_3$	10.1	−61	42
	C$_2$H$_5$	10.7	−50	42
	CH(CH$_3$)$_2$	10.1	−62	42
	C(CH$_3$)$_3$	9.3	−80	41, 42
	CH$_2$CH$_2$CH$_3$	10.2	−60	42
	CH$_2$CH(CH$_3$)$_2$	9.7	−70	42
	CH$_2$C(CH$_3$)$_3$	<9	<−100	42
	CH$_2$C$_6$H$_5$	10.8	−45	41, 42
32, R = CH$_3$, n = 3	CH$_2$C$_6$H$_5$	9.9	−65	41
33, R = CH$_3$	CH$_3$	9.7	−68	42
	C$_2$H$_5$	10.0	−62	42
	CH(CH$_3$)$_2$	10.2	−57	42
	C(CH$_3$)$_3$	9.1	−82	42
	CH$_2$C$_6$H$_5$	11.1	−40	42

33, R = CH$_2$CH$_2$OH	CH$_2$C$_6$H$_5$	12.1	−15	43
	CH$_2$CH$_2$OH	11.0	−40	43
33, R = CH$_2$CON(CH$_3$)$_2$	CH$_2$C$_6$H$_5$	12.8	−5	43
	CH$_2$CH$_2$OH	12.2	−15	43
47	CH$_2$C$_6$H$_5$	12.2	−35	50
68	CH$_2$C$_6$H$_5$	13.1	0	42
	(R)-CH(CH$_3$)C$_6$H$_5$	12.6	−5	42
69, n = 4	CH$_2$C$_6$H$_5$	9.5	−70	43
n = 5	CH$_2$C$_6$H$_5$	11.4	−35	43
n = 6	CH$_2$C$_6$H$_5$	10.4	−60	43
n = 7	CH$_2$C$_6$H$_5$	10.1	−60	43
n = 8	CH$_2$C$_6$H$_5$	9.5	−70	43
70, n = 3	CH$_2$C$_6$H$_5$	<9	<−80	43
n = 4	CH$_2$C$_6$H$_5$	10.4	−55	43
n = 5	CH$_2$C$_6$H$_5$	10.7	−55	43
n = 6	CH$_2$C$_6$H$_5$	10.0	−65	43
n = 7	CH$_2$C$_6$H$_5$	9.4	−80	43
n = 8	CH$_2$C$_6$H$_5$	<<9	<−80	43

actions between cation and crown ether destabilize the complex. In compounds represented by the general structure 71 the structure units A and B had the following effect on $\Delta G_d^{\neq}$ for benzylammonium thiocyanate complexes (Table 20):

1. For B = o-C$_6$H$_4$ $\Delta G_d^{\neq}$ decreases in the order

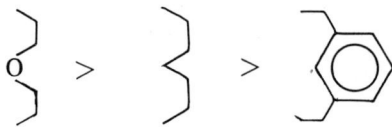

Since relative values of $\Delta G_d^{\neq}$ are directly related to relative free energies of binding (42, 74), these kinetic measurements reveal only a small difference (1.7 kcal mole^{-1}) in free energies of binding between the CH$_2$CH$_2$OCH$_2$CH$_2$ and CH$_2$CH$_2$CH$_2$CH$_2$CH$_2$ units. For the same structural change in t-BuNH$_3$·SCN complexes of crown ethers, Cram and co-workers, (39) observed a difference of 4.4 kcal mole^{-1} in the free energy of binding. Furthermore of the crown ethers the 1,3-xylyl compound was thermodynamically more stable than the 1,5-pentyl compound (39), in contrast to the diazo compounds, where the 1,3-xylyl compound was less stable than the 1,5-pentyl compound.

2. For A = 1,3-xylyl or 1,5-pentyl $\Delta G_d^{\neq}$ decreases when the o-phenylene unit is replaced by an ethylene unit. Again this order is reversed in t-BuNH$_3$·SCN complexes of crown ethers (39).

The different requirements for complex formation with crown ethers on the one hand and diazo compounds on the other has been explained by Sutherland and co-workers (43) on the basis of a two-point binding model (Figure 14). An X-ray analysis of the crystal structure of the C$_6$H$_5$CH$_2$NH$_3$·SCN complex of compound 32 showed that the cation forms only two hydrogen bonds with the

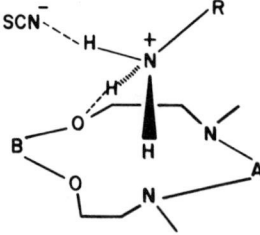

Figure 14 Most likely conformation of RNH$_3$·SCN complexes of diazacrown ethers.

ligand, the third being formed with the anion (102). A similar observation was made with crown ethers: the X-ray structure of the 1:1 complex of *tert*-butylamine and 1,3-xylyl-18-crown-5 having an intraannular CO_2H substituent (14, R^1 = H, R^2 = CO_2H, n = 3) showed that two hydrogen bonds were formed with crown ether oxygens, while the third was formed with the carboxylate anion (134). A structure like that shown in Figure 14 might be quite general in those cases in which either the binding sites of the ligand are not properly located for the formation of a complex through three hydrogen bonds, or the anion is such a strong hydrogen-bonding acceptor that ion separation does not occur.

REFERENCES

1. C. J. Pedersen, *J. Am. Chem. Soc.,* **89**, 2495 (1967).
2. C. J. Pedersen, *J. Am. Chem. Soc.,* **89**, 7017 (1967).
3. R. M. Izatt, D. J. Eatough, and J. J. Christensen, *Struct. Bonding,* **16**, 161 (1973).
4. C. L. Liotta and H. P. Harris, *J. Am. Chem. Soc.,* **96**, 2250 (1974).
5. C. L. Liotta, E. E. Grisdale, and H. P. Hopkins, *Tetrahedron Lett.,* **1975**, 4205.
6. D. J. Sam and H. E. Simmons, *J. Am. Chem. Soc.,* **96**, 2252 (1974).
7. D. J. Cram and J. M. Cram, *Science,* **183**, 803 (1974).
8. D. J. Cram, R. C. Helgeson, L. R. Sousa, J. M. Timko, M. Newcomb, P. Moreau, F. de Jong, G. W. Gokel, D. H. Hoffman, L. A. Domeier, S. C. Peacock, K. Madan, and L. Kaplan, *Pure Appl. Chem.,* **43**, 327 (1975).
9. Y. Chao and D. J. Cram, *J. Am. Chem. Soc.,* **98**, 1015 (1976).
10. T. J. van Bergen and R. M. Kellogg, *J. Chem. Soc., Chem. Commun.,* **1976**, 964.
11. British Patent 785.229 (October 23, 1957).
12. J. Dale, G. Borgen, and K. Daasvatn, *Acta Chem. Scand.,* **B28**, 378 (1974).
13. J. Dale and K. Daasvatn, *J. Chem. Soc., Chem. Commun.,* **1976**, 295.
14. J. Cooper and P. H. Plesch, *J. Chem. Soc., Chem. Commun.,* **1974**, 1017.
15. J. M. Lehn, *Struct. Bonding,* **16**, 1 (1973).
16. F. Vögtle and E. Weber, *Angew. Chem.,* **86**, 126 (1974).
17. H. Feuer and J. Hooz, in *The Chemistry of the Ether Linkage,* (S. Patai, Ed.), Interscience, London, 1967, p. 445.
18. Ref. 17, p. 462.
19. D. N. Reinhoudt, R. T. Gray, C. J. Smit, and I. Veenstra, *Tetrahedron,* **32**, 1161 (1976).
20. C. J. Pedersen, *Aldrichim. Acta,* **4**, 1 (1971).
21. C. J. Pedersen and H. K. Frensdorff, *Angew. Chem.,* **84**, 16 (1972).
22. F. Vögtle and P. Neumann, *Chem. Ztg.,* **97**, 600 (1973).
23. G. W. Gokel and H. D. Durst, *Synthesis,* **1976**, 168.
24. J. S. Bradshaw and J. Y. K. Hui, *J. Heterocycl. Chem.,* **11**, 649 (1974).
25. J. J. Christensen, D. J. Eatough, and R. M. Izatt, *Chem. Rev.,* **74**, 351 (1974).
26. D. Midgley, *Chem. Soc. Rev.,* **4**, 549 (1975).

27. F. Vögtle and M. Zuber, *Tetrahedron Lett.,* **1972**, 561.
28. E. Weber and F. Vögtle, *Chem. Ber.,* **109**, 1803 (1976).
29. L. R. Sousa and J. M. Larson, *J. Am. Chem. Soc.,* **99**, 307 (1977).
30. S. E. Drewes and B. G. Riphagen, *J. Chem. Soc., Perkin I,* **1974**, 323.
31. D. N. Reinhoudt and R. T. Gray, *Tetrahedron Lett.,* **1975**, 2105.
32. R. T. Gray, D. N. Reinhoudt, C. J. Smit, and I. Veenstra, *Rec. Trav. Chim. Pays-Bas,* **95**, 258 (1976).
33. M. Newcomb and D. J. Cram, *J. Am. Chem. Soc.,* **97**, 1257 (1975).
34. K. E. Koenig, R. C. Helgeson, and D. J. Cram, *J. Am. Chem. Soc.,* **98**, 4018 (1976).
35. M. A. McKervey and D. L. Mulholland, *J. Chem. Soc., Chem. Commun.,* **1977**, 438.
36. D. N. Reinhoudt and F. de Jong, unpublished results.
37. K. Frensch and F. Vögtle, *Tetrahedron Lett.,* **1977**, 2573.
38. J. M. Timko and D. J. Cram, *J. Am. Chem. Soc.,* **96**, 7159 (1974).
39. J. M. Timko, S. S. Moore, D. M. Walba, P. C. Hiberty, and D. J. Cram, *J. Am. Chem. Soc.,* **99**, 4207 (1977).
40. F. Wudl and F. Gaeta, *J. Chem. Soc., Chem. Commun.,* **1972**, 107.
41. S. J. Leigh and I. O. Sutherland, *J. Chem. Soc., Chem. Commun.,* **1975**, 414.
42. L. C. Hodgkinson, S. J. Leigh, and I. O. Sutherland, *J. Chem. Soc., Chem. Commun.,* **1976**, 639.
43. L. C. Hodgkinson, S. J. Leigh, and I. O. Sutherland, *J. Chem. Soc., Chem. Commun.,* **1976**, 640.
44. W. Wehner and F. Vögtle, *Tetrahedron Lett.,* **1976**, 2603.
45. G. R. Newkome and J. M. Robinson, *J. Chem. Soc., Chem. Commun.,* **1973**, 831.
46. M. Newcomb, G. W. Gokel, and D. J. Cram, *J. Am. Chem. Soc.,* **96**, 6810 (1974).
47. J. M. Timko, R. C. Helgeson, M. Newcomb, G. W. Gokel, and D. J. Cram, *J. Am. Chem. Soc.,* **96**, 7097 (1974).
48. G. W. Gokel, J. M. Timko, and D. J. Cram, *J. Chem. Soc., Chem. Commun.,* **1975**, 444.
49. J. M. Girodeau, J. M. Lehn, and J. P. Sauvage, *Angew. Chem.,* **87**, 813 (1975).
50. D. A. Laidler and J. F. Stoddart, *J. Chem. Soc., Chem. Commun.,* **1976**, 979.
51. G. R. Newkome, G. L. McClure, J. B. Simpson, and F. Danesh-Khoshboo, *J. Am. Chem. Soc.,* **97**, 3232 (1975).
52. G. R. Newkome, A. Nayak, G. L. McLure, F. Danesh-Khoshboo, and J. B. Simpson, *J. Org. Chem.,* **42**, 1500 (1977).
53. D. E. Fenton, D. H. Cook, and I. W. Nowell, *J. Chem. Soc., Chem. Commun.,* **1977**, 274.
54. E. Weber and F. Vögtle, *Tetrahedron Lett.,* **1975**, 2415.
55. F. Vögtle and R. Nätscher, *Chem. Ber.,* **109**, 994 (1976).
56. E. Weber, W. Wieder, and F. Vögtle, *Chem. Ber.,* **109**, 1002 (1976).
57. J. Dale and P. O. Kristiansen, *Acta Chem. Scand.,* **26**, 1471 (1972).
58. R. N. Greene, *Tetrahedron Lett.,* **1972**, 1793.
59. D. H. Busch, *Helv. Chim. Acta, Fasc. Extr. Alfred Werner,* **1967**, 174.
60. L. F. Lindoy, *Chem. Soc. Rev.,* **1975**, 421.
61. J. E. Richman and T. Atkins, *J. Am. Chem. Soc.,* **96**, 2268 (1974).
62. I. Tabuski, H. Okino, and Y. Kuroda, *Tetrahedron Lett.,* **1976**, 4339.

63. B. L. Shaw, *J. Am. Chem. Soc.,* **97**, 3856 (1975).
64. F. L. Cook, T. Caruso, M. Byrne, C. W. Bowers, D. H. Speck, and C. L. Liotta, *Tetrahedron Lett.,* **1974**, 4029.
65. R. G. Ackman, W. H. Brown, and G. F. Wright, *J. Org. Chem.,* **20**, 1147 (1955).
66. M. Chastrette and F. Chastrette, *J. Chem. Soc., Chem. Commun.,* **1973**, 534.
67. A. J. Rest, S. A. Smith, and I. D. Tyler, *Inorg. Chim. Acta,* **16**, L1 (1976).
68. D. N. Reinhoudt, R. T. Gray, F. de Jong, and C. J. Smit, *Tetrahedron Lett.,* **33**, 563 (1977).
69. N. S. Poonia, *J. Inorg. Nucl. Chem.,* **37**, 1855 (1975).
70. D. Live and S. I. Chan, *J. Am. Chem. Soc.,* **98**, 3769 (1976).
71. F. de Jong, D. N. Reinhoudt, and C. J. Smit, *Tetrahedron Lett.,* **1976**, 1371.
72. F. de Jong, D. N. Reinhoudt, and C. J. Smit, *Tetrahedron Lett.,* **1976**, 1375.
73. F. de Jong, D. N. Reinhoudt, C. J. Smit, and R. Huis, *Tetrahedron Lett.,* **1976**, 4783.
74. F. de Jong, D. N. Reinhoudt, and R. Huis, *Tetrahedron Lett.,* **1977**, 3985.
75. D. Leibfritz, *Tetrahedron Lett.,* **1974**, 4125.
76. D. A. Jaeger and R. R. Whitney, *J. Org. Chem.,* **40**, 92 (1975).
77. R. T. Gray, D. N. Reinhoudt, K. Spaargaren, and J. F. de Bruijn, *J. Chem. Soc., Perkin II,* **1977**, 206.
78. C. J. Pedersen, *J. Am. Chem. Soc.,* **92**, 391 (1970).
79. H. K. Frensdorff, *J. Am. Chem. Soc.,* **93**, 4684 (1971).
80. R. T. Gray and D. N. Reinhoudt, *Tetrahedron Lett.,* **1975**, 2109.
81. D. E. Fenton, M. Mercer, N. S. Poonia, and M. R. Truter, *J. Chem. Soc., Chem. Commun.,* **1972**, 66.
82. R. M. Izatt, R. E. Terry, B. L. Haymore, L. D. Hansen, N. K. Dalley, A. G. Avondet, and J. J. Christensen, *J. Am. Chem. Soc.,* **98**, 7620 (1976).
83. R. M. Izatt, D. P. Nelson, J. H. Rytting, B. L. Haymore, and J. J. Christensen, *J. Am. Chem. Soc.,* **93**, 1619 (1971).
84. S. Moore, T. L. Tarnowski, M. Newcomb, and D. J. Cram, *J. Am. Chem. Soc.,* **99**, 6398 (1977).
85. M. A. McKervey and D. L. Mulholland, *J. Chem. Soc., Chem. Commun.,* **1977**, 438.
86. N. S. Poonia, *J. Inorg. Nucl. Chem.,* **37**, 1859 (1975).
87. A. Pullman and A. M. Armbruster, *Chem. Phys. Lett.,* **36**, 558 (1975).
88. S. N. Vinogradov and R. H. Linnell, *Hydrogen Bonding,* Van Nostrand-Reinhold, New York, 1970.
89. Ref. 88, p. 121.
90. A. Kon-no and A. Kitahari, *J. Colloid Interface Sci.,* **34**, 221 (1970).
91. F. L. Slijko and R. S. Drago, *J. Am. Chem. Soc.,* **95**, 6935 (1973).
92. N. Nae and J. Jagur-Grodzinski, *J. Am. Chem. Soc.,* **99**, 489 (1977).
93. T. Nakabayashi, S. Kawamura, T. Horii, and M. Hamada, *Chem. Lett.,* 869 (1976).
94. R. D. Gree, *Hydrogen Bonding by C-H Groups,* Macmillan, London, 1974.
95. G. W. Gokel, D. J. Cram, C. L. Liotta, H. P. Harris, and F. L. Cook, *J. Org. Chem.,* **39**, 2445 (1974).
96. U.S. Patent 3997562 (December 14, 1976).
97. A. el. Basyomy, J. Klimes, A. Knöchel, J. Oehler, and G. Rudolph, *Z. Naturforsch.,*

B31, 1192 (1976).
98. R. D. McLachlan, *Spectrochim. Acta,* A30, 2153 (1974).
99. R. Kaufmann, A. Knöchel, J. Kopf, J. Oehler, and G. Rudolph, *Chem. Ber.,* 110, 2249 (1977).
100. I. Goldberg, *Acta Crystallogr.,* B31, 754 (1975).
101. I. Goldberg, *Acta Crystallogr.,* B33, 472 (1977).
102. N. A. Bailey and S. Chidlow, cited in Ref. 43.
103. F. P. Boer, M. A. Neuman, F. P. van Remoortere, and E. C. Steiner, *Inorg. Chem.,* 13, 2826 (1974).
104. F. P. van Remoortere and F. P. Boer, *Inorg. Chem.,* 13, 2071 (1974).
105. B. Metz, D. Moras, and R. Weiss, *J. Chem. Soc., Chem. Commun.,* 1970, 217.
106. P. D. Chadwick and S. Poonia, *Acta Crystallogr.,* B33, 197 (1977).
107. B. Metz, J. M. Rosalky, and R. Weiss, *J. Chem. Soc., Chem. Commun.,* 1976, 533.
108. M. A. Bush and M. R. Truter, *J. Chem. Soc. B,* 1971, 1440.
109. M. A. Bush and M. R. Truter, *J. Chem. Soc., Chem. Commun.,* 1970, 1439.
110. B. Metz, D. Moras, and R. Weiss, *J. Am. Chem. Soc.,* 93, 1806 (1971).
111. D. G. Parsons, M. R. Truter, and J. N. Wingfield, *Inorg. Chim. Acta,* 14, 45 (1975).
112. G. Bombieri, G. De Paoli, A. Cassol, and R. Imirzi, *Inorg. Chim. Acta,* 18, L23 (1976).
113. M. A. Neuman, E. C. Steiner, F. P. van Remoortere, and F. P. Boer, *Inorg. Chim. Acta,* 14, 734 (1975).
114. T. Iwachido, M. Kimura, and K. Tôei, *Chem. Lett.,* 1976, 1101.
115. H. L. Yeager and B. Kratochvil, *Can. J. Chem.,* 53,,3448 (1975).
116. M. Newcomb, J. M. Timko, D. M. Walba, and D. J. Cram, *J. Am. Chem. Soc.,* 99, 6392 (1977).
117. M. Newcomb, S. S. Moore, and D. J. Cram, *J. Am. Chem. Soc.,* 99, 6405 (1977).
118. W. D. Curtis, D. A. Laidler, and J. F. Stoddart, *J. Chem. Soc., Chem. Commun.,* 1975, 833.
119. D. A. Laidler and J. F. Stoddart, *J. Chem. Soc., Chem. Commun.,* 1977, 481.
120. P. B. Chock and E. O. Titus, *Prog. Inorg. Chem.,* 18, 287 (1973).
121. T. Funck, F. Eggers, and E. Grell, *Chimia,* 12, 637 (1972).
122. M. Shporer, H. Zemel, and Z. Luz, *FEBS Lett.,* 40, 357 (1974).
123. D. Haynes, *FEBS Lett.,* 20, 221 (1972).
124. G. W. Liesegang, M. M. Farrow, N. Purdie, and E. M. Eyring, *J. Am. Chem. Soc.,* 98, 6905 (1976).
125. G. W. Liesegang, M. M. Farrow, F. Aree Vazquez, N. Purdie, and E. M. Eyring, *J. Am. Chem. Soc.,* 99, 3240 (1977).
126. M. Shporer and Z. Luz, *J. Am. Chem. Soc.,* 97, 665 (1975).
127. E. Shchori, J. Jagur-Grodzinski, Z. Luz, and M. Shporer, *J. Am. Chem. Soc.,* 93, 7133 (1971).
128. E. Shchori, J. Jagur-Grodzinski, and M. Shporer, *J. Am. Chem. Soc.,* 95, 3842 (1973).
129. P. B. Chock, *Proc. Natl. Acad. Sci. USA,* 69, 1939 (1972).
130. Y. Kobuke, K. Hanji, K. Horiguchi, M. Asada, Y. Nakayama, and J. Furukawa, *J. Am. Chem. Soc.,* 98, 7414 (1976).

References

131. E. V. Dehmbov, *Angew. Chem.*, **89**, 521 (1977).
132. I. O. Sutherland, *Ann. Rep. NMR Spectrosc.*, **4**, 71 (1971).
133. F. de Jong, D. N. Reinhoudt, and R. Huis, to be published.
134. I. Goldberg, *Acta Crystallogr.*, **B31**, 2592 (1975).

CHAPTER FIVE
IONOPHORES
BIOLOGICAL TRANSPORT MEDIATORS

S. LINDENBAUM, J. H. RYTTING, and L. A. STERNSON

Department of Pharmaceutical Chemistry
University of Kansas
Lawrence, Kansas

1 Introduction 220
   1.1 Brief History, 220
   1.2 Structural Features, 221
   1.3 Chemical Applications, 224
   1.4 Biological Applications, 225
2 Complexation 228
   2.1 Neutral Ionophores, 229
   2.2 Charged Ionophores, 231
   2.3 Cation Selectivity, 237
3 Theory of Membrane-Transport Phenomena 239
   3.1 Theoretical Model for Carrier-Mediated Permeation, 241
   3.2 The Role of Ion-Transport Mediators in Energy Coupling, 245
References 250

1 INTRODUCTION

Ionophores are organic molecules that form specific complexes with metal cations and certain hydrophilic organic cations, rendering them lipophilic and providing a means for their transport across apolar barriers. Thus their importance rests with their ability to selectively transport cations across lipid membranes.

Biological applications of ionophores have been the subjects of several reviews (1-5, 37). The contents of some of these are summarized in this article. However, the major emphasis of this chapter is on the factors contributing to the specific ion permeability of membranes mediated by ionophores. This includes discussions of complexation and membrane transport phenomena.

1.1 Brief History

The unique transporting properties of ionophores were first recognized when it was shown that valinomycin, a Streptomyces fermentation product (6), induces highly selective energy-dependent uptake of potassium ion (but not sodium ion) by mitochondria (7, 8), thus interfering with cellular processes. Pressman (9) showed that ionophores (valinomycin and nigericin) act as mobile cation carriers that are capable of transporting ions (i.e., controlling a dynamic process) through a lipid phase. This property suggested that ionophores must be capable of forming reversible, lipid-soluble complexes with these cations. Harned and co-workers (10) first demonstrated the ability of ionophores to complex with alkali metal cations, by observing that in the presence of nigericin, such cations are soluble in hexane. The isolation of ionophores from biological systems has recently been reviewed by Shamoo and Ryan (11). They present evidence for (a) the isolation of cation-specific ionophores from $Na^+ + K^+$-ATPase, (b) the isolation of a Ca^{2+}-dependent ionophore from sarcoplasmic reticulum $Ca^{++} + Mg^{++}$-ATPase, and (c) the occurrence of an ionophoric protein associated with the cholinergic receptor (12).

The ion selectivity of ionophores varies. The ionophore A-23187 displays a high affinity for divalent cations, but does not effectively complex with monovalent cations (13). The high specificity of this ionophore is contrasted by the broad-spectrum complexing ability of the carboxylic acid ionophore, X-537A. In apolar media X-537A forms 1:1 (2) or 2:1 (14, 15) complexes with monovalent cations and also forms neutral (2:1) complexes with divalent cations (16, 17) [most notably Ca^{2+}]. This complexation enables the conductance of these ions across lipid barriers (2, 3, 18). Among the cations that are amenable to such complexation (2, 19) and transport (2) are the protonated phenethylamines (particularly catecholamines). X-537A initiates many physiological responses that are apparently caused by its ability to facilitate Ca^{2+} ion and cate-

1 Introduction

cholamine transport across lipid membranes. These properties are only now being exploited to elicit pharmacologically and therapeutically desired responses.

Hotchkiss (20) found that one of the gramicidins suppresses the synthesis of ATP in bacterial and animal systems without interfering with electron transport. Since that time a large family of molecules has been found that uncouple ATP-requiring processes without affecting electron flow. Most recently Green (21) has postulated a role of ionophores in dissociating electron flow from the synthesis of ATP in mitochondria, thus acting as uncouplers of oxidative phosphorylation processes.

1.2 Structural Features

Ionophores are structurally diverse molecules with certain common characteristics. They have a large hydrocarbon backbone incorporating functional groups (e.g., oxygen-containing moieties) with electron-donating ability. The neutral oxygen atoms complex with cations through ion-dipole interactions, resulting in a perturbation of the ionophore conformation. The cation is centrally located, and the polar nucleophilic moieties are oriented toward the interior of the resulting complex, creating a hydrophilic, electron-rich environment; the hydrocarbon backbone is exposed to the exterior. The cation is thereby encased in a complex having a lipophilic exterior, conferring lipid solubility on it. Complexation appears to take place by concerted exchange of the hydration shell of the cation with the ionophoric nucleophilic system (1). The stability of the complex ($\Delta G_{\text{complex}}$) is described by Eq. 1 as the sum of several energy terms (3, 4, 22, 23):

$$\Delta G_{\text{complex}} = \Delta G_{\text{dehydration}} + \Delta G_{\text{conform}} - \Delta G_{\text{ligand}} \qquad (1)$$

(*a*) the energy required to desolvate the cation ($\Delta G_{\text{dehydration}}$), (*b*) the energy required to convert the conformation of the ionophore from its free form to the form assumed for complexation ($\Delta G_{\text{conform}}$), and (*c*) the energy released in cation-ionophore solvation as a result of the ligand-formation process (ΔG_{ligand}). Thus variations in the value of these terms for various cations result in the ability of ionophores to discriminate among different cations.

For ionophores with flexible backbones cation solvation by ionophores is primarily governed by the charge density of the ion (3, 4) in a manner similar to ion solvation by normal solvents. Complexation by such ligands is relatively nonspecific. For ionophores with inflexible backbones, however, the complexation energy is dependent on the ionic radius of the participating ions (7, 24, 25). This dependency introduces a much greater degree of specificity into the complexation process, since a "form fit" determines the net free energy difference between ion dehydration ($\Delta G_{\text{dehydration}}$) and ionophoric solvation (ΔG_{ligand}).

Ionophores can be classified on the basis of structural features, which in turn determine the mechanism by which they facilitate ion conductance, as (*a*) neutral, (*b*) charged, and (*c*) channel-forming ionophores. Neutral ionophores contain no ionizable functional groups. The complex that forms between such compounds and metal ions results from ion-dipole interactions when the cation encounters an apolar medium containing ionophore. The water of solvation

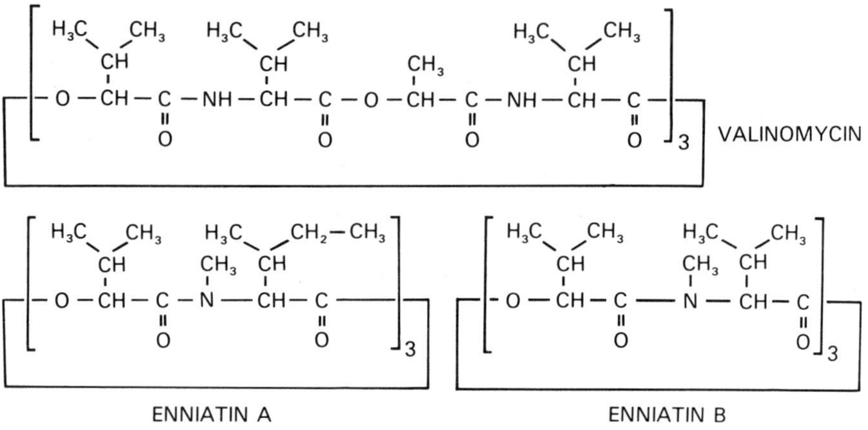

Figure 1 Structure of depsipeptide antibiotics.

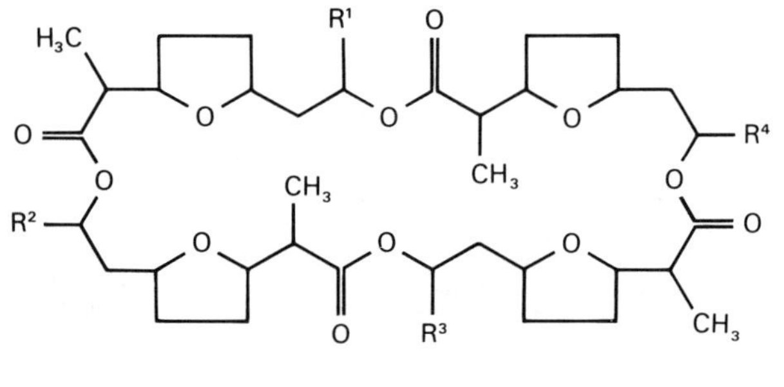

$R^1 = R^2 = R^3 = R^4 = CH_3$		NONACTIN
$R^1 = R^2 = R^3 = CH_3$	$R^4 = C_2H_5$	MONACTIN
$R^1 = R^3 = CH_3$	$R^2 = R^4 = C_2H_5$	DINACTIN
$R^1 = CH_3$	$R^2 = R^3 = R^4 = C_2H_5$	TRINACTIN
$R^1 = R^2 = R^3 = R^4 = C_2H_5$		TETRANACTIN

Figure 2 Structure of macrotetralides.

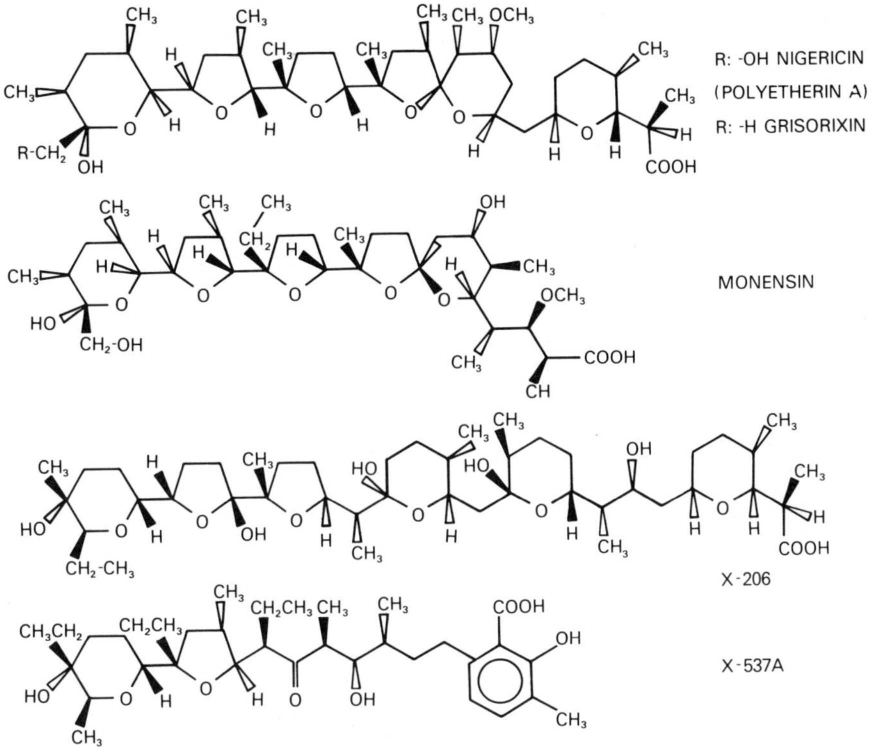

1 Introduction

Figure 3 Structure of the antibiotics of the nigericin group.

about the cation is replaced by oxygens (e.g., six carbonyl ester oxygens in valinomycin) of the ionophore in a concerted reaction. The resulting complex assumes the charge of the cation. When the complex then encounters a hydrophilic environment (e.g., at the other side of the membrane), decomplexation occurs as the reverse reaction of complexation. Among the ionophores included in this group are valinomycin (Figure 1), macrotetralide nactins (Figure 2) (produced by Actinomyces strains) (29), synthetic polyethers (28, 29), and certain cyclic polypeptides [cyclo-(L-N-Me-Ala-Sar-)$_3$ (30), cyclo-(L-Pro-Gly-)$_3$ (31)].

The charged ionophores (Figure 3) are open-chain compounds that have oxygen-containing heterocyclic aromatic rings incorporated in their structure with a terminal charged head group (carboxylic acid) and a tail capable of participating in hydrogen bonding [usually a hydroxyl group or an amino ($\overset{R}{\underset{R}{\diagdown}}$N–H) tail in the case of A-23187 (1)]. These molecules also contain additional functional groups for cation complexation, including ether, carbonyl, and

hydroxyl moieties. The carboxylic acid group must be ionized for ionophoric activities (3). In the protonated form the ionophore conducts hydrogen ions across apolar barriers. Complexation with cations requires ionophore cyclization involving heat-to-tail interaction, through hydrogen bonding. The cyclic structures are stabilized by the rigidity of the heterocyclic rings. In most cases ion-pair formation appears to account for initiation of complexation (10, 31), followed by cyclization of the ionophore and stabilization of the complex through ion-dipole interactions with oxygen-containing substitutents on the ionophore. In some instances, however, electrostatic interaction of the metal ion with the carboxylate anion does not appear to contribute to complexation (e.g., monensin) (32). The structures of the complexes are such that there is internal compensation of charge (3).

Channel-forming ionophores differ from other ionophores in that they only complex with metal ions in the presence of a membrane. This class of ionophores is most notably represented by the gramicidins (e.g., gramicidin A, which is a pentadecapeptide). These molecules dimerize in the lipid membrane to form channels 25-30 Å long and about 4 Å in diameter, which act as ion-conducting channels through the membrane (33-36). The interior of these channels is lined with a series of oxygen-containing functionalities that form an electron-rich field which stabilizes cations.

1.3 Chemical Applications

The ability of certain macromolecules to complex with various cationic species has found use in several types of chemical systems. Ionophores, specifically crown ethers, have found use in organic synthesis. These compounds have been used to dissolve reactive inorganic salts in low-polarity solvents. The coordination of the cation with ionophore results in a large increase in the activity coefficient of the anion and hence in its reactivity (37). Crown ethers have been used to solubilize potassium hydroxide, permanganate, and borohydride (38).

1 Introduction

[Structure 2: binaphthyl with O-CH₂-CH₂ and O-(CH₂CH₂O)ₙ substituents, and R group]

Chiral polyethers (e.g., **2**) have been used to resolve racemic mixtures of optically active cationic species (39, 40). In such cases the resulting complexes arising from interactions of enantiomers have different physical properties (i.e., diastereomeric products) that facilitate separation.

The selectivity of neutral carrier complexes toward alkali metal ions has been utilized in the design of ion-selective electrodes. Potassium-ion-selective electrodes are commercially available which employ a valinomycin-containing membrane separating the test solution from a reference solution of constant potassium ion activity (41, 42). The high selectivity of valinomycin for K^+ over Na^+ makes this K^+-responsive electrode a significant improvement over the K^+ glass membrane electrode (Beckman 39049) (selectivity coefficient K_{K^+,Na^+} = 0.01 for the glass electrode; for the valinomycin membrane electrode K_{K^+,Na^+} = 2.5 × 10^{-4}) (41, 43). Other, less favorable, K^+-ion-selective electrodes have been described using nonactin and monactin as the neutral carrier present in the membrane (44). The effectiveness of the neutral carrier electrode is based on the ability of certain macrocyclic antibiotics to specifically complex with particular ions. Attempts to synthesize artificial carriers for use in Ca^{2+}-ion-selective electrodes has only met with moderate success to date based on measured selectivity coefficients: $K_{Ca,K} = 0.1; K_{Ca,Na} = 0.01; K_{Ca,Mg} = 3.3 \times 10^{-5}$ (45).

1.4 Biological Applications

Much of the current work with ionophores involves the carboxylic acid derivatives X-537A and A-23187. The physiological effects associated with X-537A (1, 3, 46) and A-23187 (13) in large part result from their ability to complex with and transport Ca^{2+} ion and catecholamines across lipid membranes. Both X-537A and A-23187 induce muscle contraction by release of Ca^{2+} from vesicles associated with sarcoplasmic reticulum (47-49), which in turn activates ATPase activity of myosin.

The myocardial effects of X-537A have recently been reviewed by Pressman (2, 3). The effects are for the most part typical of an adrenergic agent, and may result from the ability of this ionophore to facilitate membrane transport of norepinephrine. Ionotropic effects and the observed drop in peripheral resistance associated with administration of X-537A were prevented by β-adrenergic

Table 1 Some Physiological Secretory Responses Associated with Ionophores

Ionophore	Ca^{2+} Dependency	Secretory Process	Ref.
X-537A	Yes	Release of taurine, glycine, and γ-amino-butyrate in chick retina	54
X-537A	Yes	Acetylcholine release	56, 57
X-537A	No	Catecholamine transport	2, 57, 58
X-537A	No	Release of histamine from mast cells	59
A-23187	Yes	Release of histamine from mast cells	60
X-537A; A-23187	Yes	Serotonin transport	62
A-23187	Yes	Fluid secretion from fly salivary glands	63
A-23187	Yes	Release of digestive enzymes	64
X-537A	Yes	Release of insulin from pancreas	65

X-537A	Yes	Vasopressin release from rat neurohypophysis	66
A-23187	Yes	Thyroxin release from thyroid	67
X-537A; A-23187	Yes	Rupture of cortical granules in sea urchin eggs and erection of fertilization membranes	68
A-23187	No	α-Amylase and lactate dehydrogenase release from dissociated pancreatic acinar cells	69
X-537A; A-23187	a	Release of endogenous amino acids (asparate, glutamate, glutamine, alanine, taurine, GABA, and glycine) from rat visual cortex	70
A-23187	Yes	Release of platelet cyclic 3':5'-nucleotide phosphodiesterase	71

[a] X-537A and A-23187 caused the non-Ca^{2+}-ion-dependent release of glutamate, taurine, and GABA. With X-537A, Ca^{2+} ion stimulated the ionophoric release of glutamate, asparate, taurine, and GABA. With A-23187, Ca^{2+} ion caused additional release of glutamate, taurine, GABA, and glycine.

blockers (e.g., propranolol). The ionophore also decreased coronary resistance, resulting in a tenfold increase in blood flow through coronary arteries. This effect was not prevented by adrenergic blockers and thus appears to operate via a different mechanism (3, 50).

Topical administration of X-537A or A-23187 produces a significant (about 25%) increase in intraocular pressure, which may result from Ca^{2+} ion uptake followed by aqueous humor production (51).

These ionophores (X-537A and A-23187) also stimulate the release of neurotransmitters from storage sites. This effect is in most cases triggered by ionophore-mediated transport of Ca^{2+} ion. Neurotransmitters that are affected by ionophores are listed in Table 1; they include acetylcholine, catecholamines, serotonin, and histamine. In addition these ionophores stimulate other physiological responses via stimulation of secretory processes; these too are presented in Table 1. In most cases Ca^{2+} ion transport by ionophore is responsible for initiation of these responses.

Ionophores [valinomycin, macrotetralide actins, enniatins, gramacidins, and antamanide (3)] have been shown by direct isotopic flux measurements (52)

$$\begin{bmatrix} \text{L-Ala-L-Phe-L-Phe-L-Pro-L-Pro} \\ \text{L-Pro-L-Pro-L-Val-L-Phe-L-Phe} \end{bmatrix}$$

3

to increase mitochondrial membrane permeability to potassium ion (21, 24, 53). This property results in the ability of these compounds to uncouple mitochondrial oxidative phosphorylation. The increased permeability to potassium ion results in energy-consuming nonproductive cyclical cation transport (5, 21, 53). The energy source for K^+ transport is a high-energy intermediate in the oxidative phosphorylation sequence [but not ATP (8)]. The uncoupling effect appears to result from dissipation of this energized intermediate.

Thus the biological responses associated with ionophores involve either their calcium ion and catecholamine transporting ability or their ability to interfere with mitochondrial oxidative phosphorylation by increasing membrane permeability to K^+ ion.

2 COMPLEXATION

The ability of ionophores to mediate the transport of metal cations and some hydrophilic organic cations is often attributed, at least in part, to their complexation tendencies with these cations. Ionophores have been shown to bind with a wide variety of cations, including various amines, and in a few cases with anions. Christensen and co-workers (72) suggest that the following factors determine the formation and stability of ion-ionophore complexes: (*a*) the type(s) of binding sites in the ring, (*b*) the number of binding sites in the ring, (*c*) the relative sizes of the ion and the macrocyclic cavity of the macrocyclic compounds, (*d*) the physical placement of the binding sites, (*e*) steric hindrance in the ring, (*f*) the solvent and extent of solvation of the ion and the binding sites, and (*g*) the electrical charge of the ion. Simon and Morf (4) have indicated that for many 1:1 complexes of monovalent metal cations with carrier antibiotics, (*a*) the metal ion is coordinated by five to eight oxygen atoms, (*b*) the external surface of the complex is lipophilic, (*c*) the alkali metal cation is not hydrated, and (*d*) the carrier molecules have sufficient flexibility to permit stepwise substitution of the solvent molecules, allowing for a low activation energy and fast reaction rates for complexation. These properties make it possible for carrier molecules to have a high degree of selectivity in binding various cations.

In this section we review recent studies of complexation by ionophores, but do not attempt to repeat those data that are presented in other reviews. A variety of methods have been used to study complexation of ionophores. These are thoroughly reviewed by Ovchinnikov and co-workers (37).

2.1 Neutral Ionophores

The neutral ionophores include valinomycin, enniatins, macrotetralide nactins, various synthetic polyethers, cryptates, and some cyclic hexapeptides and decapeptides. Valinomycin has probably been studied more than other neutral antibiotic ionophore because of its early discovery and its ability to discriminate so effectively between Na^+ and K^+. The structures of these substances in solution, both in the uncomplexed state and as complexes, have been determined by a variety of methods (37). These studies have shown considerable differences between the complexes and uncomplexed structures, providing effective methods for measuring the nature and extent of complexation. With K^+ ion valinomycin has been found to coordinate with six oxygen atoms with an approximately octahedral distribution (4). K^+ ion coordinates with eight oxygens in a cubic arrangement with nonactin and with eight oxygens in enniatin B. Ag^+ has been found to coordinate with five oxygen atoms in nigericin and grisorixin and with six oxygen atoms in monensin. Simon and Morf (4) report a number of complex-formation constants for alkali metal ion interaction with a number of carrier antibiotics in methanol and ethanol. The complexation constants reported range from about 100 to 1,000,000.

A considerable literature exists describing complexation studies between cations and synthetic macrocyclic polyethers and related compounds. In general the uncomplexed polyethers appear to have relatively unordered conformations

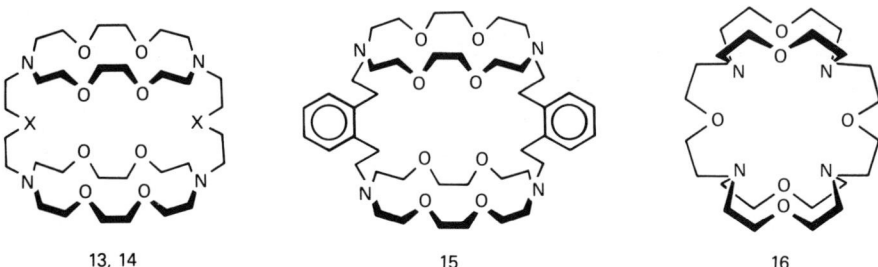

Figure 4 Structures of several cryptate ligands: (**4**) $m = 0$; $n = 1$ [2.1.1]; (**5**) $m = 1$; $n = 0$ [2.2.1]; (**6**) $m = n = 1$ [2.2.2]; (**7**) $m = 1$; $n = 2$ [3.2.2]; (**8**) $m = 2$; $n = 1$ [3.3.2] (**9**) $m = n = 2$ [3.3.3]; (**10**) $m = 1$, (third bridge = $-(CH_2)_8$-[2.2.C_8]; (**11**) X = CH_3; (**12**) X = $CH_2CH_2OCH_2CH_2OCH_3$; (**13** X = CH_2; (**14**) X = 0.

compared with the ordered conformations observed in their metal complexes. Typically the uncomplexed cyclic polyethers have some of their oxygen atoms pointed outward and the rings have an elliptical shape (73).

Metal-cyclic polyether complexation has been classified by Dalley (73) into four groups based on the relative position of the cation to the donor atoms of the polyether. First, the cations fit well in the cavity of the ligand. Second, the cation is too large to fit in the ligand cavity. Third, the cation is smaller than the ligand cavity, but complexation still occurs either by the ligand wrapping around the cation or by two cations sharing the cavity. Fourth, the cation coordinates with only some of the available donor atoms. Examples of each of these situations have been reported (73). An extensive review (37) of structures of both complexed and uncomplexed ionophores has been published. A thorough compilation of association constants and related thermodynamic data has also appeared (72).

Lehn and co-workers (74-76) have synthesized a number of macrobicyclic ligands or cryptate compounds (see Figure 4 for examples) and studied their complexation properties with alkali and alkaline earth cations. In these cases the metal ion is enclosed in the bicyclic ligand, and coordination occurs to all the nitrogen and oxygen atoms (73). The conformation changes slightly with various cations to accommodate larger or smaller ions. Stability constants and selectivities for a number of these cryptate-metal ion complexes in water and hydroalcoholic solvents have been reported (74-76), and several are included in Table 2.

A few investigators have studied the complexation of ammonium and alkylammonium ions with cyclic polyethers. Izatt and co-workers (77) have reported the thermodynamic quantities (log K, ΔH, and ΔS values) for the complexation of 18-crown-6 with several inorganic, aliphatic, and aromatic ammonium ions in methanol. These values are given in Table 3. Chain length appears to have little if any effect on the thermodynamics of binding. As the number of hydrogen atoms available for hydrogen bonding decreases, both the enthalpy change and equilibrium constant decrease markedly. This decrease is consistent with the suggestion (78) that complexation with alkylammonium ions involves hydrogen bonding between the protons on the ammonium ion and the cyclic polyether oxygen atoms. The reduction in the enthalpy change would also be expected with fewer hydrogen atoms available for hydrogen bonding. Increased chain branching decreased the complex stability, probably because of steric factors. Ortho substitution on aromatic amines was also found to decrease the complexation constants. Again changes in ΔH seem to dominate.

Cram and co-workers (79-81) have examined the interaction of *tert*-butyl ammonium thiocyanate with several cyclic polyethers and their derivatives in chloroform. They observed large differences in equilibrium constants among the various ligands. The most stable complexes were formed with the cyclic polyether, 18-crown-6.

2.2 Charged Ionophores

The class of charged ionophores includes nigericin, grisorixin, dianemycin (X-206), X-537A, A-23187, and monensin. This class of ionophores is characterized by a carboxylic acid head group and a tail (usually a hydroxyl group or an amino group) capable of hydrogen-bonding interactions. The backbones are linear, containing heterocyclic rings with a variety of oxygen-containing functional groups such as ethers, carboxyl, hydroxyl, and carbonyl. To function effectively the carboxylic acid group must be ionized, and the final complex often will carry no charge (3).

Complexation with cations typically involves cyclization of the ionophore through hydrogen bonding of the head and tail. The resulting cyclic structures are stabilized by the rigid heterocyclic rings. In most cases ion-pair formation through electrostatic interaction of the cation with the carboxylate ion appears to initiate the complexation. Detailed examples and discussions of the structural features of these ionophores and their various complexes are given in Ref. 37.

X-537A in particular has received a great deal of attention, and the thermodynamics and kinetics for metal ion binding have been studied in some detail

Table 2 Stability Constants (log K) of Alkali and Alkaline-Earth Metal Cations with Cryptates at 25° C[a]

Ligand[b]	Solvent	Log K_s with Cation[c]							
		Li^+	Na^+	K^+	Rb^+	Cs^+	Mg^{2+}	Ca^{2+}	Sr^{2+}	Ba^{2+}
4	W	5.5	3.2	<2.0	<2.0	<2.0	2.5 ± 0.3	2.50	<2.0	<2.0
	M/W	7.58	6.08	2.26	<2.0	<2.0	4.0 ± 0.8	4.34	2.90	<2.0
	M	>6.0	6.1	2.3	1.9	<2.0	—	—	—	—
5	W	2.50	5.40	3.95	2.55	<2.0	<2.0	6.95	7.35	6.30
	M/W	4.18	8.84	7.45	5.80	3.90	<2.0	9.61	10.65	9.70
	M	>5.0	>8.0	>7.0	>6.0	~5.0	—	—	—	—
6	W	<2.0	3.9	5.4	4.35	<2.0	<2.0	4.4	8.0	9.5
	M/W	1.8	7.21	9.75	8.40	3.54	<2.0	7.60	11.5	12 ± 0.7
	M	2.6	>8.0	>7.0	>6.0	4.4	—	—	—	—
7	W	<2.0	1.65	2.2	2.05	2.0	<2.0	~2.0	3.4	6.0
	M/W	<2.0	4.57	7.0	7.30	7.0	<2.0	4.74	7.06	10.40
	M	2.3	4.8	>7.0	>6.0	>6.0	—	—	—	—
8	W	<2.0	<2.0	<2.0	<0.7	<2.0	<2.0	~2.0	~2.0	3.65
	M	—	3.2	6.0	6.15	>6.0	—	—	—	—
9	W	<2.0	<2.0	<2.0	<0.5	<2.0	<2.0	<2.0	<2.0	—
	M	—	2.7	5.4	5.7	5.9	—	—	—	—

10	M/W	—	3.00	4.35	—	—	—	—	—
	M	<2.0	3.5	5.2	3.4	2.7	—	—	<2.0
11	M/W	—	3.26	4.38	—	—	4.4	—	6.7
	M	—	3.7	5.3	4.3	—	—	6.1	—
12	M/W	—	3.35	4.80	—	—	—	—	—
13[d]	M/W(1)	—	3.2	4.0	3.5	3.5	—	—	6.7
	(2)	—	1.5	3.2	3.0	2.5	—	—	6.3
14[d]	M/W(1)	—	3.6	4.8	3.7	4.4	4.0	5.5	4.4
	(2)	—	3.2	3.9	3.3	3.0	—	5.5	5.9
	W(1)	—	<1.5	~1.5	~1.5	—	—	3.5	6
15[d]	M/W(1)	—	3.0	3.6	3.0	—	3.6	4.9	8.0
	(2)	—	2.9	2.7	2.8	—	—	5	—
16	W	—	1.7	1.1	1.0	1.45	6.53	6.97	—
	M	—	4.5	5.85	6.2	<6.0	—	—	—

[a] W refers to water, M/W refers to 95% methanol–5% water mixtures, and M refers to methanol.
[b] The numbers for the ligand correspond to the numbered structure in Figure 4.
[c] The values reported for ligands **4-12** are taken from Ref. 75. Those for ligands **13-16** are from Ref. 76.
[d] (1) refers to the 1:1 complex; (2) refers to the stepwise stability constant for the 2:1 complex (i.e., $K_2 = [M_2L]/[ML][M]$ where M refers to the cation concentration and L to the ligand concentration).

Table 3 Log K, ΔH, and ΔS Values for the 1:1 Reaction of Several Amine Cations with 18-Crown-6 in CH_3OH at 25°C as a Function of Several Structural Parameters of the Cations (77)

Cation	Log K	ΔH (kcal mole^{-1})	ΔS (cal deg^{-1} mole^{-1})
		Chain Length	
NH_4^+	4.21	−9.27	−11.8
$CH_3NH_3^+$	4.25	−10.71	−16.5
$CH_3CH_2NH_3^+$	3.99	−10.65	−17.5
$CH_3CH_2CH_2NH_3^+$	3.97	−10.06	−15.6
		Hydrogen Bonding Capability	
$CH_3CH_2NH_3^+$	3.99	−10.65	−17.5
$(CH_3)_2NH_2^+$	1.76	−6.67	−14.3
$(CH_3)_3NH^+$		No heat of reaction observed	

		Steric Bulk	
$CH_3CH_2CH_2NH_3^+$	3.97	−10.06	−15.6
$(CH_3)_2CHNH_3^+$	3.56	−9.65	−16.1
$(CH_3)_3CNH_3^+$	2.90	−7.76	−12.8
		Steric Tie back	
$\underset{CH_2}{\overset{CH_2}{\triangle}}CHNH_3^+$	3.90	−10.12	−16.1
$(CH_3)_2CHNH_3^+$	3.56	−9.65	−16.1
		Aromatic Substitution	
ϕNH_3^+	3.80	−9.54	−14.6
$m,m\text{-}(CH_3)_2\phi NH_3^+$	3.74	−9.07	−13.3
$o\text{-}(CH_3)\phi NH_3^+$	2.86	−7.59	−12.4
$o,o\text{-}(CH_3)_2\phi NH_3^+$	2.00	−5.65	−9.8

(82-84). Stability constants for a number of uni- and divalent metal ions in hexane and methanol have been reported (82-84). Although A-23187 has been demonstrated (13, 85) to be effective in transporting divalent cations across bilogical membranes, the thermodynamic properties associated with complexation have not been studied in detail.

X-537 A has also been shown to complex with and transport biogenic amines such as norepinephrine and epinephrine across vesicle membranes (2). Equilibrium constants for complexation in a solvent mixture containing 70% toluene and 30% n-butanol yielded values of 163 and 9.8 for norepinephrine and epinephrine, respectively (2). More recently Kafka and Holz (86) reported that X-537A also carried dopamine across lipid bimolecular membranes. Although A-23187 is an effective carrier for Ca^{2+}, it appears to be ineffective as an ionophore for dopamine.

The solution conformation of X-537A complexes has been the subject of several investigations. CD measurements (84, 87) are consistent with the open-straight-chain conformation in polar solvents. The closed ring structure predominates in nonpolar hydrocarbon solvents and for the metal-ion-complexed species. These data also support a proposed mechanism of transport (88) by which the carrier metal ion complex is formed at the membrane-solution interface, is stable in a nonpolar medium and can transport metal ions as complexes through the lipid membrane, and dissociates to release the metal ion at the membrane-polar solution interface. The ionophore remains in the lipid membrane, in its lipophilic circular structure, and may transport back through the membrane. The ion-carrying ability of these substances therefore depends on strong complexation in the lipid membrane phase with decreasing stability of the complex in more polar media. Support for this hypothesis is found in a recent complexation study by Lindenbaum and co-workers (19).

Complexation constants for several biogenic amines with X-537A were measured in pure methanol, 2-octanol, and isooctane (19). Complexation of these amines did not appear to occur in methanol, in contrast with metal ions, which have been shown to complex with X-537A in methanol. The results of this study are shown in Table 4. Solvent polarity is seen to be very important in the binding of epinephrine and norepinephrine in that the complexation constants for these amines decrease by nearly 2 orders of mganitude in 2-octanol compared to isooctane. The binding of amphetamine, phenylephrine, and dopamine were less affected by solvent polarity.

Westley and co-workers (89) have found that a number of primary biogenic amines form crystalline complexes with X-537A, and used this property to resolve optical isomers of some of the amines. Similar results were obtained by Cram and co-workers (90-92) using synthetic crown ethers. Shen and Patel (93) have examined the complexation between X-537A and several amines in chloroform solution using proton NMR. They concluded that the X-537A backbone

Table 4 Complexation Constants (1 mole^{-1}) for X-537A
(lasalocid) with Biogenic Amines at 37°C (19)

Amine	Solvent	
	Isooctane	Octan-2-ol
Epinephrine	1.6×10^5	6.1×10^3
Norepinephrine	3.4×10^5	4.7×10^3
Amphetamine	2.1×10^5	3.6×10^4
Phenylephrine	6.2×10^3	2.7×10^3
Dopamine	1.1×10^4	2.2×10^3

conformation and primary amine binding rates in these complexes were similar to those found for alkali and alkaline earth complexes for X-537A in solution. They also found that nonpolar interactions between X-537A and the biogenic amine side chain enhance the stability of the complex in chloroform.

2.3 Cation Selectivity

One of the most striking features of many of the ionophores is their ability to complex selectively with various cations. This ability has provided a basis for some ion-selective electrodes as well as preferential transport of certain cations across membranes.

The relative selectivities of various ionophores toward alkali and alkaline earth metal ions have been reviewed by Pressman (3) and Ovchinnikov and co-workers (37). A summary based on these reviews is shown in Table 5.

A number of factors influence cation selectivity. For an inorganic cation the following factors have been suggested (77): charge, diameter (size), type or electronic structure, and associated ions or counterions. With respect to organic cations the type of ion (e.g., ammonium, diazonium, pyridinium, sulfonium), the possibilities for hydrogen bonding (number of sites and availability), electronic effects, steric considerations (e.g., type, location, and size of substituents), and isomeric form (e.g., cis or trans, optical) all have been shown (77) to influence the binding to macrocyclic compounds. With regard to the ligand, cavity size, donor atom type and number, ring conformation, and ring substituents are all important. The reaction environment can also have a significant influence, particularly with respect to solvent type and polarity as well as temperature.

Size relationships appear to be very important, particularly with respect to the

Table 5 Cation Selectivities of Several Ionophores (3, 37)

Ionophore	Selectivity Sequence
Neutral ionophores	
Valinomycin	Rb > K > Cs > Ag > Tl >> NH_4 > Na > Li
	Ba > Ca > Sr > Mg
Enniatin A	K > Rb ~ Na > Cs >> Li
Enniatin B	Rb > K > Cs > Na >> Li
	Ca > Ba > Sr > Mg
Beauvericin	Rh > Cs > K >> Na > Li
Nonactin	NH_4 > K ~ Rb > Cs > Na
Monactin	NH_4 > K > Rb > Cs > Na > Ba
Dinactin	NH_4 > K ~ Rb > Cs > Na > Ba
Trinactin	NH_4 > Rb > Na > Cs
Cryptate 211	Li > Na > K ~ Rb ~ Cs
	Ca > Sr ~ Ba
Cryptate 221	Ag > Tl > Na > K > Li ~ Rb > Cs
	Sr > Ca > Ba > Mg
Cryptate 222	Ag > Tl > K > Rb > Na > Cs ~ Li
	Ba > Sr > Ca > Mg
Antanamide	Na > Li > Tl > K > Rb > Cs
Dicyclohexyl 19-crown-6	Ag > K > Rb > NH_4 > Cs > Na > Li
	Ba > Sr > Mg > Ca
Dicyclohexyl 14-crown-4	Na > K > Cs
15-Crown-5[a]	Tl > Ag > Cs ~ K ~ Na ~ Rb
18-Crown-6[a]	Tl > K > Rb ~ Ag > Cs > Na
18-Crown-6[b]	Pb > Ba > Sr > K > Rb > Cs ~ Na
Benzo-15-crown-5[b]	Pb > Na ~ Rb > K
Dibenzo-24-crown-8[b]	Rb ~ Cs ~ K > Na
Dibenzo-27-crown-9[b]	K > Na ~ Cs
2,6-Dioxo-18-crown-6[c]	Ba > K > Na
2,4-Dioxo-19-crown-6[c]	K > Na > Ba
Charged ionophore	
Monensin	Na >> K > Rb > Li > Cs
Nigericin	K > Rb > Na > Cs >> Li
Dianemycin	Na > K > Rb ~ Cs > Li
X-206	K > Rb > Na > Cs > Li
X-537A	Cs > Rb ~ K > Na > Li
	Ba > Sr > Ca > Mg
A-23187	Li > Na > K
	Mn > Ca > Mg > Sr > Ba

Table 5. Continued

Ionophore	Selectivity Sequence
Channel-forming ionophores	
Gramicidin A	$H^- > Cs \sim Rb > NH_4 > K > Na > Li$
Alamethecin	$K > Rb > Cs > Na$
Monazomycin	$Cs > Rb > K > Na > Li$
Polyene antibiotics	Low selectivity

[a] Reference 94; solvent, water.
[b] Reference 95; solvent, 70% methanol in water.
[c] Reference 96; solvent, methanol.

rigid synthetic cyclic polyethers. In general optimum stability occurs when the cation diameter is approximately the same as the diameter of the ligand cavity (77). The sensitivity of equilibrium constants to such a diameter ratio appears to be greater for the bivalent cations compared with univalent cations. Even greater correspondence of cavity size to complex stability is observed for the selectivity of macrobicyclic (cryptand) ligands for alkali metal ions. For example, cryptand [2.1.1] (cavity diameter = 1.66 Å) has a maximum complexation constant for Li^+ having an ionic diameter of 1.48 Å, while [2.2.1] (cavity diameter = 2.2 Å) prefers Na^+ (ionic diameter = 2.04 Å) and cryptand [2.2.2] (cavity diameter = 2.8 Å) has a maximum stability constant with K^+ (ionic diameter = 2.46 Å). The larger macrobicyclic compounds exhibit less pronounced selectivities, probably because of increased ligand flexibility.

Generally as one goes to less polar solvents, the complexation constants increase. In most cases, however, the relative selectivity sequences do not change. More complete discussions of cation selectivity are available in the literature (3, 37, 77).

The complexation tendencies and unusual cation selectivities shown by ionophores can be modified by structural and environmental changes on the ionophore. These properties should lead to a number of possibilities in the design of drugs and other biologically important compounds.

3 THEORY OF MEMBRANE-TRANSPORT PHENOMENA

Considerable evidence has been cited for the currently accepted mechanism by which ionophores transport ions across biological membranes (88). According to this model the ionophore at one membrane-solution interface envelops an ion so as to form an uncharged neutral complex with all polar or hydrophilic moieties

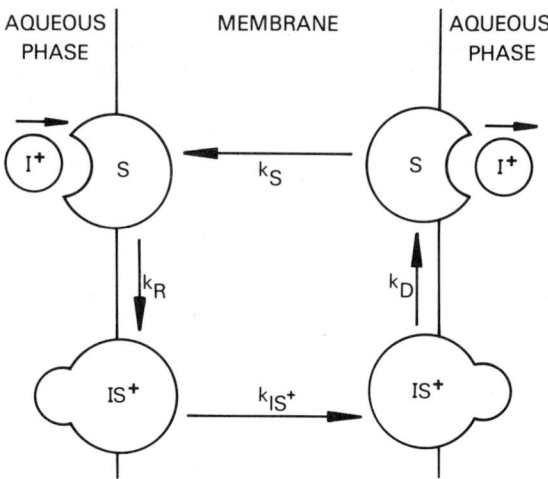

Figure 5 Diagram of carrier-mediated ion translocation across a membrane. The rate constants k_R and k_D are the rates of ion (I^+) - carrier complex formation and dissociation, respectively, and the constants K_{IS^+} and k_S represent the rates of transfer of the ion-carrier complex (IS^+) and the free carrier(s), respectively, across the membrane. Reproduced, with permission, from the *Annual Review of Biophysics and Bioengineering,* Volume 4. © (1975) by Annual Reviews Inc.

of the carrier and ion "hidden" on the inside of the complex, so that the exterior of the ionophore-ion complex is lipophilic. This complex is free to traverse the membrane by random walk. At the other interface dissociation of the complex may occur and the ion may be released to the aqueous phase. The free carrier in its "protected" lipophilic configuration diffuses back through the membrane to repeat the process. This model is illustrated in Figure 5, which was taken from the review of McLaughlin and Eisenberg (88).

The following evidence has been cited for this model of ion translocation (88).

1. A linear relationship is observed between the conductance of a black lipid membrane and the carrier concentration, consistent with the formation of a lipid-soluble 1:1 complex (97-101).
2. The conductance across the membrane increases linearly with the concentration of the selectively carried ion (99, 101).
3. Black lipid membranes are selective to cations in the presence of ionophores, and a Nernstian membrane potential is observed (103).
4. Valinomycin and nonactin, for example, enhance the conductance of thick (1000 Å) lipid membranes (98), providing evidence that these ionophores function as carriers rather than forming gramicidin like pores (3, 34).

5. These ionophores solubilize cations in bulk nonpolar organic phases (1, 2, 19).
6. The selectivity ratios for alkali metal ions are virtually identical, whether determined by conductance ratio, permeability ratio, or partition coefficient ratios. Thus for any pair of cations I and J

$$\frac{G(I)}{G(J)} = \frac{P_I}{P_J} = \frac{K(I)}{K(J)} \tag{2}$$

where G, P, and K refer to conductance, permeability, and partition coefficients for the equilibrium extraction of an ion into a bulk organic phase (102).

The permeability ratio is obtained from the transmembrane zero current potential described by the Goldman-Hodgkin-Katz equation (88):

$$V_0 = \frac{RT}{F} \ln \frac{a_i' + (P_j/P_i)\, a_j'}{a_i'' + (P_j/P_i)\, a_j''} \tag{3}$$

where V_0 is the zero current membrane potential, (') and ('') refer to the two aqueous phases separated by an ionophore-doped membrane, and the activities of the ions on the two sides of the membrane are indicated by a_i', a_i'', a_j', and a_j''.

3.1 Theoretical Model for Carrier-Mediated Permeation

Equation 3 has been shown to hold for carrier-mediated permeation in bilayers (102-104) with the restriction that the complexation reactions are assumed to be rapid compared to the translocation of the complex. This restriction is referred to by Eisenman and co-workers as the *equilibrium domain*. Under these conditions the permeability ratios of Eq. 3 are true constants, that is,

$$\left(\frac{P_j}{P_i}\right)_{eq} = \text{Const}$$

In the equilibrium domain this ratio is also equal to the ratio of the equilibrium constants for ion extraction into a lipid bilayer ($\overline{K}_j/\overline{K}_i$) or for extraction into model organic solvents (K_j/K_i) (22, 102).

Membrane potentials across isolectin bilayers in alkali chloride-potassium nitrate aqueous mixtures in the presence of macrotetralide actins (nonactin, monactin, dinactin, and trinactin) were shown to obey Eq. 3 with constant values of (P_j/P_i) (104).

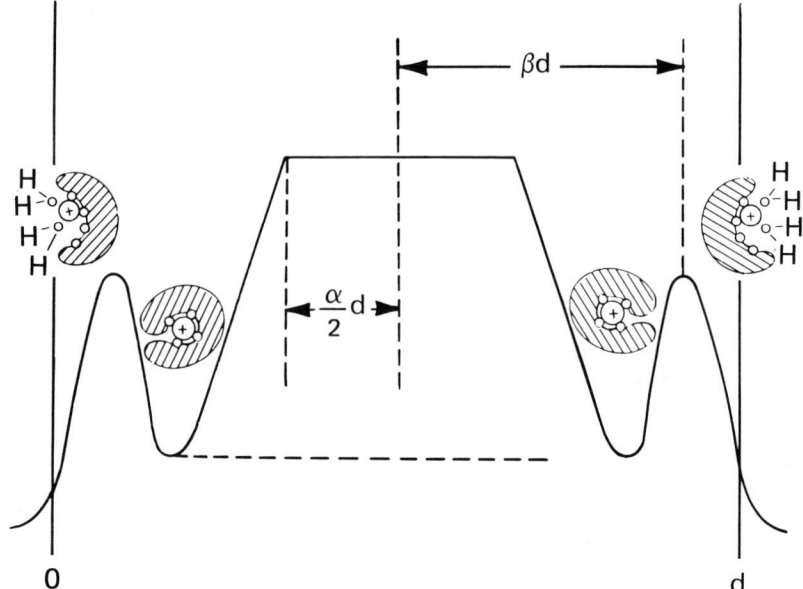

Figure 6 Schmatic diagram of the energy profile for the "extended model" of Ciani (23) and Krasne and Eisenman (109) for the carrier-mediated transport of ions across a lipid membrane. The diagram shows that the equilibrium position of the ion-carrier complex is displaced from the interfaces by some fraction of the membrane thickness of d. The free energy of the ion-carrier complex in the central portion of the membrane is approximated by a trapezoid. The hypothetical conformations of the ion-carrier complex show the transition state for the complexation reaction at the interfaces, and the "equilibrium state" of the complex within the membrane. [Adapted from Refs. 109, 123.)]

Other systems involving different lipid bilayer membranes and more strongly complexing ionophores do not obey this simple equation with constant values of (P_j/P_i). Examples of the more general "kinetic domain" behavior have been demonstrated in studies by Läuger, Stark, and co-workers (101, 105-108). In the kinetic domain (Figure 6) the rates of complex formation k_R and dissociation k_D are comparable to the rates of membrane transport of the carrier-ion complex k_{IS} and the uncomplexed carrier k_S (see Figure 5). In this case $(P_j/P_i)_{eq}$ of Eq. 3 must be replaced by a variable term $(P_j/P_i)_{app}$.

According to the Eyring single barrier model introduced by Läuger and Stark (105-108), the apparent permeability ratio $(P_j/P_i)_{app}$ is given by

$$\left(\frac{P_j}{P_i}\right)_{app} = \left(\frac{P_j}{P_i}\right)_{eq} \left(\frac{1 + 2\, W_i \cosh \phi/2}{1 + 2\, W_j \cosh \phi/2}\right) \quad (4)$$

where $\phi = FV_0/RT$, R is the gas constant, T is the absolute temperature, and W_i and W_j are the ratios of the rate constants for translocation of the complex across the membrane to the rate constant for the dissociation of the complex. According to the notation of Figure 5, $W_i = k_{IS^+}/k_D$. In the limit, as W_i becomes small, that is, when the rate of translocation is slow compared to the dissociation of the complex, $(P_j/P_i)_{app}$ becomes equal to $(P_j/P_i)_{eq}$. The equilibrium domain according to this model is that situation for which the rate-limiting step for membrane transport is the diffusion of the complex across the lipid bilayer.

Ciani and co-workers (99) have measured transmembrane potentials for glycerol dioleate bilayers separating aqueous solutions of alkali halide salts in the presence of trinactin. The transmembrane potentials as a function of aqueous salt concentration were in good agreement with Eq. 3 and 4. The "kinetic" term of Eq. 3 was required to rationalize the data, suggesting that the interfacial barriers are substantial and contribute appreciably to the overall rate of ion permeation of the lipid bilayer (102).

An extended model based on a trapezoidal diffusion barrier has been developed by Ciani and co-workers (22, 103). A schematic diagram of this model is shown in Figure 6. This model is similar to that derived earlier by Hall, Mead, and Szabo (110) and by Hladky (111, 112). This model leads to the following equation for the apparent permeability ratio:

$$\left(\frac{P_j}{P_i}\right)_{app} = \left(\frac{P_j}{P_i}\right)_{eq} \frac{1 + 2W_i\phi_0 \dfrac{\cosh(\beta_i \phi_0)}{\sinh[(\bar{\alpha}_i/2)\phi_0]}}{1 + 2W_j\phi_0 \dfrac{\cosh(\beta_j\phi_0)}{\sinh[(\bar{\alpha}_j/2)\phi_0]}} \quad (5)$$

The terms α and β are best visualized by reference to Figure 6. Here β is the fractional distance from the "reaction plane" to the midpoint of the membrane, and α is the fractional width of the plateau of the diffusion barrier (109) $\phi_0 = FV_0/RT$, that is, the transmembrane potential at zero current in F/RT units. Equation 5 may be simplified to read

$$\left(\frac{P_j}{P_i}\right)_{app} = \left(\frac{P_j}{P_i}\right)_{eq} \frac{1 + 2W_i[f(\phi_0)]}{1 + 2W_j[f(\phi_0)]} \quad (6)$$

where

$$f(\phi_0) = \phi_0 \frac{\cosh\beta\phi_0}{\sinh(\alpha/2)\phi_0} \quad (7)$$

As in the case of the simpler model (Eq. 3), if W_i and W_j are small, the condition defining the equilibrium domain $(P_j/P_i)_{app}$ also becomes equal to $(P_j/P_i)_{eq}$. In this situation the equilibrium permeability ratio is given by

$$\left(\frac{P_j}{P_i}\right)_{eq} = \frac{\overline{K}_j}{\overline{K}_i} \frac{\overline{A}_{js}^*}{\overline{A}_{is}^*} \tag{8}$$

where K_i and K_j are the equilibrium constants for the heterogeneous interfacial association reactions, and $\overline{A}_{is}^*$ and $\overline{A}_{js}^*$ are the rate constants for the translocation of the ion-carrier complex across the membrane (23, 109). In the other extreme, that is, the conditions under which the kinetic terms of Eq. 4 and 5 are large compared to 1, these equations reduce to

$$\left(\frac{P_i}{P_j}\right)_{app} = \left(\frac{P_j}{P_i}\right)_{eq} \frac{W_i}{W_j} \tag{9}$$

and

$$\left(\frac{P_j}{P_i}\right)_{app} = \frac{\overline{K}_j^F}{\overline{K}_i^F} \tag{10}$$

where $\overline{K}_i^F$ and $\overline{K}_j^F$ are the rate constants of the heterogeneous reaction describing the formation of the ion-carrier complexes (23). In this case the apparent permeability ratio reflects only the ratio of the rate constants for the complexation of ions i and j by the carrier; the contribution of a diffusion barrier to ion selectivity is negligible.

These different limiting cases of selectivity in ion permeation may be visualized by referring to Figure 6, which has been adapted from the paper of Krasne and Eisenman (109).

The hypothetical conformation of the incipient carrier-ion complex is depicted as occurring near the peak, just within the membrane solution interface. The selectivity resulting from the complex formation step is determined by the ratio $(\overline{K}_j^F/\overline{K}_i^F)$. The "fully formed" complex which occurs at the minimum energy position is the equilibrium state. The selectivity contribution attributed to differences in the diffusion rate of these equilibrium state complexes is given by $(\overline{K}_j/\overline{K}_i)$.

The extended model has been applied to the determination of the relative equilibrium and kinetic contributions to the selectivity of the carrier-induced permeability across membranes for the macrotetralide actin-ion interaction (109). It was found that a change in lipid composition altered the relative contributions of the "kinetic" and "equilibrium" components to the selectivity.

Increased methylation of the ionophore increases the "kinetic" contribution by decreasing the total constant for the dissociation of the ion-carrier complex. Whereas the permeability ratio and the ratios W_i and W_j vary with the structure of the carrier and the ion as expected, the parameters α and β remain virtually constant for all ions and carriers studied. These parameters represent the width of the diffusion barrier and the distance of the reaction plane for complex formation from the center of the membrane respectively. The fact that a good fit to the data can be obtained with this theory when these parameters are held constant provides encouraging support for the physical interpretation of the parameters in the equation as depicted in Figure 6.

On the basis of the correlations obtained with this theory, it appears that this model for carrier-mediated permeation will be a valuable tool for further study of the effect of varying the molecular structure of the ionophores, the properties of the lipid environment, and the nature of the ionic species on ion permeability.

3.2 The Role of Ion-Transport Mediators in Energy Coupling

Energy coupling is the mechanism by which energy-producing reactions or sequences of reactions provide the energy for a driven chemical reaction. The detailed molecular mechanism for oxidative phosphorylation, for example, in which the energy from the respiratory chain is coupled to the synthesis of ATP, has not been worked out. However, a great deal of information has been accumulated concerning the factors that control the electron-transport chain and the coupled synthesis of ATP (113).

D. E. Green and co-workers have proposed a theory of energy coupling based on the paired moving charges (PMC) model (114). They have enthusiastically promoted this model and the role of identified, and as yet unidentified, ionophores in mitochondrial energy-coupling processes (115).

The simplest form of energy coupling according to the PMC model depends on the interaction of a negatively charged species in the driving chemical reaction (energy source) with a positively charged species in the driven reaction. This situation, in which the positive and negative charges are simultaneously moving in the same direction, is referred to as *symport coupling* (118, 121, 122). An example of symport coupling is the coupling of electron transfer in an electron-transfer complex in the exergonic center on one side of a membrane with the ionophore-mediated transport of K^+, the driven chemical reaction, on the other side of the membrane. The separation of electron from proton is paired to and dependent on the ionophore-mediated separation of cation from anion. An alternative coupling process may also be visualized with which transmembrane flow of negative charge is coupled to the transfer of negatively charged species in the opposite direction (*antiport coupling*). The distinction between symport and antiport coupling is diagrammed in Figure 7, from the

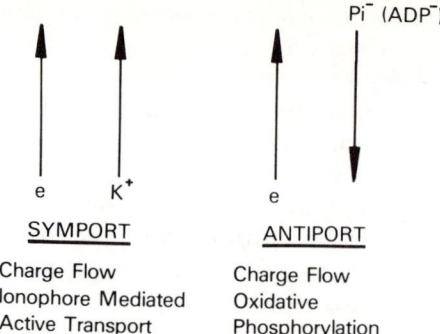

Figure 7 Symport versus antiport energy coupling (114). Reproduced, with permission, from *Ann. NY Acad. Sci.*, Volume 264. © 1975 by the New York Academy of Sciences.

review by Green (114). The example shown in Figure 7 for antiport coupling is the oxidative phosphorylation process, in which the movement of an electron is coupled to the removal of a proton from ADP and the movement of negatively charged Pi^- and ADP^-. This would appear to require antiport coupling. Kemeny, however, has shown that all energy coupling must be symport (121, 122). Green and Reible (118, 119) have reconciled the requirements of oxidative phosphorylation and the principle that the coupling must be symport by suggesting that antiport coupling may be reduced to two symport charge flows by means of a linkage system of two circulating positive charges. These positive charges consist of metal ions complexed by neutral ionophore molecules rendering them capable of diffusion in a lipid membrane environment. This linkage system is illustrated in Figure 8 (117).

An alternative to an ionophore linkage system to transform antiport into symport charge flow has recently been suggested (117, 123). If the moving charged species are the Mg^{2+} complexes of Pi^- and ADP^-, that is, $[Mg^{2+}...Pi^-]$ and $[Mg^{2+}....ADP^-]$, then these *positively* charged ion pairs can move in

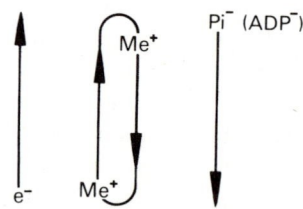

Figure 8 Linkage system for mediating antiport energy coupling (114). Reproduced with permission from *Ann. NY Acad. Sci.* Volume 264. © 1975 by the New York Academy of Sciences.

concert with an electron flow without violating the requirement for symport coupling.

Further evidence that helps to define the nature of energy coupling comes from studies with uncouplers of oxidative phosphorylation. These uncouplers include a wide variety of structures. Some of the compounds that have been studied most extensively include 2,4-dinitrophenol, carbonylcyanide p-trifluoromethoxyphenylhydrazone (FCCP), carbonylcyanide m-chlorophenylhydrazone (m-ClCCP), 5-chloro-3-t-butyl-2′-chloro-4′-nitrosalicylanilide (S13), 3,5-di-t-butyl-4-hydroxy-benzylidinemalononitrile (SF6847), and 4,5,6,7-tetrachloro-2-trifluoromethylbenzimidazole (TTFB) (128).

According to the Mitchell model of mitochondrial energy coupling, a gradient of H^+ ions across the mitochondrial inner membrane serves as a means of coupling the electron flow from the respiratory chain to the synthesis of ATP (113, 126, 127). An intact, proton impermeable membrane is an essential requirement of the Mitchell hypothesis. According to the Mitchell chemiosmotic coupling theory, it is the function of the electron carriers of the respiratory chain to serve as active transport agents for H^+ ions across the inner mitochondrial membrane. The free energy stored in this gradient is then available at the active ATPase site to promote the dehydration of ADP and Pi to yield ATP. This dehydration is accomplished by the high external H^+ concentration, which pulls OH^- in the outward direction, and the relatively high OH^- concentration on the inner side of the membrane, which pulls protons inward. The intact mitochondrial membrane is essential for this process to maintain the energy source inherent in the pH gradient (113). The function of uncoupling agents is to render the membrane permeable to protons and to collapse the pH gradient required for energy coupling. It has been shown, in agreement with the requirements of the model, that the electron-transport chain can pump hydrogen ions outward, and that ATP formation is accompanied by movement of hydrogen ions in the opposite direction. Experiments by Racker and co-workers (125) have also shown that oxidative phosphorylation can be reconstituted with a proton pump (i.e., an active transport system for H^+ ions) substituted for respiratory chain components. Considerable evidence was also provided to support the thesis that uncouplers of oxidative phosphorylation were protonophores (127, 128) in every case examined, and that rendering the membrane permeable to protons is essential to uncoupling action.

Recently Green (115) and Kessler and co-workers (128, 129) have called attention to the fact that uncouplers abolish not only protonic gradients, but also cationic gradients (53, 131). It was also shown by Montal and co-workers (53) that uncoupler action could be duplicated by the combination of nigericin and valinomycin in the presence of potassium ion. The uncoupling action in this case could be explained as follows: electron flow, in the presence of valinomycin and potassium ion, is accompanied by the inward flow of the charged K^+VAL

complex (symport coupling). In the presence of nigericin NIG^- the potassium ion is carried out as the neutral NIG^-K^+ complex. The function of the uncoupler (in this case the valinomycin-nigericin combination) is to substitute the inward flow of K^+ ions for the coupled process leading to ATP synthesis. It has been suggested that "uncoupling" may be an unfortunate choice of terminology (28). This example shows that what is in fact involved is the substitution of one coupled process for another. Since the action of uncouplers in every case examined appears to be identical to the valinomycin-nigericin combination, it was assumed that the uncoupling action requires a synergism between uncoupler, intrinsic mitochondrial ionophore, and cations. To prove that this "cyclical cation transport hypothesis" was valid, it was necessary to demonstrate that (a) uncouplers are in every case ionophores, (b) cations are required for ionophore action, and (c) for maximum uncoupling the concentration of uncoupler must be at least equal to the concentration of electron-transfer species.

In a series of convincing experiments Kessler and co-workers have validated each of these conditions. It was shown (129) for a variety of uncoupling agents that ionophore activity, with respect to univalent and divalent cations, can be demonstrated in each case. It was further shown that this ionophore activity requires the presence of an electrogenic ionophore (e.g., nigericin). Uncoupler, electrogenic ionophore, and univalent cation were shown to form a 1:1:1 complex. Similarly it was shown that for the cyclical transport of Ca^{2+} the molar ratio of uncoupler, cation, and electrogenic ionophore is 1:1:2 (130). The requirement that cations be present for uncoupler action was demonstrated by depleting coupled mitochondria of Ca^{2+} and Mg^{2+}. These cation-depleted preparations required the addition of $CaCl_2$ or $MgCl_2$ to induce uncoupler response.

The approximate equivalence of electron-transfer complex to uncoupler concentration for maximum uncoupling action was demonstrated in beef heart mitochondria, lending further validity to the proposed hypothesis.

Since uncoupler action requires the presence of an electrogenic ionophore, it must be assumed that mitochondria contain intrinsic electrogenic ionophores. This has been demonstrated by Blondin and co-workers (131, 133), who have isolated peptidic ionophores from beef heart mitochondria. They have also shown that these ionophores increase the transport of K^+ ion in the presence of uncoupler.

These results have led Green (115) and Kessler's group (129) to a formulation of uncoupler-induced cyclical transport of univalent and divalent cations, based on the paired moving charge model. This formulation is illustrated in Figure 9. The successful correlation of the experimental evidence with the model provides evidence that cyclical cation transport takes precedence over all other coupled processes and that uncoupling proceeds with the formation of a complex involving uncoupler (U^-), electrogenic ionophore (e.g., valinomycin), and cation (K^+).

3 Theory of Membrane-Transport Phenomena 249

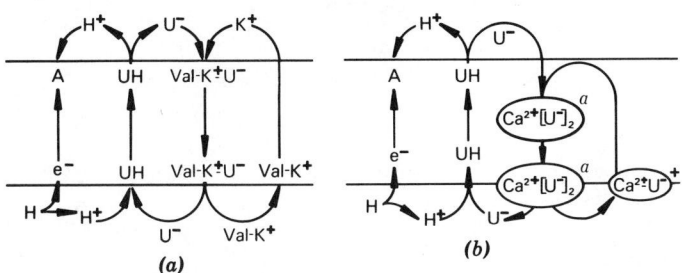

Figure 9 (*a*) Cyclical transport of K$^+$ mediated by the combination of uncoupler (U$^-$) and a neutral electrogenic ionophore such as valinomycin (VAL). The movement of the electron (*e*$^-$) through the electron-transfer complex is paired to the movement of valinomycin-K$^+$ (Val-K$^+$) through the membrane. The charge separation of H into *e*$^-$ and H$^+$ in the electron transfer complex is also paired to the charge separation of Val-K$^+$-U$^-$ into Val-K$^+$ and U$^-$. At the terminus of its trajectory the electron is transferred to the final acceptor in the complex (A), and this transfer is simultaneous with the uptake of a proton. Charge elimination in the electron-transfer chain is paired to charge elimination of Val-K$^+$ by combination with U$^-$. The protonated form of U$^-$ is represented as UH. The diagram shows how the rate of oxidation of substrate by the electron transfer complex can be maximized by drawing off the protons as fast as these are released on one side of the membrane, and feeding in the protons required on the other side of the membrane. Cyclical transport of K$^+$ mediated by the combination of electrogenic ionophore and uncoupler can accomplish this maximization of the rate of oxidation of substrates in the electron-transfer complex. (*b*) Cyclic transport of Ca^{2+} mediated by the combination of uncoupler (U$^-$) and electrogenic ionophore active on divalent metal cations (O). The principle for uncoupler-mediated cyclical transport is identical whether the cation is monovalent or divalent. The new feature is that the uncoupler participating in the electrogenic as well as in the nonelectrogenic ionophore is 1+ in the electrogenic step and 0 in the nonelectrogenic step. Ref. Reproduced with permission from R. J. Kessler, H. Vandezande, C. A. Tyson, G. A. Blondin, J. Garfield, P. Glasser and D. E. Green, Proc. Natl. Acad. Sci. USA, **74**, 2241 (1977) Ref. (129).

If this mechanism for uncoupling is correct, important conclusions may also be drawn concerning the nature of respiratory control and energy coupling. Green (115) thus defines respiratory control as that state in which electron flow cannot be coupled to the flow of some positively charged species, and concludes that energy coupling always involves participation of an electrogenic ionophore. Cyclical cation transport, however, would eliminate all protonic and cationic gradients. This requirement would be in direct conflict with the chemiosmotic model, which requires a proton gradient for driving coupled processes. This conflict is still not resolved. It is clear that further evidence will be required to determine the exact mechanisms invovled in energy coupling. Current evidence suggests that ionophore action is involved in coupled processes and that uncouplers generally possess ionophore (and protonophore) properties.

REFERENCES

1. B. C. Pressman, *Fed. Proc., Fed. Am. Soc. Exp. Biol.,* **32**, 1698 (1973).
2. B. C. Pressman and N. T. deGuzman, *Ann. NY Acad. Sci.,* **264**, 373 (1975).
3. B. C. Pressman, *Ann. Rev. Biochem.,* **45**, 501 (1976).
4. W. Simon and W. E. Morf, in *Membranes—A Series of Advances,* Vol. 2 (G. Eisenman, Ed.), Marcel-Dekker, New York, 1973, p. 329.
5. G. A. Blondin and D. E. Green, *Chem. Eng. News,* **53**, 26 (1975).
6. H. Brockmann and G. Schmidt-Kastner, *Chem. Ber.,* **88**, 57 (1955).
7. W. McMurray and R. W. Begg, *Arch. Biochem. Biophys.,* **84**, 546 (1959).
8. C. Moore and B. C. Pressman, *Biochem. Biophys. Res. Commun.,* **15**, 562 (1964).
9. B. C. Pressman, E. J. Harris, W. S. Jagger, and J. H. Johnson, *Proc. Natl. Acad. Sci.,* **58**, 1949 (1967).
10. R. L. Harned, P. H. Hidy, C. J. Corum, and K. L. Jones, *Antibiot. Chemother.,* **1**, 594 (1951).
11. A. E. Shamoo and T. E. Ryan, *Ann. NY Acad. Sci.,* **264**, 83 (1975).
12. A. E. Shamoo, T. E. Ryan, P. S. Stewart, and D. H. MacLennan, *J. Biol. Chem.,* **251**, 4147 (1976).
13. P. W. Reed and H. A. Lardy, *J. Biol. Chem.,* **247**, 6970 (1972).
14. D. J. Patel and C. Shen, *Proc. Natl. Acad. Sci.,* **73**, 1786 (1976).
15. C. Shen and D. J. Patel, *Proc. Natl. Acad. Sci.,* **73**, 4277 (1976).
16. S. M. Johnson, J. Herrin, S. J. Liu, and I. C. Paul, *J. Am. Chem. Soc.,* **92**, 4428 (1970).
17. M. J. O. Anteunis, *Bioorg. Chem.,* **6**, 1 (1977).
18. R. G. Johnson and A. Scarpa, *FEBS Lett.,* **47**, 117 (1974).
19. S. Lindenbaum, L. Sternson, and S. Rippel, *J. Chem. Soc., Chem. Commun.,* 268 (1977).
20. R. D. Hotchkiss, *Adv. Enzymol.,* **4**, 153 (1944).
21. D. E. Green, *Trends Biochem. Sci.,* **2**, 113 (1977).
22. G. Eisenman, S. Ciani, and G. Szabo, *Fed. Proc.,* **27**, 1289 (1968).
23. S. Ciani, *J. Membr. Biol.,* **30**, 45 (1976).
24. B. C. Pressman, *Proc. Natl. Acad. Sci.,* **53**, 1076 (1965).
25. B. C. Pressman, *Fed. Proc., Fed. Am. Soc. Exp. Biol.,* **27**, 1283 (1968).
26. V. T. Ivanov, I. A. Laine, N. D. Abdullaev, L. B. Senyavina, E. M. Popov, Yu. A. Ovchinnikov, and M. M. Shemyakin, *Biochem. Biophys. Res. Commun.,* **34**, 803 (1969); M. Pinkerton, L. K. Steinrafu, and P. Dawkins, *Biochem. Biophys. Res. Commun.,* **35**, 512 (1969).
27. J. Beck, H. Gerlach, V. Prelog, and W. Yoser, *Helv. Chim. Acta,* **45**, 620 (1962).
28. H. K. Frensdorff, *J. Am. Chem. Soc.,* **93**, 600 (1971).
29. J. J. Christensen, J. O. Hill, and R. M. Izatt, *Science,* **174**, 459 (1971).
30. M. M. Shemyakin, Yu. A. Ovchinnikov, V. I. Ivanov, V. K. Antonov, E. I. Vinogradova, A. M. Shkrob, G. G. Malenkov, A. V. Evstratov, I. D. Ryabova, I. A. Laine, and E. I. Melnik, *J. Membr. Biol.,* **1**, 402 (1969).

References

31. C. M. Deber, D. A. Torchia, S. Wong, and E. R. Blout, *Proc. Natl. Acad. Sci.*, **69**, 1825 (1972).
32. L. K. Steinrauf, M. Pinkerton, and J. W. Chamberlin, *Biochem. Biophys. Res. Commun.*, **33**, 29 (1968).
33. D. W. Urry, M. C. Goodall, J. D. Glickson, and D. F. Mayers, *Proc. Natl. Acad. Sci.*, **68**, 1907 (1971).
34. D. Neubert and A. L. Lehninger, *Biochem. Biophys. Acta,* **62**, 556 (1962).
35. W. R. Veatch, R. Mathies, M. Eisenberg, and L. Stryer, *J. Mol. Biol.,* **99**, 75 (1975).
36. R. Sarges and B. Witkop, *J. Am. Chem. Soc.,* **86**, 1862 (1964); **87**, 2011 (1963).
37. Yu. A. Ovchinnikov, V. T. Ivanov, and A. M. Shkrob, in *Membrane Active Complexones,* Vol. 12. (B.B.A. Library, Ed.), Elsevier, New York, 1974.
38. C. L. Liotta and H. P. Harris, *J. Am. Chem. Soc.,* **96**, 2250 (1974).
39. R. C. Helgeson, J. M. Timko, and D. J. Cram, *J. Am. Chem. Soc.,* **95**, 3021, 3023 (1973).
40. E. P. Kyba, K. Koga, L. R. Sousa, M. Siegel, and D. J. Cram, *J. Am. Chem. Soc.,* **95**, 2692 (1973).
41. L. A. R. Pioda, V. Stankova, and W. Simon, *Anal. Lett.,* **2**, 665 (1969).
42. J. Pick, K. Toth, E. Pungor, M. Vasak, and W. Simon, *Anal. Chim. Acta,* **64**, 477 (1973).
43. Z. Stefanac and W. Simon, *Chimia,* **20**, 436 (1966).
44. Z. Stefanac and W. Simon, *Microchem. J.,* **12**, 125 (1967).
45. D. Ammann, E. Pretsch, and W. Simon, *Tetrahedron Lett.,* 2473 (1972).
46. B. C. Pressman, in *The Role of Membranes in Metabolic Regulation,* (M. A. Mehlman and R. W. Hanson, Eds.), Academic Press, New York, 1972, p. 149.
47. A. H. Caswell and B. C. Pressman, *Biochem. Biophys. Res. Commun.,* **49**, 292 (1972).
48. M. L. Entman, P. C. Gillette, E. T. Wallick, B. C. Pressman, and A. Schwartz, *Biochem. Biophys. Res. Commun.,* **48**, 847 (1972).
49. A. Scarpa, J. Baloassare, and G. Inesi, *J. Gen. Physiol.,* **60**, 735 (1972).
50. M. W. Osborne, J. J. Wenger, and M. T. Zanko, *J. Pharmacol. Exp. Ther.,* **200**, 195 (1977).
51. S. M. Podos, *Invest. Ophthalmol.,* **15**, 851 (1976).
52. E. J. Harris, G. Catlin, and B. C. Pressman, *Biochemistry,* **6**, 1360 (1967).
53. M. Montal, B. Lee, and C. P. Chance, *J. Membr. Biol.,* **2**, 201, 234 (1970).
54. H. Pasantes-Morales, R. Salceda, and A. Gomez-Puyou, *Biochem. Biophys. Res. Commun.,* **58**, 847 (1974).
55. H. Kita and W. Van der Kloot, *Nature* **250**, 658 (1974).
56. H. Kita and W. Van der Kloot, *Bioscience,* **24**, 13 (1974).
57. M. Schadt and G. Haeusler, *J. Membr. Biol.,* **18**, 277 (1974).
58. R. Holz, *Biochim. Biophys. Acta,* **345**, 138 (1975).
59. J. C. Foreman, J. L. Monger, and B. D. Gomperts, *Nature,* **245**, 249 (1973).
60. W. W. Douglas, *Biochem. Symp.,* **39**, 1 (1974).
61. P. Massini and E. E. Luscher, *Biochim. Biophys. Acta,* **372**, 109 (1974).
62. M. B. Feinstein and C. Fraser, *J. Gen. Physiol.,* **66**, 561 (1975).

63. W. T. Prince, T. H. Rassmussen, and M. J. Berridge, *Biochim. Biophys. Acta,* **329**, 98 (1973).
64. I. S. Eimerl, N. Savion, O. Heichal, and Z. Selinger, *J. Biol. Chem.,* **249**, 3991 (1974).
65. G. W. G. Sharp, C. Wollhein, W. A. Muller, A. Gutzeit, P. A. Trueheart, B. Blondel, L. Orci, and A. E. Reynolds, *Fed. Proc., Fed. Am. Soc. Exp. Biol.,* **34**, 1537 (1975).
66. Y. Nakazato and W. W. Douglas, *Nature,* **249**, 479 (1974).
67. G. Grenier, J. Van Strio, and J. E. Dumont, *FEBS Lett.,* **49**, 96 (1974).
68. R. A. Steinhardt and D. Epel, *Proc. Natl. Acad. Sci.,* **71**, 915 (1974).
69. D. E. Chandler and J. A. Williams, *J. Membr. Biol.,* **32**, 201 (1977).
70. G. G. S. Collins, *J. Neurochem.,* **28**, 461 (1977).
71. H. Hidaka and T. Asano, *J. Biol. Chem.,* **251**, 7508 (1976).
72. J. J. Christensen, D. J. Eatough, and R. M. Izatt, *Chem. Rev.,* **74**, 351 (1974).
73. N. K. Dalley, in *Synthetic Multidentate Macrocyclic Compounds,* (R. M. Izatt and J. J. Christensen, Eds.), Academic Press, New York, in press.
74. J. M. Lehn, *Struct. Bonding,* **16**, 1 (1973).
75. J. M. Lehn and J. P. Sauvage, *J. Am. Chem. Soc.,* **97**, 6700 (1975).
76. J. M. Lehn and J. Simon, *Helv. Chim. Acta,* **60**, 141 (1977).
77. R. M. Izatt, J. D. Lamb, D. J. Eatough, J. J. Christensen, and J. H. Rytting, in *Drug Design,* Vol. 8 (E. J. Ariens Ed.), Academic Press, New York, in press; R. M. Izatt, N. E. Izatt, B. E. Rossiter, J. J. Christensen, and B. L. Haymore, *Science,* **199**, 994 (1978).
78. I. Goldberg, *Acta Crystallogr.,* **B31**, 2592 (1975).
79. D. J. Cram, R. C. Helgeson, L. R. Sousa, J. M. Timko, M. Newcomb, P. Moreau, F. DeJong, G. W. Gokel, D. H. Hoffman, L. A. Domeier, S. C. Peacock, K. Madan, and L. Kaplan, *Pure Appl. Chem.,* **43**, 327 (1975).
80. E. P. Kyba, R. C. Helgeson, K. Madan, G. W. Gokel, T. L. Tarnowski, S. S. Moore, and D. J. Cram, *J. Am. Chem. Soc.,* **99**, 2564 (1977).
81. J. M. Timko, S. S. Moore, D. M. Walba, P. C. Hiberty, and D. J. Cram, *J. Am. Chem. Soc.,* **99**, 4207 (1977).
82. H. Degani, H. L. Friedman, G. Navon, and E. M. Kosower, *J. Chem. Soc., Chem. Commun.,* **1973**, 431.
83. H. Degani and H. L. Friedman, *Biochemistry,* **13**, 5022 (1974).
84. H. Degani, R. M. D. Hamilton, and H. L. Friedman, *Biophys. Chem.,* **4**, 363 (1976).
85. D. R. Pfeiffer, P. W. Reed, and H. A. Lardy, *Biochemistry,* **19**, 4007 (1974).
86. M. S. Kafka and R. W. Holz, *Biochim. Biophys. Acta,* **426**, 31 (1976).
87. S. R. Alpha and A. H. Brady, *J. Am. Chem. Soc.,* **95**, 7043 (1973).
88. S. McLaughlin and M. Eisenberg, *Ann. Rev. Biophys. Bioeng.,* **4**, 355 (1975).
89. J. W. Westley, R. H. Evans, Jr., and J. F. Blount, *J. Am. Chem. Soc.,* **99**, 6057 (1977).
90. R. C. Hegelson, J. M. Timko, P. Moreau, S. C. Peacock, J. M. Mater and D. I. Cram, *J. Amer. Chem. Soc.,* **96**, 6762 (1974).
91. M. Newcomb and D. J. Cram, *J. Am. Chem. Soc.,* **97**, 1257 (1975).
92. S. C. Peacock and D. J. Cram, *J. Chem. Soc., Chem. Commun.,* **1976**, 282.
93. C. Shen and D. J. Patel, *Proc. Natl. Acad. Sci. USA,* **74**, 4734 (1977).

References

94. R. M. Izatt, R. E. Terry, B. L. Haymore, L. D. Hansen, N. K. Dalley, A. G. Avondet, and J. J. Christensen, *J. Am. Chem. Soc.*, 98, 7620 (1976).
95. R. M. Izatt, R. E. Terry, D. P. Nelson, Y. Chan, D. J. Eatough, J. S. Bradshaw, L. D. Hansen, and J. J. Christensen, *J. Am. Chem. Soc.*, 98, 7626 (1976).
96. R. M. Izatt, J. D. Lamb, G. E. Maas, R. E. Asay, J. S. Bradshaw, and J. J. Christensen, *J. Am. Chem. Soc.*, 99, 2365 (1977).
97. D. A. Haydon and S. B. Hladky, *Q. Rev. Biophys.*, 5, 187 (1972).
98. G. Eisenman, G. Szabo, G. Ciani, S. McLaughlin, and S. Krasne, in *Progress in Surface and Membrane Science*, Vol. 6 (J. Danielli, M. Rosenberg, and D. Cadenhead, Eds.), Academic Press, New York, 1973, p. 139.
99. S. M. Ciani, G. Eisenman, R. Laprade, and G. Szabo, in *Membranes—A Series of Advances*, Vol. 2 (G. Eisenman, Ed.), Marcel-Dekker, New York, 1973, p. 61.
100. G. Szabo, G. Eisenman, R. Laprade, S. M. Ciani, and S. Krasne, *ibid.*, Vol. 2, 1973, p. 179.
101. G. Stark and R. Benz, *J. Membr. Biol.*, 5, 133 (1971).
102. G. Eisenman, S. Krasne, and S. Ciani, *Ann. NY Acad. Sci.*, 264, 34 (1975).
103. S. M. Ciani, G. Eisenman, and G. Szabo, *J. Membr. Biol.*, 1, 1 (1969).
104. G. Szabo, G. Eisenman, and S. Ciani, *J. Membr. Biol.*, 1, 346 (1969).
105. P. Läuger and G. Stark, *Biochim. Biophys. Acta*, 211, 458 (1970).
106. G. Stark, B. Ketterer, R. Benz, and P. Läuger, *Biophys. J.*, 11, 339 (1973).
107. R. Benz, G. Stark, K. Sanko, and P. Läuger, *J. Membr. Biol.*, 14, 339 (1973).
108. P. Läuger, *Science*, 178, 24 (1972).
109. S. Krasne and G. Eisenman, *J. Membr. Biol.*, 30, 1 (1976).
110. J. E. Hall, C. A. Mead, and G. Szabo, *J. Membr. Biol.*, 11, 75 (1973).
111. S. B. Hladky, *Biochim. Biophys. Acta*, 352, 71 (1974).
112. S. B. Hladky, *Biochim. Biophys. Acta*, 375, 327 (1975).
113. A. L. Lehninger, *Biochemistry*, 2nd ed., Worth, New York, 1975.
114. D. E. Green, *Ann. NY Acad. Sci.*, 264, 61 (1975).
115. D. E. Green, *Trends Biochem. Sci.*, 2, 113 (1977).
116. D. E. Green, *Biochim. Biophys. Acta*, 346, 27 (1974).
117. D. E. Green, *Ann. NY Acad. Sci.*, 227, 6 (1974).
118. D. E. Green and S. Reible, *Proc. Natl. Acad. Sci. USA*, 71, 4850 (1974).
119. D. E. Green and S. Reible, *Proc. Natl. Acad. Sci. USA*, 72, 253 (1975).
120. D. E. Green, G. Blondin, R. Kessler, and J. H. Southard, *Proc. Natl. Acad. Sci. USA*, 72, 896 (1975).
121. G. Kemeny, *Proc. Natl. Acad. Sci. USA*, 71, 3064 (1974).
122. G. Kemeny, *Proc. Natl. Acad. Sci. USA*, 71, 3669 (1974).
123. D. E. Green, in *The Structural Basis of Membrane Function*, (Y. Hatefi and L. Djavadi-Ohaniance, Eds.), Academic, New York, 1976, p. 241.
124. P. Mitchell, *Biol. Rev. Cambridge Philos. Soc.*, 41, 445 (1966).
125. E. Racker, A. F. Knowles, and E. Eytan, *Ann. NY Acad. Sci.*, 264, 17 (1975).
126. P. Ting, D. F. Wilson, and B. Chance, *Arch. Biochem. Biophys.*, 141, 141 (1970).

127. V. P. Skulachev, A. A. Sharaf, and E. A. Lieberman, *Nature,* **216**, 718 (1967).
128. R. J. Kessler, C. A. Tyson, and D. E. Green, *Proc. Natl. Acad. Sci., USA,* **73**, 3141 (1976).
129. R. J. Kessler, H. Vande Zande, C. A. Tyson, G. A. Blondin, J. Fairfield, P. Glasser, and D. E. Green, *Proc. Natl. Acad. Sci. USA,* , **74**, 2241 (1977).
130. H. Komai, D. R. Hunter, J. H. Southard, R. A. Haworth, and D. E. Green, *Biochem. Biophys. Res. Commun.,* **69**, 695 (1976).
131. G. A. Blondin, A. F. DeCastro, and A. E. Senior, *Biochem. Biophys. Res. Commun.,* **43**, 28 (1971).
132. G. A. Blondin, R. J. Kessler, and D. E. Green, *Proc. Natl. Acad. Sci. USA,* **74**, 3667 (1977).
133. G. A. Blondin, *Ann. NY Acad. Sci.,* **264**, 98 (1975).

AUTHOR INDEX

Numbers in regular type are page numbers that indicate that each author of a given contribution is referred to although his name is not cited in the text. Italicized numbers show the page on which the complete reference is listed. More than one complete reference for a given author may be found on the same page.

Abdullaev, N.D., 140, 187, *154*
Abragam, A., 92, *113*
Ackman, R.G., 117, *215*
Acrivos, J.V., 66, *110*
Adamson, A.W., 12, 15, 30, *58*
Adolphson, D.G., 86, *112*
Adrian, W., 53, *60*
Agtarap, A., 121, 140, 142, *151*
Akazaki, H., 140, *153*
Alder, B.J., 66, *110*
Alkins, T., 177, *214*
Alleaume, M., 121, 140, 142, *151*, *153*
Alpatova, P.M., 77, *112*
Alpha, S.R., 236, *252*
Ammann, D., 2, 18, 21, 23, 25, 29, 30, 35, 36, 37, *57*, *58*, *59*, *60*, 118, 147, *149*, 225, *251*
Anderegg, G., 64, *108*, 138, *152*
Ando, K., 140, *153*
Andreoli, T.E., 118, 140, *150*, *154*
Andrews, C.W., 64, 65, 75, 83, 87, 90, 92, 103, 105, *108*, 138, 139, *153*
Anteunis, M.J.O., 220, *250*
Antonov, V.K., 119, 140, 142, *150*, 223, *250*
Arkhipova, S.F., 140, *154*
Armatis, F.J., 86, *112*
Armbruster, A.M., 189, *215*
Arnett, E.M., 28, *59*
Arvanitis, S., 38, 39, 40, *60*
Asada, M., 201, 209, *216*
Asano, T., 227, *252*
Asay, R.E., 239, *253*
Avondet, A.G., 64, *109*, 188, *215*, 239, *253*

Bailey, N.A., 195, 213, *216*
Balashova, T.L., 140, *154*
Baldeschwieler, J.D., 140, *154*
Baloassare, J., 225, *251*
Banthia, A., 128, *152*
Bar-Eli, K., 70, 72, 73, 76, *111*, *112*
Barnett, B.L., 64, 65, 75, 83, 84, 85, *108*, 138, *153*
Baron, B., 77, *112*
Basyomy, A. el, 191, *215*
Baumann, E.W., 37, *59*
Becker, E., 66, *110*
Beckman, A., 90, *113*
Bedekovic, D., 2, *57*
Begg, R.W., 220, 221, *250*
Belousova, M.I., 64, *109*
Benz, R., 240, 242, *253*
Bergen, T.J. van, 158, 174, *213*
Berger, J., 140, *154*
Berlinova, I.V., 64, 69, *109*
Berridge, M.J., 226, *252*
Berthod, H., 53, *61*
Bhagwat, V.W., 128, 132, 137, *152*
Bissig, R., 2, 21, 23, 25, 30, 36, *57*, *59*
Bittar, E.E., 116, 137, *149*
Bittman, R., 2, *57*
Blasius, E., 53, *60*
Blondel, B., 226, *252*
Blondin, G.A., 220, 228, 247, 248, 249, *250*, *254*
Bloor, E.G., 90, *113*
Blount, J.F., 121, 140, 142, *151*, *153*, 223, 224, 236, *251*, *252*
Boaz, H.E., 140, *153*
Bockrath, B., 81, *112*

Bodner, R.L., 90, *113*
Böklen, K.D., 90, *113*
Boer, F.P., 134, 136, *152*, 195, *216*
Boileau, S., 78, 79, 80, *112*
Boller, A., 140, *153*
Bombieri, G., 195, *216*
Borgen, G., 159, *213*
Borowitz, G.B., 2, *57*
Borowitz, I.J., 2, 21, 23, *57*
Bourgoin, M., 124, 125, 126, 145, *152*
Bowers, C.W., 177, 190, *215*
Bradshaw, J.S., 64, *109*, 159, *213*, 239, *253*
Brady, A.H., 236, *252*
Bright, D., 65, 82, *109*, 119, 123, 124, 133, *150*
Brockmann, H., 140, *153*, 220, *250*
Brooks, J.M., 77, *112*
Browall, K.W., 68, 69, 71, 75, *111*
Brown, W.H., 177, *215*
Buck, R.P., 35, 44, *59*, *60*
Büchi, R., 23, 25, *59*
Burgermeister, W., 142, *154*
Büsch, D.H., 145, *155*, 177, *214*
Bush, M.A., 65, 82, *110*, 119, 123, 124, 125, 132, 133, *150*, *152*, 195, *216*
Byrne, M., 177, 190, *215*
Bystrov, V.F., 140, *154*

Cabbiness, D.K., 12, *58*, 130, 134, *152*
Cafasso, F., 64, 68, *109*
Cahen, Y.M., 64, 90, 91, 99, *109*, *112*, 138, *152*
Callear, A.B., 12, *58*
Calvin, M., 12, *58*
Carafoli, E., 137, *152*
Caruso, T., 177, 190, *215*
Cassol, A., 195, *216*
Caswell, A.H., 225, *251*
Catlin, G., 118, 143, *149*, 228, *251*
Catterall, R., 68, 70, 72, 73, 81, *111*
Cavallone, F., 53, *61*
Ceraso, J.M., 64, 65, 75, 79, 81, 83, 90, 92, 93, 94, 99, 102, 103, 105, *108*, *109*, *112*, 138, *152*, *153*
Chabanel, M., 124, 132, *152*
Chadwick, P.D., 195, *216*
Chamberlain, J.W., 121, 140, 142, *151*, 224, *251*

Chan, S.I., 90, *112*, 140, *154*, 179, 186, 190, *215*
Chan, Y., 64, *109*, 239, *253*
Chance, B., 247, *253*
Chance, C.P., 226, 228, 247, *251*
Chandler, D.E., 227, *252*
Chao, Y., 158, *213*
Chappell, J.B., 145, *155*
Chastrette, F., 177, *215*
Chastrette, M., 177, *215*
Cheney, J., 119, 138, *151*, *152*
Chidlow, S., 195, 213, *216*
Chinson, G.W., 64, *109*
Chmielowiec, J., 53, *61*
Chock, P.B., 93, *113*, 143, *155*, 201, *216*
Christensen, J.J., 64, 80, 96, *108*, *109*, *112*, 116, 118, 119, 120, 121, 125, 132, 140, 143, 144, 147, *149*, *150*, *152*, 158, 159, 181, 188, 195, *213*, *215*, 223, 229, 230, 231, 234, 237, 239, *250*, *252*, *253*
Christian, G.D., 19, *59*
Chumburidz, T.S., 140, *154*
Ciani, G., 240, *253*
Ciani, S., 18, 44, 45, 46, 47, 50, 51, 52, *58*, *60*, 118, 124, 140, 142, 143, *149*, *152*, 221, 241, 242, 243, 244, *250*, *253*
Cimerman, Z., 2, 30, 36, *57*
Cisar, A., 86, *112*
Clement, D., 139, *153*
Clementi, E., 8, 9, 11, 53, *57*, *58*, *61*
Cockrell, R.S., 118, *149*
Cohen, M.H., 64, *107*
Collet, A., 78, 79, *112*
Collins, G.G.S., 227, *252*
Cook, D.H., 174, *214*
Cook, F.L., 177, 190, 191, *215*
Cook, P.C., 118, 140, *150*, *154*
Coolen, R.B., 64, 79, 81, *108*
Cooper, J., 159, *213*
Copeland, D.A., 66, 67, 77, *110*, *111*
Corbett, J.D., 86, *112*
Corum, C.J., 140, *153*, 220, 224, *250*
Cox, B.G., 64, *109*
Cradwick, P.D., 132, 133, 137, *152*
Cram, D.J., 2, 53, *57*, *60*, 158, 159, 164, 165, 166, 167, 168, 169, 170, 171, 172, 173, 177, 180, 183, 184, 186, 188, 189,

191, 195, 196, 199, 200, 201, *213,*
214, 215, 216, 225, 231, 236, *251,*
252
Cram, J.M., 169, *213*
Crofts, A.R., 145, *155*
Curtis, W.D., 200, 201, *216*
Czerwinski, E.W., 121, 140, *151*

Daasvatn, K., 159, *213*
DaGue, M.G., 64, 65, 86, 87, 88, 89, *108*
Dainton, F.S., 64, *109*
Dale, J., 121, *151*, 159, 177, *213, 214*
Dalley, N.K., 64, *109*, 188, *215*, 230, 239, *252, 253*
Dalton, L.R., 70, 72, *111*
Damay, P., 66, *110*
Damm, F., 139, *153*
Danesh-Khoshboo, F., 174, *214*
Das, T.P., 64, 90, *107, 113*
Dawkins, P., 121, 125, 142, *151*
DeBacker, M.G., 64, 66, 75, 79, 81, *108, 110, 111*
Deber, C.M., 223, 224, *251*
DeBruijn, J.F., 180, *215*
DeCastro, A.F., 247, 248, *254*
Degani, H., 236, *252*
DeGazman, N.T., 220, 236, 241, *250*
Dehayes, L.J., 145, *155*
Dehmbov, E.V., 201, *217*
DeJong, F., 2, *57,* 158, 159, 164, 165, 166, 169, 178, 179, 180, 184, 185, 186, 190, 191, 193, 195, 196, 199, 200, 202, 204, 212, *213, 214, 215, 217,* 231, *252*
Delahay, P., 77, *112*
Demortier, A., 66, *110*
Dennhardt, R., 35, *60*
DePaoli, G., 195, *216*
Deverell, C., 90, *113*
DeVries, J.X., 140, *153*
Dewald, R.R., 67, 68, 69, 71, 75, 77, 86, *111, 112*
DeWitt, R., 53, 61, 90, 91, *113*
Diaddaria, L.L., 30, *59*
Diakiw, V., 2, *57*
Diebler, H., 143, *155*
Dietrich, B., 2, 30, *57, 59,* 64, 90, 93, 95, *108,* 119, 138, *151, 152*
Dobler, H., 121, 125, 142, 145, *151*

Dobler, M., 23, 26, *59,* 65, 82, *110,* 123, 124, 125, 133, 142, 143, *151, 152, 155*
Dohner, R.E., 35, *60*
Domeier, L.A., 2, *57,* 158, 159, 169, 186, 195, 200, *213,* 231, *252*
Dorfman, L.M., 70, 75, 81, *111, 112*
Dotsevi, G., 53, *60*
Douglas, W.W., 226, 227, *251, 252*
Down, J.L., 64, *109*
Drago, R.S., 191, *215*
Drewes, S.E., 163, *214*
Dullenkopf, W.D., 86, *112*
Dumont, J.E., 227, *252*
Dunitz, J.D., 65, 82, *110,* 121, 123, 124, 125, 142, 143, 145, *151, 152*
Durst, H.D., 159, 177, 201, *213*
Duval, E., 66, *110*
Dye, J.L., 64, 65, 66, 67, 68, 69, 70, 71, 72, 73, 74, 75, 76, 78, 79, 81, 83, 84, 85, 86, 87, 88, 89, 90, 91, 92, 93, 94, 97, 98, 99, 100, 101, 102, 103, 104, 105, *107, 108, 109, 110, 111, 112, 113,* 138, 139, *152, 153*
Džidić, I., 8, 9, *57*

Eatough, D.J., 64, 80, 96, *108, 109, 112,* 116, 118, 119, 120, 121, 125, 143, 147, *149,* 158, 159, 181, 188, 195, *213,* 229, 230, 231, 234, 237, 239, *252, 253*
Edwards, P.P., 70, 72, 86, *111, 112*
Efremov, E.S., 140, *154*
Eggers, F., 119, 140, 142, *150,* 201, *216*
Eigen, M., 143, *155*
Eimerl, I.S., 226, *252*
Eisenberg, M., 224, 236, 239, 240, 241, *251, 252*
Eisenman, G., 15, 16, 18, 35, 44, 45, 46, 47, 50, 51, 52, *58, 59, 60,* 118, 119, 124, 140, 142, 143, 147, *149, 150,* *152, 155,* 221, 240, 241, 242, 243, 244, *250, 253*
El Haj, B., 80, 112
Elke, D., 90, *113*
Elkins, D., 22, 24, *59*
Entman, M.L., 225, *251*
Epel, D., 227, *252*
Erlich, R.H., 90, *113*
Estrada, O.S., 38, *60,* 140, *154*

Evans, R.H., Jr., 236, *252*
Evstralov, E.V., 140, *154*
Evstratov, A.V., 119, 140, 142, *150*, 223, 250
Eyal, E., 18, *58*, 118, 119, *150*
Eyre, J.A., 75, *111*
Eyring, E.M., 64, *109*
Eytan, E., 247, *253*

Faesel, J., 140, *153*
Fairfield, J., 247, 248, 249, *254*
Farrow, M.M., 64, *109*
Faulstich, H., 140, 142, *154*
Fedarko, M.C., 140, *154*
Feinstein, M.B., 226, *251*
Feng, D.F., 67, *111*
Fenton, D.E., 119, 123, 124, 125, 133, *150*, 174, 186, *214*, *215*
Feuer, H., 159, *213*
Ffromov, E.S., 140, *154*
Fiedler, U., 2, 21, 30, 36, *57*, *59*, 137, *152*
Fleming, I., 12, *58*
Fletcher, J.W., 70, 71, 72, 77, 81, *111*
Flygare, W.H., 90, *113*
Foreman, J.C., 226, *251*
Fraga, S., 90, 102, *113*
Frant, M.S., 19, *59*
Fraser, C., 226, *251*
Freed, S., 66, *110*
Frensch, K., 167, 174, 176, 182, *214*
Frensdorff, H.K., 30, *59*, 64, 100, *108*, 118, 119, 123, 124, 125, 126, 127, 128, 133, 143, 144, 145, *149*, *150*, 159, 182, *213*, *215*, 223, 248, *250*
Friedenberg, A., 81, *112*
Friedman, H.L., 236, *252*
Fruh, P.U., 119, 140, *151*
Fueki, K., 67, *111*
Funch, Th., 119, 140, 142, *150*
Funck, R.J.J., 14, *58*
Funck, T., 201, *216*
Furukawa, J., 201, 209, *216*

Gabor, G., 76, *112*
Gachon, P., 140, *153*
Gaeta, F., 170, 177, *214*
Gavlas, J.F., 81, *112*
Gazzotti, P., 137, *152*

Gerlach, H., 140, *153*
Gillette, P.C., 225, *251*
Girodeau, J.M., 172, 173, 177, *214*
Glasser, P., 247, 248, 249, *254*
Glickson, J.D., 224, *251*
Gokel, G.W., 2, *57*, 158, 159, 169, 172, 173, 177, 180, 186, 191, 195, 200, 201, *213*, *214*, *215*, 231, *252*
Gold, M., 66, *110*
Goldberg, I., 193, 213, *216*, *217*, 231, *252*
Goldberg, M.W., 140, *154*
Golden, S., 66, 74, 75, *110*, *111*
Gomperts, B.D., 226, *251*
Goodall, M.C., 224, *251*
Goodman, M., 140, *154*
Gorman, M., 140, *153*
Goudeau, J., 86, *112*
Gould, R.F., 64, *108*
Graf, E., 139, *153*
Graven, S.N., 38, *60*, 140, 143, *154*, *155*
Gray, R.T., 159, 161, 162, 163, 164, 165, 167, 168, 169, 177, 178, 179, 180, 184, 185, 191, *213*, *214*, *215*
Gree, R.D., 191, *215*
Green, D.E., 211, 220, 228, 245, 246, 247, 248, 249, *250*, *253*, *254*
Greenberg, M.S., 90, *113*
Greene, R.N., 121, *151*, 177, *214*
Grell, E., 119, 140, 142, *150*, 201, *216*
Gremmo, N., 68, 70, 77, *111*, *112*
Grenier, G., 227, *252*
Gresh, N., 53, *61*
Grisdale, E.E., 158, 179, *213*
Grodzinski, J.J., 121, 138, *151*
Grossmann, P., 53, *60*
Güggi, M., 2, 21, 23, 30, 31, 36, 37, *57*, *59*, 137, *152*
Guttman, C., 66, 74, *110*
Gutzeit, A., 226, *252*

Haeusler, G., 226, *251*
Hall, J.E., 46, *60*, 243, *253*
Hamada, M., 191, *215*
Hamill, R.L., 140, *153*
Hamilton, R.M.D., 236, *252*
Hancock, R.D., 12, 15, *58*, 138, *153*
Hanji, K., 201, 209, *216*
Hansch, C., 22, 24, *59*

Hansen, E.M., 70, 72, *111*
Hansen, L.D., 64, *109*, 188, *215*, 239, *253*
Harder, A., 86, *112*
Harned, R.L., 140, *153*, 220, 224, *250*
Harris, E.J., 118, 140, 143, *149*, *150*, *154*, 220, *250*, 228, *251*
Harris, H.P., 158, 179, 181, 191, *213*, *215*, 224, *251*
Harter, P.H., 140, *153*
Haworth, R.A., 248, *254*
Haydon, D.A., 50, *60*, 240, *253*
Haymore, B.L., 64, 96, *108*, *109*, 119, 121, 140, 143, 144, *150*, 188, *215*, 231, 234, 237, 239, *252*, *253*
Haynes, D.H., 93, *113*, 118, 119, 140, 142, 143, *149*, *150*, *154*, 201, *216*
Heftman, E., 128, *152*
Hegelson, R.C., 236, *252*
Heichal, O., 226, *252*
Helgeson, R.C., 2, *57*, 158, 159, 164, 165, 169, 170, 171, 177, 180, 183, 186, 188, 189, 195, 200, *214*, 225, 231, *251*, *252*
Henco, K., 64, 91, *109*
Herceg, B., 65, *110*
Herlem, M., 90, *113*
Herrin, J., 140, *153*, 220, *250*
Hertz, H.G., 92, *113*
Hiberly, P.C., 167, 168, 177, 180, 189, 196, 199, 200, 212, *214*, 231, *252*
Hickel, D., 121, 140, 142, *151*, *153*
Hidaka, H., 227, *252*
Hidy, P.H., 220, 224, *250*
Higgens, C.E., 140, *153*
Hill, J.O., 64, *108*, 143, *150*, 223, *250*
Hinz, F.P., 130, 134, *152*
Hirano, S., 140, *153*
Hladky, S.B., 45, *60*, 240, 243, *253*
Hodgkinson, L.C., 170, 177, 180, 209, 210, 212, *214*
Hofer, I., 140, *153*
Hoffman, D.H., 2, *57*, 158, 159, 169, 186, 195, 200, *213*, 231, *252*
Hofmeister, D.W., 90, *113*
Holz, R.W., 226, 236, *251*, *252*
Hooz, J., 159, *213*
Hopkins, H.P., 158, 179, *213*
Horie, Y., 67, 68, *111*
Horiguchi, K., 201, 209, *216*

Horii, T., 191, *215*
Hotchkiss, R.D., 221, *250*
Hourdakis, A., 90, *113*
Hughes, D.L., 124, 126, 130, 133, *152*
Hui, J.Y.K., 159, *213*
Huis, R., 180, 199, 202, 204, 212, *215*, *217*
Hunter, D.R., 248, *254*
Hurley, I., 70, 72, 74, *111*
Huster, E., 66, *110*
Hutchison, C.A., Jr., 66, *110*

Ikenberry, D., 90, *113*
Illgenfritz, G., 143, *155*
Imirzi, R., 195, *216*
Inesi, G., 225, *251*
Ivanov, V.T., 2, *57*, 119, 140, 142, *150*, *151*, *154*, 220, 223, 224, 229, 230, 231, 237, 238, 239, *250*, *251*
Iwachido, T., 195, *216*
Izatt, N.E., 231, 234, 239, *252*
Izatt, R.M., 64, 80, 96, *108*, *109*, *112*, 116, 118, 119, 120, 121, 125, 132, 140, 143, 144, 147, *149*, *150*, *152*, 158, 159, 181, 188, 195, *213*, *215*, 223, 229, 230, 231, 234, 237, 239, *250*, *252*, *253*

Jaeger, D.A., 180, *215*
Jagger, E.J., 118, 140, 143, *150*
Jagger, W.S., 220, *250*
Jagur-Grodzinski, J., 64, 80, 90, 93, *109*, *112*, 191, 201, 205, *215*, *216*
Jakubetz, W., 53, *61*
Janacek, K., 116, 137, 143, *149*
Janzen, K.P., 53, *60*
Jepson, B.E., 53, *61*
Johnson, J.H., 118, 140, 143, *150*, 220, *250*
Johnson, L.F., 140, *154*
Johnson, R.G., 220, *250*
Johnson, S.M., 140, *153*, 220, *250*
Jolly, W.L., 64, 66, *107*, *110*
Jones, K.L., 140, *153*, 220, 224, *250*
Jones, T.E., 30, *59*
Jordan, P., 2, *57*, 140, *154*
Jortner, J., 64, 66, 77, *108*, *110*
Jou, F.Y., 70, *111*
Justice, J.C., 80, *112*

Kaempf, B., 78, 79, *112*

Kafka, M.S., 236, *252*
Kaplan, J., 66, *110*
Kaplan, L., 2, *57*, 121, 134, 142, 143, 146, *151*, 158, 159, 169, 186, 195, 200, *213*, 231, *252*
Karle, I.L., 121, 134, 142, 143, 146, *151*
Kauffman, E., 64, 96, 97, *109*, 138, *152*
Kawamura, S., 191, *215*
Kebarle, P., 8, 9, *57*
Keller-Schierlein, W., 140, *153*, *154*
Kellogg, R.M., 158, 174, *213*
Kemeny, G., 245, 246, *253*
Kergomard, A., 140, *153*
Kessler, M., 2, *57*
Kessler, R.J., 247, 248, 249, *254*
Kestner, N.R., 64, 66, 67, 77, *108*, *110*, *111*
Ketterer, B., 242, *253*
Kevan, L., 67, *111*
Kidd, R.G., 90, *113*
Kilbourn, B.T., 121, 125, 142, 145, *151*
Kimura, M., 195, *216*
Kintzinger, J.P., 64, 90, 91, *109*, *113*
Kirsch, N.N.L., 14, 15, 16, 18, 23, *58*, 137, *152*
Kistenmacher, H., 8, 9, 53, *57*
Kita, H., 226, *251*
Kitahari, A., 190, *215*
Kittel, C., 66, *110*
Klaulke, G., 53, *60*
Klimes, J., 191, *215*
Knipe, A.C., 64, 107, *108*, *109*
Knöchel, A., 191, *215*
Knowles, A.F., 247, *253*
Kobuke, Y., 201, 209, *216*
Koenig, K.E., 164, 165, 169, 170, 177, 183, 188, 189, *214*
Koga, K., 225, *251*
Komai, H., 248, *254*
Kon-no, A., 190, *215*
Konz, W., 140, *153*
Koryta, J., 35, *60*
Kosower, E.M., 236, *252*
Kotyk, A., 116, 137, 143, *149*
Kowalsky, A., 93, *113*, 140, *154*
Koyama, H., 140, *153*
Krajewski, J., 121, 125, 142, 143, *151*
Krasne, S., 44, 45, 46, 47, 50, 51, 52, *60*, 147, *155*, 240, 241, 242, 243, 244, *253*

Kratochvil, B., 196, *216*
Kraus, C.A., 64, 75, *107*, *112*
Kristiansen, P.O., 121, *151*, 177, *214*
Kubota, T., 140, *153*
Kuo, K.H., 50, *60*
Kuroda, Y., 177, *214*
Kyba, E.P., 225, 231, *251*, *252*

Läuger, P., 44, 46, 47, 49, 50, *60*, 242, *253*
Lagowski, J.J., 64, *108*
Laidler, D.A., 172, 173, 180, 199, 200, 201, 211, *214*, *216*
Laidler, K.J., 8, *57*
Laine, I.A., 140, *154*, 223, *250*
Lal, S., 19, *59*
Lamb, J.D., 231, 234, 237, 239, *252*, *253*
Land, R.H., 67, *110*
Landers, J.S., 64, 90, 94, 99, 102, *109*
Laprade, R., 44, 46, 50, 51, 52, *60*, 240, 243, *253*
Lardy, H.A., 38, *60*, 119, 140, 143, *150*, *154*, *155*, 220, 225, 236, *250*, *252*
Larson, J.M., 162, 166, *214*
Lee, B., 226, 228, 247, *251*
Lehn, J.M., 2, 12, 13, 15, 23, 30, *57*, *58*, 64, 65, 78, 79, 86, 87, 88, 89, 90, 91, 93, 95, 96, 97, *108*, *109*, *112*, *113*, 116, 119, 120, 121, 125, 126, 130, 131, 135, 138, 139, 145, 147, *149*, *151*, *152*, *153*, 159, 171, 172, 173, 177, 196, *213*, *214*, 230, 231, 233, *252*
Lehninger, A.L., 224, 240, 245, 247, *251*, *253*
Leibfritz, D., 180, *215*
Leigh, S.J., 170, 177, 179, 180, 209, 210, 212, *214*
Lelieur, J.P., 66, *110*
Leo, A., 22, 24, *59*
Lepoutre, G., 64, 66, *108*, *110*
Lev, A.A., 19, 41, 50, *59*
Levanon, H., 81, *112*
Levins, R.J., 37, *59*
Lewis, J., 64, *109*
Lieberman, E.A., 247, *254*
Liesegang, G.W., 64, *109*
Lin, W., 2, *57*
Lindenbaum, S., 220, 236, 237, 241, *250*
Lindoy, L.F., 177, *214*

Lindquist, R.H., 66, *110*
Linnell, R.H., 190, *215*
Linschitz, H., 70, 72, *111*
Liotta, C.L., 158, 177, 179, 181, 190, 191, *213*, *215*, 224, *251*
Lipkind, G.M., 140, *154*
Liu, L., 90, 91, 93, 97, *113*, 140, *153*, 220, *250*
Live, D., 90, *112*, 179, 186, 190, *215*
Logan, J., 67, *111*
Lok, M.T., 64, 65, 70, 71, 75, 78, 79, 81, 83, *108*, *111*, *153*
Long, L.D., 79, *112*
Lorscheider, R., 53, *60*
Loyola, V.M., 64, 95, *109*
Luben, G., 140, *153*
Lugo, R., 77, *112*
Lutz, W.K., 119, 140, *151*
Luz, Z., 64, 90, 93, *109*, 138, *152*, 201, 205, *216*

Maas, G.E., 239, *253*
Maass, G., 64, 91, *109*
McClure, G.L., 174, *214*
McDougall, G.J., 138, *153*
McKervey, M.A., 164, 165, 166, 177, 188, 189, *214*, *215*
McLachlan, R.D., 191, *216*
McLaughlin, S., 16, *58*, 124, *152*, 236, 239, 240, 241, *252*, *253*
MacLennan, D.H., 220, *250*
McMurray, W., 220, 221, *250*
Madan, K., 2, *57*, 158, 159, 169, 186, 195, 200, *213*, 231, *252*
Maier, C.A., 140, *153*
Malenkov, G.G., 140, 142, *150*, 199, 223, *250*
Malev, V.V., 19, 41, 50, *59*
Malli, G., 90, 102, *113*
Mallinson, P.R., 119, 124, 125, 132, 133, 145, *150*
Mal'tsev, E.I., 77, *112*
Manohar, H., 132, 137, *152*
Margalit, R., 50, *60*
Margerum, D.W., 12, *58*, 130, 134, *152*
Maricano, F., 12, 15, *58*
Marius, W., 53, *61*
Martell, A.E., 12, *58*

Martin, L.Y., 145, *155*
Mass, G., 143, *155*
Matalon, S., 75, *111*
Mater, J.M., 236, *252*
Mathews, S.E., 64, 65, 83, 87, *108*, 138, 139, *153*
Mathies, R., 224, *251*
Mathieu, F., 121, 138, *151*
Matsutani, S., 140, *153*
Matsuura, N., 80, *112*
Maurer, P.G., 53, *60*
May, K., 2, *57*
Mayers, D.F., 140, *154*, 224, *251*
Mead, C.A., 47, *60*, 243, *253*
Mei, E., 64, 90, 91, 93, 97, 98, 99, 100, 101, *109*, *113*, 138, *152*
Meier, P. Ch., 12, 13, 15, 17, 18, 19, 20, 28, 35, 38, *58*, *60*, 116, 118, 119, 120, 121, 125, 126, 137, 139, 141, 143, 147, *149*
Melnik, E.I., 140, *154*, 223, *250*
Menard, M.C., 124, 132, *152*
Mercer, M., 119, 123, 124, 125, 133, *150*, 186, *215*
Mernery, P., 80, *112*
Merryman, D.J., 86, *112*
Metz, B., 13, *58*, 65, 82, 84, *110*, 121, 138, 139, *151*, *153*, 195, *216*
Midgley, D., 159, 181, *213*
Mikhaleva, I.I., 119, 140, 142, *150*, *151*
Minnich, E.R., 79, *112*
Miroshnikov, A.I., 140, *154*
Monger, J.L., 226, *251*
Montal, M., 226, 228, 247, *251*
Montavon, F., 119, *151*
Moore, B., 64, *109*
Moore, C., 2, *57*, 140, *154*, 220, 228, *250*
Moore, S., 118, 199, 200, 201, *215*
Moore, S.S., 167, 168, 177, 189, 196, 199, 200, 212, *214*, *216*, 231, *252*
Moras, D., 13, *58*, 65, 82, 83, 84, *110*, 121, 138, *151*, 195, *216*
Moreau, P., 2, *57*, 158, 159, 169, 180, 186, 195, 200, *213*, *214*, 231, 236, *252*
Morf, W.E., 2, 8, 10, 11, 12, 13, 15, 16, 17, 18, 19, 20, 21, 23, 25, 28, 30, 31, 35, 36, 38, 39, 40, 41, 42, 43, 44, 45, 46, 47, 48, 50, 51, 53, *57*, *58*, *59*, *60*, 64,

108, 116, 118, 119, 120, 121, 125, 126,
 137, 139, 140, 141, 142, 143, 146, 147,
 149, *150*, *151*, *152*, 220, 221, 229, *250*
Mortimer, C.L., 130, *152*
Moschler, H.J., 140, 142, *150*, 199
Mueller, P., 118, 140, 142, 145, *149*
Muirhead-Gould, J.S., 8, *57*
Mulholland, D.L., 164, 165, 166, 177, 188,
 189, *214*, *215*
Muller, W.A., 226, *252*
Murakami, Y., 140, *153*

Nae, N., 80, *112*, 191, *215*
Nätscher, R., 174, *214*
Nager, U., 140, *153*
Nakabayashi, T., 191, *215*
Nakamura, Y., 67, 68, *111*, 201, 209, *216*
Nakazato, Y., 227, *252*
Navon, G., 236, *252*
Nawata, Y., 140, *153*
Nayak, A., 174, *214*
Neider, F., 140, *154*
Nelson, D.P., 64, 96, *108*, *109*, 119, 121,
 140, 143, 144, *150*, 188, *215*, 239,
 253
Neubert, D., 224, 240, *251*
Neumann, P., 159, 177, 195, *213*, *216*
Neumcke, B., 44, 47, 50, *60*
Neupert-Laves, K., 23, 26, *59*
Newcomb, M., 2, *57*, 158, 159, 164, 165,
 166. 169, 172, 177, 180, 183, 184, 186,
 188, 189, 195, 199, 200, 201, *213*, *214*,
 215, *216*, 231, 236, *252*
Newkome, G.R., 171, 174, 177, *214*
Newman, M.S., 28, *59*
Newton, M.D., 66, 67, 77, *110*
Nguyen, V.B., 53, *60*
Nguyentien, T., 53, *60*
Nicely, V.A., 64, 70, 72, *108*, *111*
Nicolsky, B.P., 35, *59*
Nowell, I.W., 174, *214*

Ochrymowycz, L.A., 30, *59*
Oehler, J., 191, *215*
Oehme, M., 2, 30, 31, 36, *57*, *59*, 137, *152*
Oesch, U., 25, *59*
Ogg, R.A., Jr., 66, *110*
Ohnishi, M., 140, *154*

Oishi, H., 140, *153*
Okino, H., 177, *214*
Okutomi, T., 140, *153*
Olivier, A., 140, *154*
Ollolenghi, M., 70, 72, *111*
Orci, L., 116, *252*
O'Reilly, D.E., 66, 67, *110*
Osborne, M.W., 228, *251*
Osipov, V.V., 19, 41, 50, *59*
Osswald, H., 2, 30, 35, 36, 37, *57*, *59*, *60*
Ottenheym, H., 140, *153*
Ottewill, R.H., 12, *58*
Ottolenghi, M., 75, *111*
Ovchinnikov, Yu. A., 2, *57*, 119, 140, 142,
 150, *151*, *154*, 220, 223, 224, 229, 230,
 231, 237, 238, 239, *250*, *251*

Pache, W., 140, *154*
Panayotov, I.M., 64, 69, *109*
Papadakis, N., 64, 79, 81, *108*
Parson, D.G., 126, 130, 133, 134, 136,
 152, 195, *216*
Pastor, R.C., 66, *110*
Patel, D.J., 140, *154*, 220, 236, *250*, *252*
Paul, I.C., 140, *153*, 220, *250*
Pauling, L., 11, *57*
Peacock, S.C., 2, 57, 158, 159, 169, 186,
 195, 200, *213*, 231, 236, *252*
Pedersen, C.J., 2, *57*, 64, 100, *108*, 119,
 121, 123, 124, 125, 126, 127, 128,
 133, 145, 147, *150*, *151*, 158, 159,
 182, 186, 190, *213*, *215*
Petránek, J., 2, 36, *57*, *59*, 121, *151*
Petrov, E.S., 64, *109*
Pfeiffer, D.R., 236, *252*
Phizackerley, R.P., 65, 82, *110*, 124,
 125, 133, 142, *152*, *155*
Pick, J., 225, *251*
Pinkerton, M., 121, 125, 140, 142, *151*,
 153, 224, *251*
Pioda, L.A.R., 19, *59*, 119, 140, 142,
 150, 225, *251*
Pitzer, K.S., 66, *110*
Pizer, R., 64, 95, *109*
Plattner, P.A., 140, *153*
Plesch, P.H., 159, *213*
Podos, S.M., 228, *251*
Poonia, N.S., 117, 119, 121, 123, 124,

Author Index

125, 126, 127, 128, 130, 132, 133, 134, 135, 136, 137, 139, 145, *149*, *150*, *151*, *152*, *153*, 179, 186, 189, 195, *215*, *216*
Popkie, H., 8, 9, 53, *57*
Popov, A.I., 64, 90, 91, 93, 97, 98, 99, 100, 101, *109*, *112*, *113*, 138, 140, *152*, *154*
Portonva, S.L., 140, *154*
Prelog, V., 2, *57*
Pressman, B.C., 2, *57*, 93, *113*, 118, 119, 139, 140, 142, 143, *149*, *150*, *153*, *154*, *155*, 220, 221, 224, 225, 226, 228, 231, 236, 237, 238, 239, 240, 241, *250*, *251*
Prestegard, J.H., 140, *154*
Pretch, E., 2, 14, 15, 16, 17, 18, 19, 21, 23, 25, 29, 30, 31, 35, 36, 37, 39, 41, 44, *57*, *58*, *59*, 118, 137, 140, 147, *149*, *152*, *154*, 225, *251*
Prince, R.H., 12, *58*
Prince, W.T., 226, *252*
Prox, A., 140, *153*
Pullman, A., 53, *61*, 189, *215*
Pungor, E., 225, *251*
Purdie, N., 64, *109*

Rachin, A.I., 140, *154*
Racker, E., 247, *253*
Ramsey, N.F., 90, *113*
Randles, J.E.B., 68, 70, 77, *111*, *112*
Rassmussen, T.H., 226, *252*
Raynal, S., 78, 79, *112*
Rechnitz, G.A., 18, *58*, 118, 119, *150*
Reed, P.W., 220, 225, 236, *250*, *252*
Reible, S., 245, 246, *253*
Reinhoudt, D.N., 159, 161, 162, 163, 164, 165, 166, 167, 168, 169, 177, 178, 179, 180, 184, 185, 190, 191, 193, 196, 199, 202, 212, *215*, 204, *213*, *214*, 217
Remoortere, F.P. van, 195, *216*
Rest, A.J., 177, *215*
Reynolds, A.E., 226, *252*
Richards, R.E., 90, *113*
Richman, J.E., 177, *214*
Rigny, P., 66, *110*
Riphagen, B.G., 163, *214*

Rippel, S., 220, 236, 237, 241, *250*
Roach, E., 90, *113*
Robinson, J.M., 171, 177, *214*
Rodriguez, L.J., 64, *109*
Rorbacher, D.B., 30, *59*
Rosalky, J.M., 195, *216*
Ross, J.W., Jr., 19, *59*
Rossiter, B.E., 231, 234, 237, 239, *252*
Rozalky, J.M., 139, *153*
Rudin, D.O., 118, 140, 142, 145, *149*
Rudolph, G., 191, *215*
Rutter, A., 143, *155*
Ryabova, I.D., 223, *250*
Ryan, T.E., 220, *250*
Ryba, O., 2, 36, *57*, *59*, 121, *151*
Rynbrandt, J.D., 70, 72, *111*
Rytting, J.H., 64, 96, *108*, 119, 121, 140, 143, 144, *150*, 188, *215*, 231, 234, 237, 239, *252*

Sagawa, T., 140, *153*
Saika, A., 90, *113*
Sam, D.J., 158, *213*
Sanderson, R.T., 125, *152*
Sandifer, J.R., 35, *59*
Sanko, K., 242, *253*
Sarges, R., 224, *251*
Sasaki, A., 80, *112*
Sauvage, J.P., 2, 13, 30, *57*, *58*, 64, 90, 93, 95, 96, 97, *108*, *109*, 119, 130, 131, 135, 138, 145, *151*, *152*, 172, 173, 177, *214*, 230, 231, 232, *252*
Savion, N., 226, *252*
Sawada, M., 140, *153*
Scarpa, A., 220, 225, *250*, *251*
Schadt, M., 226, *251*
Schindewolf, U., 64, *107*, *108*
Schindler, J.G., 35, *60*
Schmid, J., 140, *153*
Schmidt-Kastner, G., 140, *153*, 220, *250*
Schneider, H., 64, *109*
Schneider, J., 2, *57*
Scholer, R., 19, *59*
Scholten, G., 53, *60*
Schué, F., 78, 79, *112*
Schuster, P., 53, *61*
Schwartz, A., 225, *251*
Schwarzenbach, G., 12, 23, *58*

Schwyzer, R., 119, 140, 142, *150*
Scordamaglia, R., 53, *61*
Scott, W.E., 140, *154*
Seddon, W.A., 70, 71, 72, 77, 81, *111*
Seiler, P., 65, 82, *110*, 121, 123, 124, *151*, *152*
Selinger, Z., 226, *252*
Senior, A.E., 247, 248, *254*
Senyavina, L.B., 140, *154*
Shamoo, A.E., 220, *250*
Sharaf, A.A., 247, *254*
Sharp, G.W.G., 226, *252*
Shatenshtein, A.I., 64, *109*
Shaw, B.L., 177, *215*
Shchori, E., 64, 80, 90, 93, *109*, *112*, 121, 138, *151*, 201, 205, *216*
Shemyakin, M.M., 119, 140, 142, 146, *150*, *153*, *154*, 223, *250*
Shen, C., 220, 236, *250*, *252*
Shepel, E.N., 140, *154*
Shih, J.S., 90, 91, *113*
Shimoji, M., 67, 68, *111*
Shiro, M., 140, *153*
Shkrob, A.M., 2, *57*, 119, 140, 142, *150*, *154*, 220, 223, 224, 229, 230, 231, 237, 238, 239, *250*, *251*
Shporer, M., 64, 90, 93, *109*, 121, 138, *151*, *152*, 201, 205, *216*
Siegel, M., 225, *251*
Sienko, M.J., 64, *108*
Simmons, H.E., 158, *213*
Simon, W., 2, 8, 10, 11, 12, 13, 14, 15, 16, 17, 18, 19, 20, 21, 23, 25, 28, 29, 30, 31, 35, 36, 37, 38, 39, 40, 41, 42, 43, 44, 50, 51, 53, *57, 58, 59, 60, 61,* 64, *108,* 116, 118, 119, 120, 121, 125, 126, 137, 139, 140, 141, 142, 143, 146, 147, *149*, *150*, *151*, *152*, *153*, *154*, 220, 225, 229, 230, 231, 233, *250*, *251*, *252*
Simpson, J.B., 174, *214*
Skulachev, V.P., 247, *254*
Slater, J., 70, 72, 81, *111*
Slichter, C.P., 90, *113*
Slijko, F.L., 191, *215*
Smid, J., 80, *112*, 124, 125, 126, 145, *152*
Smit, C.J., 159, 161, 162, 163, 164, 165, 167, 168, 169, 177, 179, 180, 184, 185, 190, 191, 193, 196, 199, 202, *213*, *214*, *215*
Smith, P.B., 64, 90, 94, 99, 102, 104, 105, *109*, *113*
Smith, S.A., 177, *215*
Sogah, Y., 53, *60*
Sonthard, J.A., 248, *254*
Sopchyshyn, F.C., 72, 81, *111*
Sousa, L.R., 2, *57*, 158, 159, 162, 166, 169, 186, 195, 200, *213*, *214*, 225, 231, *251*, *252*
Spaargaren, K., 180, *215*
Speck, D.H., 177, 190, *215*
Springourum, M., 140, *153*
Stankova, V., 19, *59*, 225, *251*
Stark, G., 46, 47, 50, *60*, 240, 242, *253*
Starr, S., 132, *152*
Stefanac, Z., 2, *57*, 119, 140, 142, *150*, *154*, 225, *251*
Steiner, E.C., 195, *216*
Steinhardt, R.A., 227, *252*
Steinrauf, L.K., 121, 125, 140, 142, *151*, *153*, 224, *251*
Sternbach, L.H., 140, *154*
Sternson, L., 220, 236, 237, 241, *250*
Stewart, P.S., 220, *250*
Stockemer, J., 53, *60*
Stoddart, J.F., 172, 173, 180, 199, 200, 201, 211, *214*, *216*
Stryer, L., 224, *251*
Stubbs, M.E., 64, 90, *109*, 119, 138, *151*, *152*
Sugarman, N., 66, *110*
Sumskaya, L.V., 140, *154*
Sundheim, B.R., 64, 68, *109*
Sutherland, I.O., 170, 177, 179, 180, 209, 210, 212, *214*
Suzuku, K., 140, *153*
Symons, M.C.R., 64, 68, 70, 71, 73, *107*, *111*
Szabo, G., 18, 44, 46, 47, 50, 51, 52, 58, *60*, 118, 124, 140, 142, 143, *149*, *152*, 221, 240, 241, 243, *250*, *253*
Szwarc, M., 79, *112*

Tabuski, L., 177, *214*
Takeda, Y., 80, *112*

Author Index

Tarnowski, T.L., 188, 189, 199, 200, 201, 214, *215*, 231, *252*
Tehan, F.J., 64, 65, 70, 71, 75, 78, 79, 81, 83, 84, 85, *108*, *111*, 138, *153*
Terry, R.E., 64, *109*, 188, *215*, 239, *253*
Thoma, A.P., 2, 38, 39, 40, 41, 42, *57*, *60*
Thompson, J.C., 64, 65, 76, *107*
Tieffenberg, M., 118, 140, *150*, *154*
Tien, H.T., 44, *60*
Timko, J.M., 2, *57*, 158, 159, 167, 168, 169, 171, 172, 173, 177, 180, 186, 189, 195, 196, 199, 200, 212, *213*, *214*, *216*, 225, 231, 236, *251*, *252*
Ting, P., 247, *253*
Tipping, J.W., 68, 70, 72, 73, *111*
Titus, E.O., 201, *216*
Tôei, K., 195, *216*
Tonelli, A.E., 140, *154*
Torchia, D.A., 223, 224, *251*
Tosteson, D.C., 118, 119, 140, 143, *150*, *154*
Toth, K., 225, *251*
Traxler, G., 140, *153*
Trueheart, P.A., 226, *252*
Truter, M.R., 65, 82, *109*, *110*, 116, 119, 120, 121, 123, 124, 125, 126, 127, 130, 132, 133, 134, 136, 139, 140, 147, *149*, *150*, *151*, *152*, 186, 195, *215*, *216*
Tsvetanov, Ch. B., 64, 69, *109*
Tümmler, B., 64, 91, *109*
Tuttle, T.R., Jr., 64, 66, 70, 72, 73, 74, 75, *109*, *110*, *111*
Twigg, M.V., 12, *58*
Tyler, I.D., 177, *215*
Tyson, C.A., 247, 248, 249, *254*

Umemoto, K., 80, *112*
Ungaro, R., 80, *112*
Urry, D.W., 140, *154*, 224, *251*

Van der Kloot, W., 226, *251*
Vande Zande, H., 247, 248, 249, *254*
Vannikov, A.V., 77, *112*
Van Strio, J., 227, *252*
Vasak, M., 140, *154*, 225, *251*
Vazquez, F.A., 64, *109*
Veatch, W.R., 224, *251*

Veenstra, I., 159, 161, 162, 163, 164, 165, 167, 168, 169, 177, 179, *214*
Velichkova, R.S., 64, 69, *109*
Veschambre, H., 140, *153*
Vinogradov, S.N., 190, *215*
Vinogradova, E.I., 223, *250*
Viullenmier, P., 137, *152*
Viviani-Nauer, A., 38, 39, 40, *60*
Vögtle, F., 2, *57*, 64, *109*, 159, 160, 166, 167, 169, 171, 172, 173, 174, 175, 176, 177, 180, 181, 182, *213*, *214*
Von Remoorkere, F.P., 134, 136, *152*
Vos, K.D., 70, 72, *111*
Vuilleumier, P., 49, 50, *60*

Wadl, F., 170, *214*
Wagemen, F., 70, *111*
Wagner, J., 139, *153*
Walba, D.M., 167, 168, 177, 180, 189, 196, 199, 200, 212, *214*, *216*, 231, *252*
Wallick, E.T., 225, *251*
Warren, S.G., 12, *58*
Weber, E., 2, *57*, 64, *109*, 159, 160, 166, 167, 169, 171, 172, 173, 174, 175, 176, 177, 180, 181, 182, *213*, *214*
Weber, H.G., 119, 140, 142, *150*
Weed, G., 132, *152*
Wehner, W., 64, *109*, 171, 172, 174, 181, *214*
Wehrli, F.W., 90, 101, *113*
Weiland, T., 140, *153*
Weiss, L., 2, 21, 23, *57*
Weiss, R., 13, *58*, 65, 82, 83, 84, *110*, 121, 138, 139, *151*, 153, 195, *216*
Weissman, S.I., 64, *109*
Wells, P.F., 28, *59*
Wenger, J.J., 228, *251*
Westley, J.W., 121, 140, 142, *151*, *153*, 236, *252*
White, R.D., 64, *109*
Whitney, R.R., 180, *215*
Wieder, W., 174, 175, 180, *214*
Wieland, T., 119, 140, 142, *150*, *154*
Wiens, T., 118, 143, *150*
Wiest, R., 121, *151*
Wiles, D.M., 64, *109*
Wilkins, R.G., 64, 95, *109*

Wilkop, B., 224, *251*
Williams, J.A., 227, *252*
Williams, R.J.P., 11, *57*, 116, 137, *149*
Wilson, D.F., 247, *253*
Wingfield, J.N., 126, 130, 133, 134, 136, *152*, 195, *216*
Winkler, R., 140, 143, 145, *154*, *155*
Wipf, H.K., 119, 140, 142, *150*, *151*, *154*
Wojtkowiak, F., 124, 132, *152*
Wollhein, C., 226, *252*
Wong, K.H., 124, 125, 126, 145, *152*
Wong, S., 223, 224, *251*
Wright, A.N., 64, *109*
Wright, G.F., 177, *215*
Wudl, F., 177, *214*
Wuhrmann, H.R., 18, *58*, 119, *150*
Wuhrmann, P., 15, 16, 17, 18, 19, 21, 31, 38, 39, 41, 42, 43, 44, 50, 53, *58*, *59*, *60*, 137, *152*
Wun, T., 2, *57*

Yadav, B.P., 139, *153*
Yeager, H.L., 196, *216*
Yemen, M.R., 64, 65, 86, 87, 88, 89, *108*

Zahner, H., 140, *154*
Zanko, M.T., 228, *251*
Zemel, H., 201, *216*
Zimmer, L.L., 30, *59*
Zintl, E., 86, *112*
Zompa, L.J., 145, *155*
Zuber, M., 160, 177, *214*
Zust, Ch. U., 119, 140, 143, *150*, *151*

SUBJECT INDEX

Absorption spectrum, of thin films, of metal-cryptand complexes, 88
   of potassium-cryptand complexes, 89
   of sodium-cryptand complex, 87
Actins, structure of, 120
Alamethecin, cation binding selectivities of, 239
Alkali metal anions, 74-75
   in amines, 67-75
   in ammonia, 65-67
   crystal structure of salt of, 85
   free energies of hydration of, 12
   NMR spectra of, 101-103
   optical spectra in films, 87-89
   in solution, 74-75
Alkali metal exchange rates, measurement by NMR of, 93-95
Alkaline earth metal ions, in ammonia, 65-67
   in ethers, 67-75
   free energies of hydration of, 12
Alkaline earth metal picrate, extraction in methylene chloride with ionophores of, 32
Alkylammonium salts, relative binding constants in $CDCl_3$ for crown ether complexes of, 194
Amine cations, log K, ΔH, and ΔS values for reaction of 18-crown-6 with, 234-235
Amines, complexation constants for ionophore X-537A with, 237
   solutions of alkali and alkaline earth metals in, 67-75
Ammonia, solutions of alkali and alkaline earth metals in, 65-67
Ammonium ion, selectivity factors for ionophores in liquid membrane electrodes complexing with, 36-37
   stability constants of ionophores in ethyl alcohol with, 14
Ammonium picrate, association constants in $CDCl_3$ for crown ethers with, 189
Anion, effect of, on association constants of crown ether complexes of alkylammonium salts in $CDCl_3$, 198-199
   on complexation in solution, 128-132
   on complexation and stoichiometry in solids, 132-133
   on synthesis of metal-crown complexes, 127-128
   role of in metal-crown complexation, 126-133
Antanamide, cation binding selectivities of, 238
   chemical formula of, 228
   stability constants for metal complexes of, 141
   structure of, 120
Antibiotics, cyclic electrically neutral, 140
   metal complexes of, 140-143
   open-chain monobasic, 140
   similarity in complexation of crowns and, 143-147
Antiport coupling, of ion-transport mediators, 245-246
   linkage system for mediating, 246
Association constants, for crown ether-ammonium picrate in $CDCl_3$, 189
   for crown ether-cation complexes in $CDCl_3$, 185
   for crown ether complexes of alkylammonium salts in $CDCl_3$, 198-199
   for metal-crown ether complexes, 183
   see also Stability constants *and* Equilibrium constants
ATPase activity, of monactin in presence of alkali metal cations, 38

Barium ion, extraction in methylene chloride with ionophores, 32
   selectivity factors for ionophores in liquid

Subject Index

membrane electrodes complexing with, 36-37
stability constant of bicyclic cryptand complexes, 182
stability constants of cryptand complexes, 131, 232-233
Beauvericin, cation binding selectivities of, 238
Benzo-15-crown-5, cation binding selectivities of, 238
  stoichiometries of K, Na, Li, and Ca with, 122
  structure of, 116
Bicyclic cryptands, 138
Bilayer membranes, 44-53
  carrier-mediated ion-transport behavior of, 44-53
  current-voltage characteristics, for dioleoyl lecithin membranes, 47
    for lecithin bilayer membranes, 49
    for phosphatidyl serine bilayer membranes, 49
  general flux equation for carrier-mediated ion transport behavior in, 45
Binding sites for cations, 8-10
Bond lengths, in sodium-cryptand crystals, 84
Bulk membranes, 39-44
  cation transference numbers, 39
  electrical current density, 39

Calcium ion, extraction in methylene chloride with ionophores of, 32
  lipophilization by crown ethers of, 184
  models of complexes of ionophores with, 27
  molecular geometry of Ca^{2+} ionophore complex, 25
  secretory responses associated with ionophores dependency on, 226-227
  selectivity factors for ionophores in liquid membrane electrodes complexing with, 36-37
  stability constants of cryptates, 131, 182, 232-233
  stoichiometries of crown ethers with, 122
  transference number through Ca^{2+}-carrier membrane, 43
  transfer numbers in neutral carrier membranes for, 40
  transport of by combination of uncomplex and electrogenic ionophore, 249
Carrier-mediated ion transport, diagram of across a membrane, 240
  energy profile for, 242
  permeability ratio, 241-244
  through membranes, 35-53
Carrier membrane electrodes, EMF response of, 41
Carriers, ion selectivity of membranes containing, 15-22
Cation selectivity of ionophores, 237-239
Cation transference numbers, in bulk membranes, 39
Cavity radius, of minimal cavity enclosed by n oxygen atoms, 11
Cavity size, stoichiometry, of large-cavity crown complexes, 135-136
  of small-cavity crown complexes, 133-135
Cesium ion, absorption spectrum of thin film of cryptand with, 88
  association constants, of crown ether complexes, 183
  of crown ether complexes in $CDCl_3$, 185
  chemical shift differences of protons in crown ether complexes, 187
  effect of temperature on solution of cryptand with, 104
  ESR spectra in various solvents of monomer, 72
  formation constants in various solvents for complex of crowns and, 98
  NMR chemical shifts in various solvents of crown complex with, 100
  NMR spectra of Ce^-, 103-105
  sandwich complexes of crown ethers with, 99-101
  selectivity factors for ionophores in liquid membrane electrodes complexing with, 36-37
  solubilities in amines and ethers of, 69
  stability constants, for antibiotic complexes, 141
    for crown ether complexes, 144
    for cryptates, 131, 182, 232-233
Charge density, effect of on metal-crown complexation, 133-137
Chemical shift differences, poly-ether

Subject Index

proton in crown ether-cation complexes, 187
Conductance, effect of crowns and cryptands on, 79-80
Complexation, ability of ionophores for, 228-239
　of alkali cations by crown ethers or cryptands, 64-65
　of crown ethers, with alkylammonium salts under anhydrous conditions, 193-195
　　in apolar solvents, 182-186
　　in polar solvents, 181-182
　effect of anion, on crown complexation in solution, 128-132
　　in crystal lattice on, 132-133
　　on metal-crown complexes, 126-133
　effect of charge density of cation on, 133-137
　similarity of crown and antibiotics, 143-147
Complexes, of metal-antibiotics, 140-143
Complexing agents, design features of membrane-active, 22-31
Conformation, of alkylammonium complex of 1,3-xylyl-18-crown-5, 194
　of metal-crown complexes, 123-126
Coordination number, of group IA and IIA cations, 11-12
　of some metal ions, 11
15-Crown-5, cation binding selectivities of, 238
18-Crown-6, cation binding selectivities of, 238
　kinetics of complexation with alkylammonium salts, 204-205
　log K, ΔH, and ΔS values for reaction of amine cations with, 234-235
　stability constants of alkali metal complexes, 144
　structure of, 116
Crown ether, association constants, in $CDCl_3$ of alkylammonium complexes of, 198-199
　in $CDCl_3$ of ammonium picrate complexes with, 189
　in $CDCl_3$ of crown ether-cation complexes, 185
　for metal complexes of, 183
　complexation, with alkylammonium salts under anhydrous conditions of, 193-195
　with alkylammonium salts in presence of water, 195
　with ammonium and alkylammonium salts, 186-213
　in apolar solvents, 182-186
　in $CDCl_3$ of water with, 190
　in polar solvents, 181-182
　control of stoichiometry of metal-amine and metal-ether solutions by, 81-82
　effect, on formation of ion pairs or monomers in metal solutions of, 80-81
　　on species present in solution of, 79-82
　free energy relationship for alkylammonium salt complexes of, 207-208, 210-211
　lipophilization of cations by, 184
　maximum synthesis yields of, 163
　metal complexes isolation from methanol-ethyl acetate of, 181
　metal solubility enhancement by, 76-78
　synthesis yields using different alkoxide cations, 161
　twisted conformation of complexes of, 185-186
　yields, obtained by cyclization reactions, 165
　　obtained by substituent interconversion, 166
　see also Ionophores and Macrocyclic polyethers
Cryptand, 221, cation binding selectivities of, 238
Cryptand, 222, cation binding selectivities of, 238
Cryptands, bond lengths in solid sodium complexes of, 84
　control of stoichiometry of metal-amine and metal-ether solutions by, 81-82
　effect, on formation of ion pairs or monomers in metal solutions of, 80-81
　　on species present in solution of, 79-82
　exclusive complexes of, 101
　metal solubility enhancement by, 76-78
　stability constant in water of metal complexes with, 182
　stability constants, for alkali and alkaline earth metal complexes, 131
　of metal cation complexes, 232-233

structure of, 118, 230
types of, 137-139
Cryptate, *see* Cryptand
Crystal lattice, effect of anion on complexation in, 132-133
Crystal structure, of alkylammonium salt complexes of diazocrown ethers, 212
 of cryptand-sodium complex, 85
 of metals with crowns and cryptands, 82-89
Current-voltage characteristics, for dioleolyl lecithin bilayer membranes, 47
 for lecithin bilayer membranes, 49
 for phosphatidyl serine bilayer membranes, 49
Cyclization reaction, yields of crown ethers obtained by, 165

Depsipeptide antibiotics, structure of, 222
Dianemycin, cation binding selectivities of, 238
Dibenzo-14-crown-8, structure of, 116
Dibenzo-18-crown-6, formation constant in various solvents of cesium complexes, 98
 stability constants of alkali metal complexes, 144
 stoichiometries of K, Na, Li, and Ca complexes, 122
 structure of, 116
Dibenzo-24-crown-8, cation binding selectivities of, 238
 stability constants of alkali metal complexes, 144
 stoichiometries of K, Na, Li, and Ca complexes, 122
Dibenzo-27-crown-9, cation binding selectivities of, 238
Dibenzo-30-crown-10, stability constants of alkali metal complexes, 144
 stoichiometries of K, Na, Li, and Ca complexes, 122
 structure of, 116
Dicyclohexano-14-crown-4, cation binding selectivities of, 238
Dicyclohexano-18-crown-6, formation constants in various solvents of cesium and, 98
 stability constants of alkali metal complexes, 144

structure of, 116
Dicyclohexano-19-crown-6, cation binding selectivities of, 238
Dicyclohexyl, *see* Dicyclohexano
Dinactin, cation binding selectivities of, 238
 structure of, 3, 222
2,4-Dioxo-19-crown-6, cation binding selectivities of, 238
2,6-Dioxo-18-crown-6, cation binding selectivities of, 238
Dipole moment, of some molecules, 10

Electrical current density, in bulk membranes, 39
Electrode assembly, schematic representation of membrane, 34
EMF response, of carrier membrane electrodes, 41
Encapsulation in solution, dynamics of, 135
Energy coupling, in ion-transport mediators, 245-249
Energy maps, for interaction of Na^+ with ionophore, 54-56
Enniatin A, cation binding selectivities of, 238
 structure of, 222
Enniatin B, cation binding selectivities of, 238
 stability constants for metal complexes of, 141
 structure of, 222
Enthalpy, values for reaction of amine cations with 18-crown-6 in methanol, 234-235
Entropy, values for reaction of amine cations with 18-crown-6 in methanol, 234-235
Equilibrium constants, values for reaction of amine cations with 18-crown-6 in methanol, 234-235
 see also Association constants *and* Stability constants
Equilibrium domain, 241
Ethers, solutions of alkali and alkaline earth metals in, 67-75
Extraction, of alkaline earth metal picrates in methylene chloride with ionophores, 32

Films, of thin solid compounds of crowns

Subject Index

and cryptands with metal ions, 86-89
Free energy, of activation of alkylammonium salt complexes of crown ethers, 207-208, 210-211
  of association of alkylammonium salt complexes of crown ethers, 207-208
  of electrostatic interaction between cationic complex and membrane solvent, 21
  of transfer of cationic complexes from water into membrane, 17

Goldman-Hodgkin-Katz equation, 241
Gramicidin A, cation binding selectivities of, 239
Grisorixin, structure of, 223

Hydration, free energies of for alkali ions and alkaline earth ions, 12
Hydrogen ion, selectivity factors for ionophores in liquid membrane electrodes complexing with, 36-37

Inclusive complexes, of cryptates, 101
Interaction energy, between binding site models and cations, 9
  change in for binding site models and cations, 9
  for interaction of Na^+ with ionophore, 54-56
Ion-Cavity-Radius concept, use in explaining stoichiometry of metal-crown systems, 125-126
Ionophore A-23187, 220, 225, 228
  cation binding selectivities of, 238
  secretory responses associated with, 226-227
  structure of, 223-224
Ionophore X-206, cation binding selectivities of, 238
  structure of, 223
Ionophore X-537A, 225, 228
  cation binding selectivities of, 238
  secretory responses associated with, 226-227
  structure of, 223
Ionophores, biological applications of, 225-228
  cation selectivity of, 237-239
  charged types complexing with cations, 231-237
  chemical applications of, 224-225
  description of, 2-8
  neutral complexing, 229-231
  physiological secretory responses associated with, 226-227
  stability constants of alkali and alkaline earth metal cations in ethyl alcohol with, 14
  structural features, 221-224
  see also Crown ethers *and* Macrocyclic polyethers
Ion-selective membrane electrodes, containing neutral carriers, 31-35
Ion selectivity, future prospects in designing ligands for, 53
  influence of ligand layer around cation in liquid membrane electrodes on, 24
  of membranes containing lipophilic neutral carriers, 15-22
  observed in polymeric membranes containing ionophores, 33
Ion transport, carrier-mediated through membranes, 35-53

Kinetics, of cation complexation as studied by alkali metal NMR spectroscopy, 89-101
  of complexation of alkylammonium salts with 18-crown-6, 204-206
  of complexation effect of solvent and complexing agent on, 96-99
  of crown ether complexation with alkylammonium salts, 201-203

Lasalocid, *see* Ionophore X-537A
Ligand encapsulation, 134
Ligands, influence on ion selectivity of liquid membrane electrodes of, 28-29
  values of lipophilicities for, 23
Lipophilicites, for various ligands, 23
Lipophilicity increments, for several structural fragments, 22
Lipophilization, of cations by crown ethers, 184
Liquid membrane electrodes, influence, of ethyleneoxide units between diphenylamide groups on ion selectivity in, 29
  of ligand on selectivity of, 28

influences of thickness of ligand layer around cation on ion selectivity of, 24
selectivity factors, for liquid-membrane electrodes containing valinomycin, 19
for neutral complexing agents in, 36-37
Lithium ion, association constants of crown ethers with, 183
effect on selectivity of reactions of 1,2-bis(bromomethyl)benzene and polyethylene glycolate, 178
lipophilization by crown ethers of, 184
molecular structure of antamanide complexes of, 142
selectivity factors for ionophores in liquid membrane electrodes complexing with, 36-37
solubilities in amines and ethers of, 68-69
stability constants, for antibiotic complexes of, 141
of crown ether complexes, 144
of cryptates, 131, 182, 232-233
stoichiometries of crown ethers with, 122
synthesis yields of crown ethers using, 161
transfer numbers in neutral carrier membranes of, 40

Macrocyclic polyethers, differentiation of Na/K and Ca/Mg with, 136-137
sandwich complexes of cesium with, 99-101
schematic representation of different metal-polyether complexes, 123
similarity in complexation of antibiotics and, 143-147
stoichiometry and conformation of metal complexes of, 123-126
see also Crown ethers and Ionophores
Macrocyclic thiapolyethers, synthesis and melting points of, 175
Macrocyclic thiazapolyethers, synthesis and melting points of, 176
Macroheterobicyclic ligands, complex formation constants for, 13
Macrotetralides, structure of, 222
Magnesium ion, extraction in methylene chloride with ionophores of, 32
selectivity factors for ionophores in liquid membrane electrodes complexing with, 36-37
stability constant of cryptates, 131, 182, 232-233
Mass spectra, used to study crown ethers, 180-181
Melting point, of crown ethers, containing 1,3-bis(methylene)benzene subunits, 167
containing 2,5-bis(methylene)furan subunits, 168
containing 2,6-bis(methylene)pyridine subunits, 172
containing 1,n-bis(methylene) aromatic subunits, 169
containing catechol subunits, 167
of macrocyclic thiapolyethers with 1,n-bis(methylene) aromatic subunits, 175
of macrocyclic thiazapolyethers with 1,n-bis(methylene) aromatic subunits, 176
Membrane-active complexing agents, design features of, 22-31
Membrane conductance, zero-current for, 50
Membrane electrode assembly, schematic representation, 34
Membrane electrodes, ion-selective based on neutral carriers, 31-35
Membrane potential, equations for, 34-35
Membranes, basic equilibrium of carriers at interfaces of, 15-16
carrier-mediated ion transport through, 35-53
containing lipophilic neutral carriers ion selectivity of, 15-22
correlation between permeabilities and kinetic parameters in glyceryl dioleate bilayers, 52
Membrane selectivity, 146
Membrane-transport, diagram of carrier mediated ion translocation across membrane, 240
energy profile for carrier mediated, 242
role of ion-transport mediators in energy coupling, 245-249
theoretical model for carrier mediated, 241-245
theory of, 239-249
Metal-ammonia compounds, solid, 75-76
Metal binding, influence of ring closure of polydentate ligands on stability constants for, 30
Metal solubility, enhancement by crowns and cryptands, 76-78

Subject Index

solvent effects, 77
Metal solutions, in amines, 67-75
   in ammonia, 65-67
Mitchell model, of mitrochondrial energy coupling, 247
Monactin, ATPase activity in presence of alkali metal cations of, 38
   cation binding selectivities of, 238
   current-voltage characteristics for phosphatidyl serine bilayer membranes containing, 49
   structure of, 3, 222
Monazomycin, cation binding selectivities of, 239
Monensin, cation binding selectivities of, 238
   stability constants for metal complexes of, 141
   structure of, 119, 223
Monomers, in metal solutions, 80-81
   spectra of in various solvents, 72

Neutral carrier membrane, selectivity between divalent and monovalent cations of, 18
   transport numbers of cations in, 40
Neutral-carrier-modified membranes, comparison of selectivities of, 51
Neutral carriers, contained in ion-selective membrane-electrodes, 31-35
Neutral complexing agents, selectivity factors for liquid membrane electrodes using, 36-37
Nicolsky equation, 35
   empirical extension of, 35
Nigericin, cation binding selectivities of, 238
   stability constants for metal complexes, 141
   structure of, 119, 223
NMR, chemical shifts in various solvents of cesium-crown complexes, 100
   studies of alkali metal spectra in metal solutions, 101-107
   to study the kinetics of crown ethers complexation with alkylammonium salts, 201-202
NMR spectra, effect of complexing agent on, 90-91
   effect of solvent on, 90-91

linewidth effects, 91-92
   measurement of exchange rates by alkali metal, 93-95
   of Na^+-cryptand-Na^- solutions in various solvents, 103
   of Rb^- and Cs^-, 103-105
   of sodium-cryptand complex, 94
   used to study crown ethers, 179-180
   at various temperatures for solutions of Na^+-crown-Na^-, 106
Nonactin, cation binding selectivities of, 238
   correlation of permeabilities and kinetic parameters in bilayers containing, 52
   selectivities of neutral-carrier-modified solvent polymeric membranes containing, 51
   stability constants for metal complexes of, 141
   structure of, 3, 222

Paper chromatography, of metal-nitropenolates, 129
Partition coefficient, 24
   for ion-selective membrane electrodes, 31
Permeabilities, correlation in bilayers of kinetic parameters with, 52
Permeability ratio, for carrier-mediated transport of ions, 241-244
Phenylethylammonium ion, transfer numbers in neutral carrier membranes of, 40
   transport and potentiometric selectivities of PVC membranes for, 42
Physiological secretory responses, associated with ionophores, 226-227
Polarizability, of some molecules, 10
Polydentate ligands, 12-15
Polyene antibiotics, cation binding selectivities of, 239
Polyetherin A, structure of, 223
Polymeric membranes, comparison of selectivities of neutral-carrier-modified membranes, 51
   ion selectivities observed for ionophores in, 33
Potassium ion, absorption spectrum of thin film of cryptate, 88-89
   association constants, of crown ether complexes in $CDCl_3$, 185
   of crown ethers with, 183

chemical shift differences of protons in crown ether complexes of, 187
effect on selectivity of reactions of 1,2-bis-(bromomethyl)benzene and polyethylene glycolates, 178
ESR spectra in various solvents of monomer, 72
lipophilization by crown ethers of, 184
maximum synthesis yields of crown ethers using, 163
model of complex of valinomycin with, 26
selectivity factors for ionophores in liquid membrane electrodes complexing with, 36-37
solubilities in amines and ethers of, 68-69
stability constants, of antibiotic complexes of, 141
of crown ether complexes, 144
of cryptates, 131, 182, 232-233
stoichiometries of crown ethers with, 122
synthesis yields of crown ethers using, 161
transfer numbers in neutral carrier membranes of, 40
transport of by combination of uncomplex and neutral ionophore, 249
Potentiometric selectivities of PVC membranes, containing enantiomer-selective carriers, 42
containing neutral Na⁺ carrier, 42
Powders, of solid compounds of crowns and cryptands with metal ions, 86-89

Ring closure, influence on stability constants of, 30
Rubidium ion, absorption spectrum of thin film of cryptates, 88
association constants, of crown ether complexes in $CDCl_3$ of, 185
of crown ethers with, 183
chemical shift differences of protons in crown ether complexes of, 187
ESR spectra in various solvents of monomer, 72
NMR spectra of Rb⁻, 103-105
selectivity factors for ionophores in liquid membrane electrodes complexing with, 36-37
solubilities in amines and ethers of, 69

stability constants, for antibiotic complexes of, 141
of crown ether complexes of, 144
for cryptates, 131, 182, 232-233

Sandwich complexes, of cesium with crown ethers, 99-101
Selectivity, influence of ligands of liquid membrane electrodes on, 28-29
of membranes, 51
of reactions of 1,2-bis(bromomethyl)-benzene and polyethylene metal glycolates, 178
Selectivity coefficient, for neutral carrier membranes toward cations, 16
Selectivity factors, dependence on polarity of membrane solvent of, 20
for liquid membrane electrodes, containing neutral complexing agents, 36-37
containing valinomycin, 19
for neutral carrier membranes toward cations, 16
Selectivity sequence, of binding of metal cation by ionophores, 238
Self-encapsulation, 134
Sensors, 137
structures of representative, 117
Sodium ion, absorption spectrum of thin film of cryptate, 87-88
association constants of crown ethers with, 183
bond lengths in crown and cryptand crystals of, 84
chemical shift differences of protons in crown ether complexes of, 187
crystal structure of cryptates, 85
effect on selectivity of reactions of 1,2-bis(bromomethyl)benzene and polyethylene glycolates, 178
energy maps for interaction of ionophore with, 54-56
lipophilization by crown ethers of, 184
maximum synthesis yields of crown ethers using, 163
NMR spectra, of cryptates, 94
at various temperatures for solution of Na⁺-crown-Na⁻, 106
in three solvents of cryptates, 103
selectivity factors for ionophores in

Subject Index

liquid membrane electrodes complexing with, 36-37
solubilities in amines and ethers of, 68
stability constants, of antibiotic complexes, 141
of crown ether complexes, 144
for cryptates, 131, 182, 232-233
stoichiometries of crown ethers with, 122
synthesis yields of crown ethers using, 161
transport and potentiometric selectivities of PVC membranes for, 42
Solid complexes, of metals with crowns and cryptands, 82-89
Solubility of alkali metals, in amines and ethers, 68-69
enhancement by cyclic polyethers, 68-69
Solvated electron, 70-72
relation between infrared absorption band and pulse-radiolysis for, 71
Stability constants, of alkali and alkaline earth metal-cryptand complexes, 131
of alkali metals-crown complexes, 144
of t-BuNH$_3$·PF$_6$ complexes of 1,3-xylyl crown ethers in CDCl$_3$, 194
of cesium-crown complexes in various solvents, 98
determination from alkali metal chemical shifts, 92-93
influence of ring closure of polydentate ligands on, 30
of ionophore X-537A with biogenic amines, 237
of ionophores with alkali and alkaline earth metal cations in ethyl alcohol, 14
of macroheterobicyclic ligands in aqueous solution, 13
of metal-antibiotic complexes, 141
of metal-bicyclic cryptate complexes in water, 182
of metal cations with cryptands, 232-233
see also Association constants *and* Equilibrium constants
Stoichiometry, of alkali and alkaline earth complexes with crown ethers, 122
effect of anion in solid phase on, 132-133
ion-cavity-radius concept to explain the stoichiometry of metal-crown systems, 125-126
of large-cavity crown complexes, 135-136
of metal-crown complexes, 123-126
of small-cavity crown complexes, 133-135
Stronium ion, extraction in methylene chloride with ionophores of, 32
selectivity factors for ionophores in liquid membrane electrodes complexing with, 36-37
stability constants for cryptates, 131, 182, 232-233
Structure, of actins, 120
of antamanide, 120
of benzo-15-crown-5, 116
of 18-crown-6, 116
of dibenzo-14-crown-8, 116
of dibenzo-18-crown-6, 116
of dibenzo-30-crown-10, 116
of dicyclohexano-18-crown-6, 116
of monensin, 119
of nigericin, 119
of several cryptands, 118, 230
of several representative sensors, 117
of valinomycin, 120
Substituent interconversion, yields of crown ethers obtained by, 166
Superchelate effect, 134
Symport coupling, of ion-transport mediators, 245-246, 248
Synthesis, of crown ethers, containing 1,3-bis(methylene)benzene subunits, 165-167
containing 2,5-bis(methylene)furan subunits, 168
containing 2,6-bis(methylene)pyridine subunits, 172
containing 1,n-bis(methylene) aromatic subunits, 169
containing catechol subunits, 167
of macrocyclic azapolyethers, 170-174
of macrocyclic polyethers, 160-170
of macrocyclic thiapolyethers, 174-176
of macrocyclic thiapolyethers with 1,n-bis(methylene) aromatic subunits, 175
of macrocyclic thiazapolyethers, 176
of macrocyclic thiazapolyethers with 1,n-bis(methylene) aromatic subunits, 176
maximum yields of crown ethers as function of cation used, 163
of metal-crown-complexes and effect of anion on, 127-128

yields of crown ethers using different alkoxide cations, 161

Templated synthesis, 176-179
Tetranactin, correlation of permeabilities and kinetic parameters in bilayers containing, 52
  selectivities of neutral-carrier-modified solvent polymeric membranes containing, 51
  structure of, 3, 222
Thermodynamics of cation complexation, effect of solvent and complexing agent on, 96-99
  as studied by alkali metal NMR spectroscopy, 89-101
Transference number, for Ca^{2+} through a Ca^{2+}-carrier membrane, 43
Transfer numbers, of cation permselective neutral carrier membranes, 40
Transport, of calcium ion, by uncoupler and electrogenic ionophore, 249
  of potassium ion, by uncoupler and neutral ionophore, 249
Transport selectivities of PVC membranes, containing enantiomer-selective carriers, 42
  containing neutral Na^+ carrier, 42
Tricyclic cryptands, 139
Trinactin, cation binding selectivities of, 238
  correlation of permeabilities and kinetic parameters in bilayers containing, 52
  selectivities of neutral-carrier-modified solvent polymeric membranes containing, 51
  structure of, 3, 222

Valinomycin, cation binding selectivities of, 238
  correlation of permeabilities and kinetic parameters in bilayers containing, 52
  current-voltage characteristics for phosphatidyl serine bilayer membranes containing, 49
  model of K^+ complex of, 26
  selectivities of neutral-carrier-modified solvent polymeric membranes containing, 51
  selectivity factors for liquid-membrane electrodes containing, 19
  stability constants for metal complexes of, 141
  structure of, 3, 120, 222
Van der Waals Radius, of some molecules, 10

Water, complexation in $CDCl_3$ of crown ethers with, 190

1,3-Xylyl crown ethers, amount complexed by alkylammonium salt, 197
  relation between chemical shift of water and water concentration for, 192
  relative binding constants for t-$BuNH_3 \cdot PF_6$ complexes of, 194
1,3-Xylyl-18-crown-5, most likely conformation of alkylammonium complex of, 194

Zero-current membrane conductance, 50
Zintl anions, in crystalline solids, 86

DATE DUE

JUL 7 1982